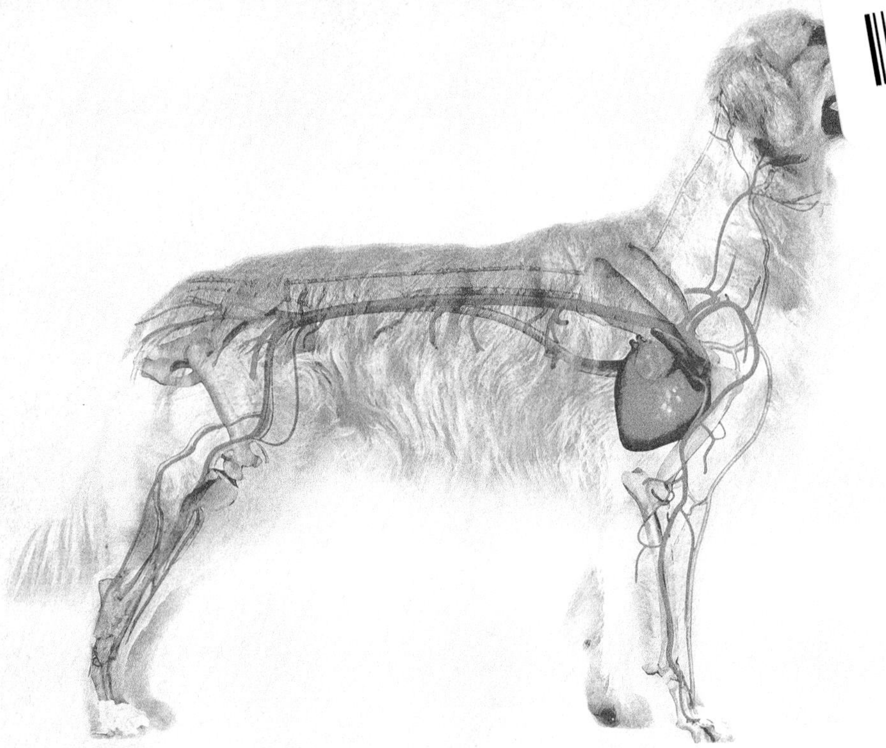

Dog Anatomy a Photo Atlas

Second Edition

Heather P. Walker

bluedoor
small publishing done big®

Chief Executive Officer: Jon K. Earl

President, College: Lucas Tomasso
President, Private Sector: Dawn Earl
Regional Manager: Greg Bartell

Business Development Coordinator: Jennifer Smith

Senior Production Manager: Connie Dayton
Production Manager: Dan Woods
Production Manager: Amber Wahl
Production Manager: Rhiannon Nelson
Assistant Production Manager: Ben Sweeney
Production Manager: Stephanie Larson
Project Manager: Peggy Li

Consulting Editors: Bruce D. Wingerd, M.S.
Suzanne S. Frucht, Ph.D.
Anna M. Kats, M.S., Florida Atlantic University
Michelle F. Cavallo, M.S., Florida Atlantic University
John F. Wiginton, Ph.D., University of Mississippi
Stephanie R. Dillon, Ph.D., Florida State University

Cover Design: Dan Woods

ISBN-13: 978-1-59984-706-1

Photographs by Wendy Savage, Chris Kennedy, and Torii Rodgers; Biomedical photographers, Department of Educational Media and Design, NC State College of Veterinary Medicine.

Illustrations by Alice MacGregor Harvey; Medical illustrator, Department of Educational Media and Design, NC State College of Veterinary Medicine.

Radiographs courtesy of Dr. Ed Smallwood, DVM and the Department of Radiology at NC State College of Veterinary Medicine.

Plastinated models courtesy of the Anatomy Lab at NC State College of Veterinary Medicine.

Published by bluedoor, LLC
10949 Bren Road East
Minneapolis, MN 55343-9613
800-979-1624
www.bluedoorpublishing.com

Printed in the United States of America.
10 9 8 7 6 5

Preface

Dog Anatomy - A Photo Atlas was created for veterinary technicians, specifically for those students who have a focus on canine training. This atlas is not meant to substitute as an anatomy textbook but rather to serve as a supplement.

The main purpose is to help the student get a better understanding of canine anatomy through visuals. This atlas contains photographs, micrographs, radiographs and schematic drawings that will help the student appreciate dog anatomy and how the anatomical structures are arranged.

Each chapter contains a combination of histological images, diagrams, illustrations, photographs and radiographs. Labeling, crosswords and multiple-choice exercises are included to reinforce material covered in each chapter.

Table of Contents

Chapter 1

Cells, Tissue and the Integument

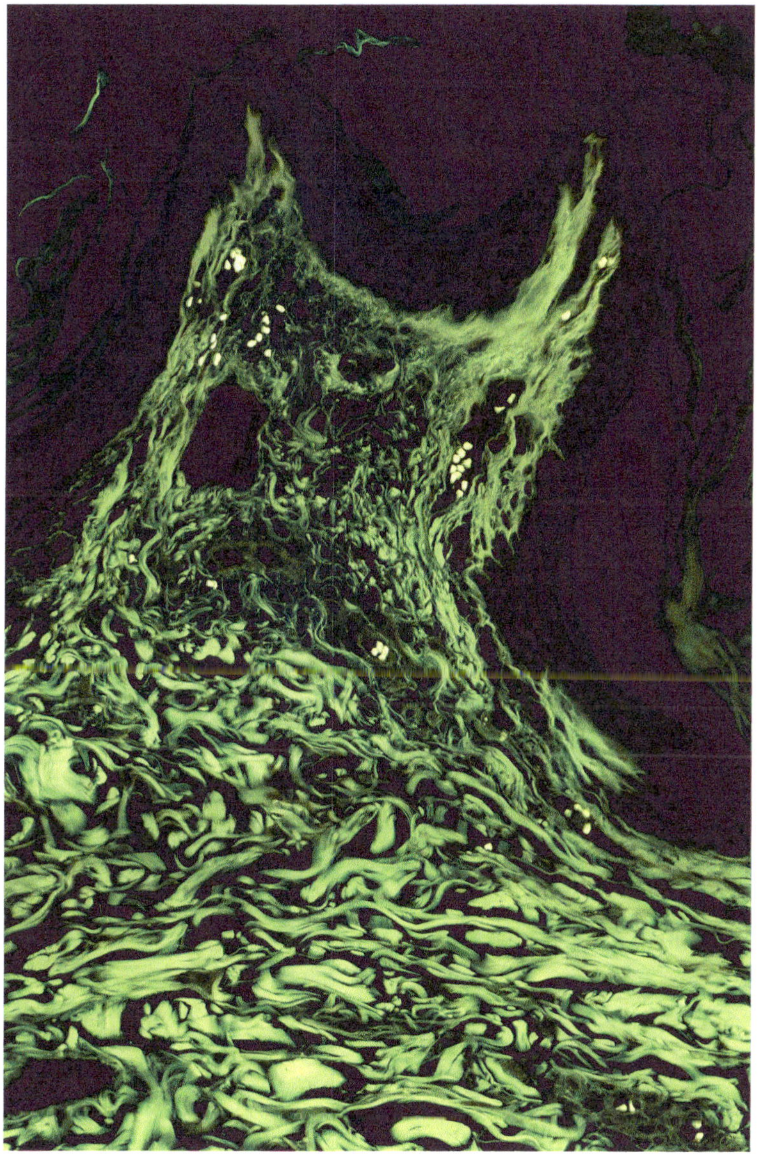

© David G. Ward

Fibrous Connective Tissue in Dermis of Skin (Autofluorescence, x380)

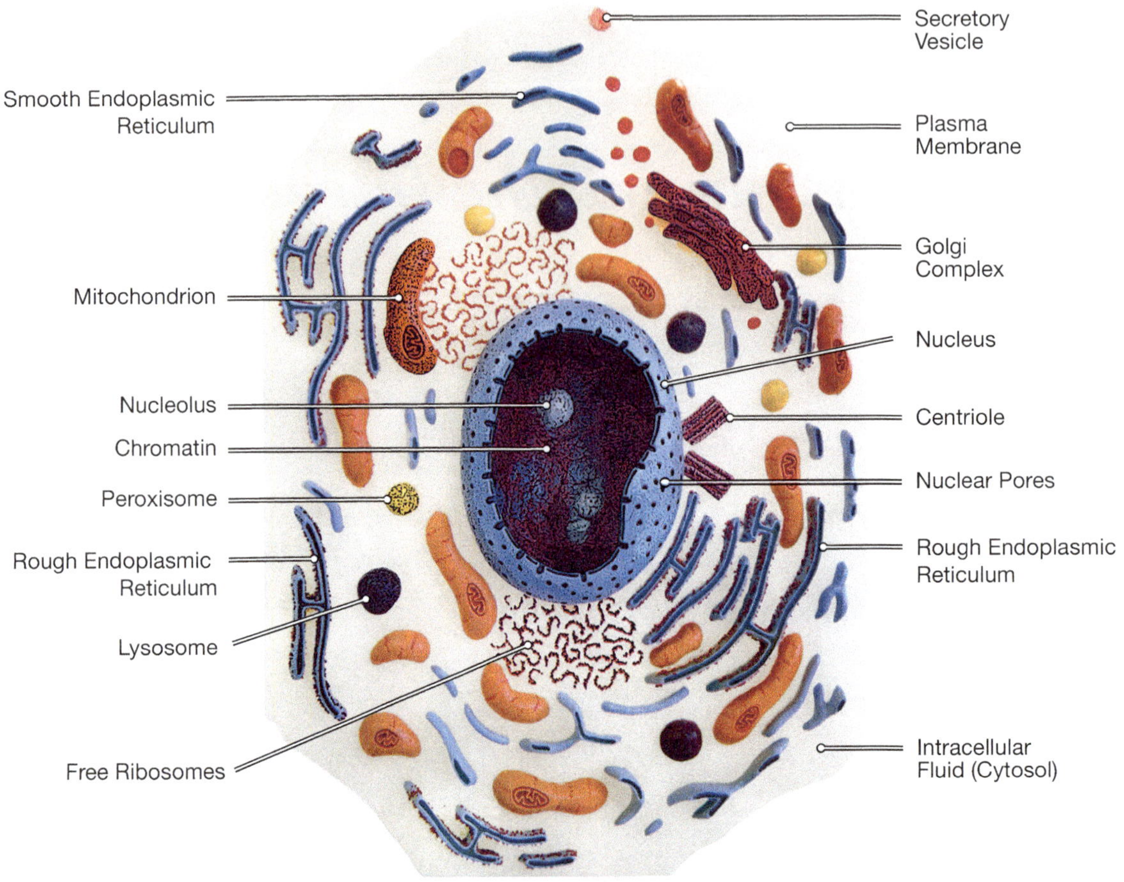

Figure 1.1: Mammalian cell model.
© David G. Ward.

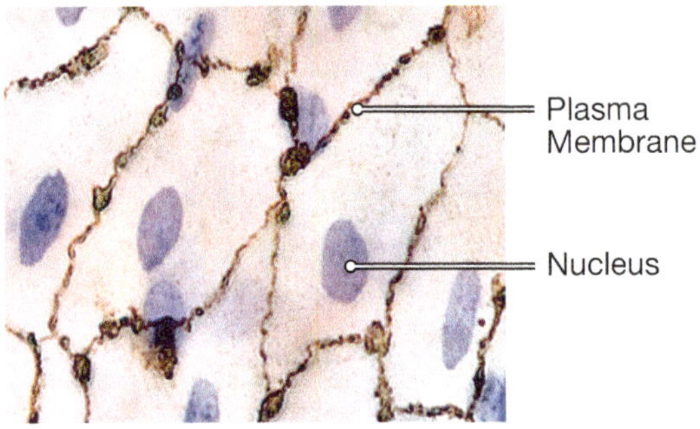

Figure 1.2: Squamous epithelium cells, mesothelium (brightfield).
© David G. Ward.

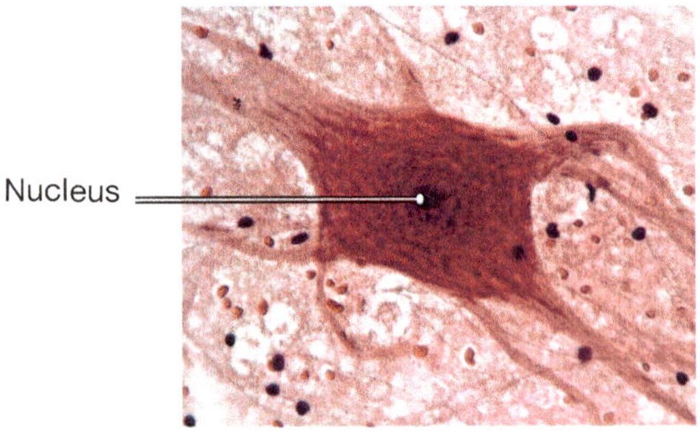

Figure 1.3: Nerve cell.
© bluedoor, LLC.

Mitosis (Cell Division)

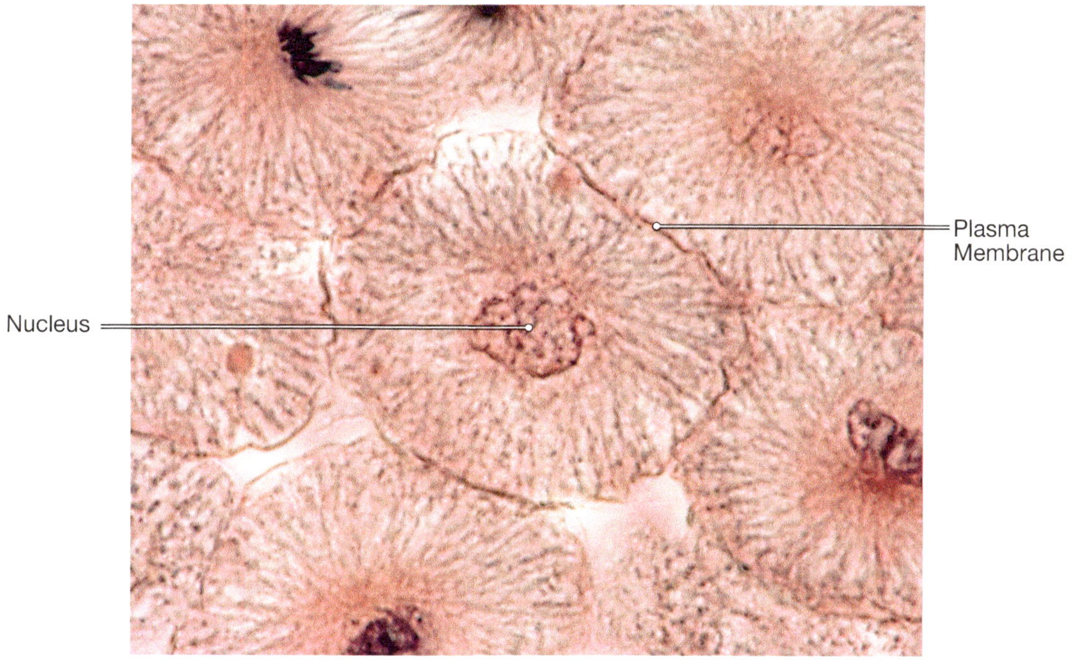

Figure 1.4: Interphase.

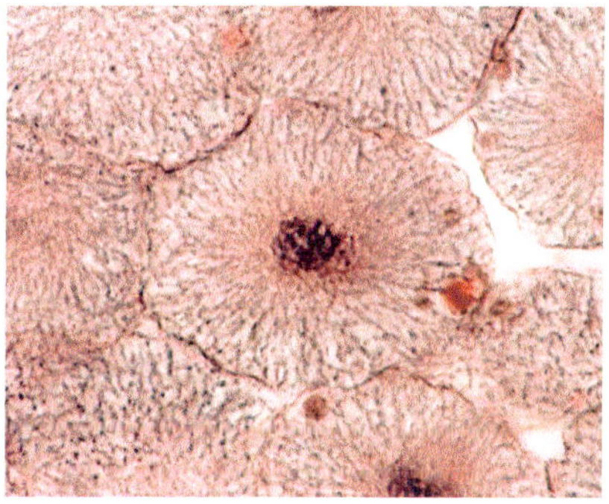

Figure 1.5: Prophase.

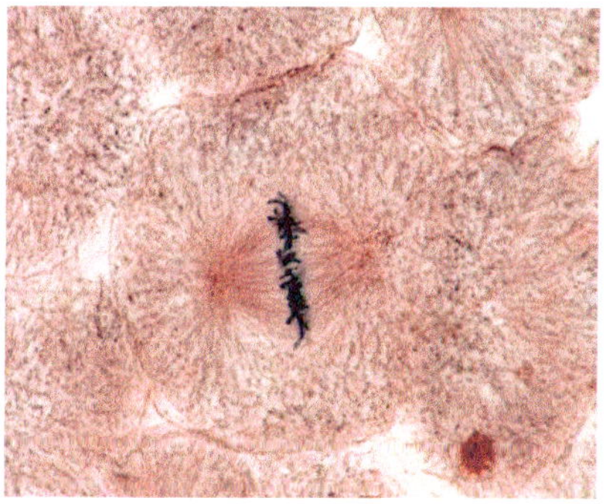

Figure 1.6: Metaphase.

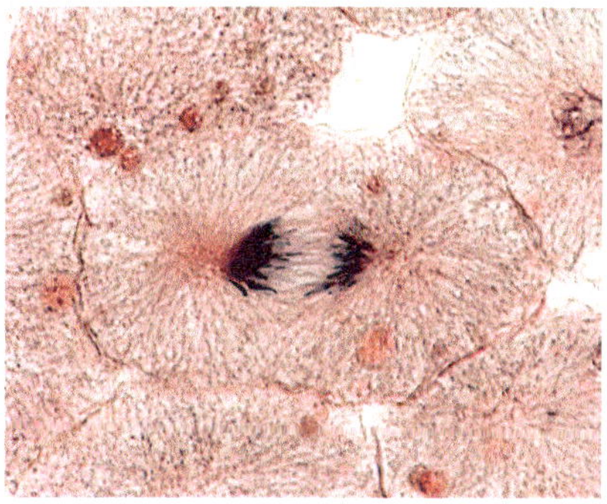

Figure 1.7: Anaphase.

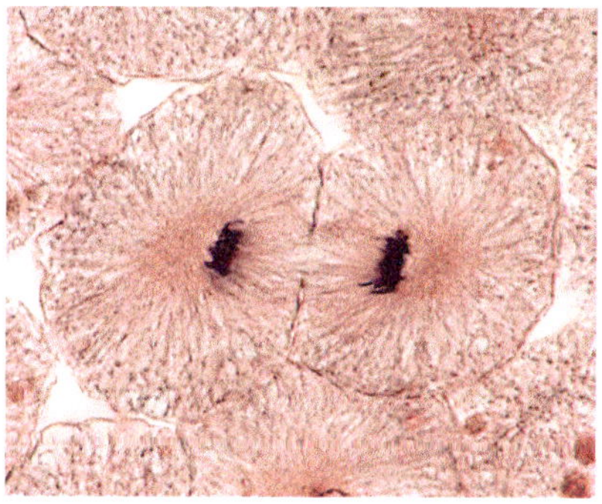

Figure 1.8: Telophase.

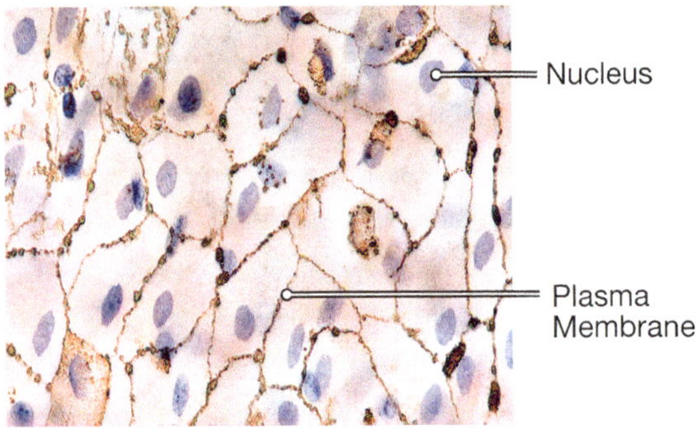

Figure 1.9: Simple squamous epithelium, mesothelium of abdominal cavity, surface view.
© David G. Ward.

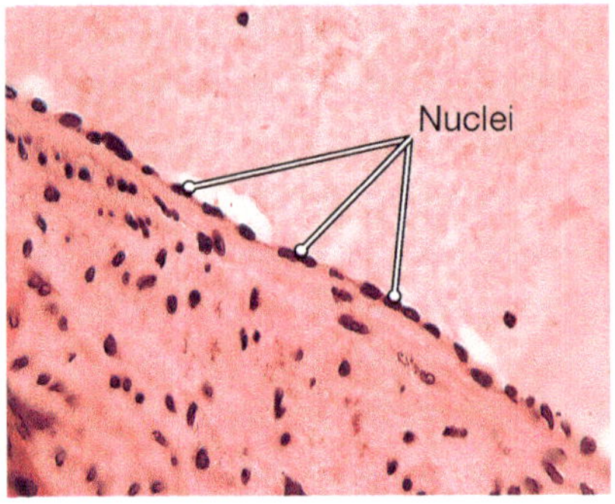

Figure 1.10: Simple squamous epithelium, endocardium of heart, side view.
© David G. Ward.

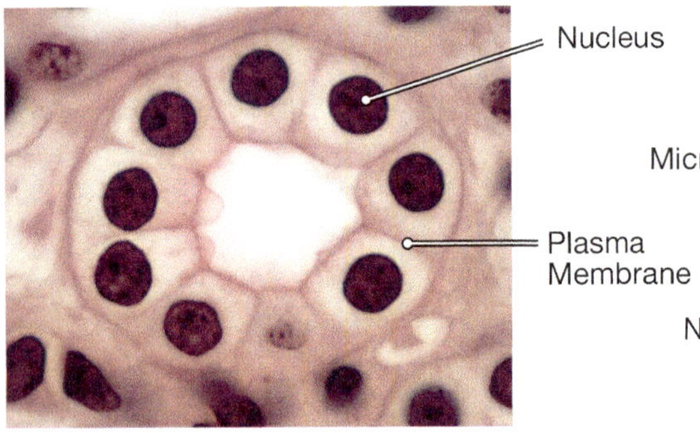

Figure 1.11: Simple cuboidal epithelium, distal tubule of kidney.
© David G. Ward.

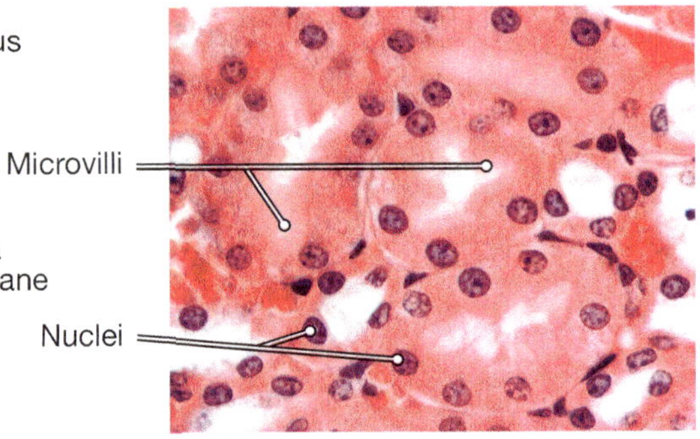

Figure 1.12: Simple cuboidal epithelium, proximal tubule of kidney.
© David G. Ward.

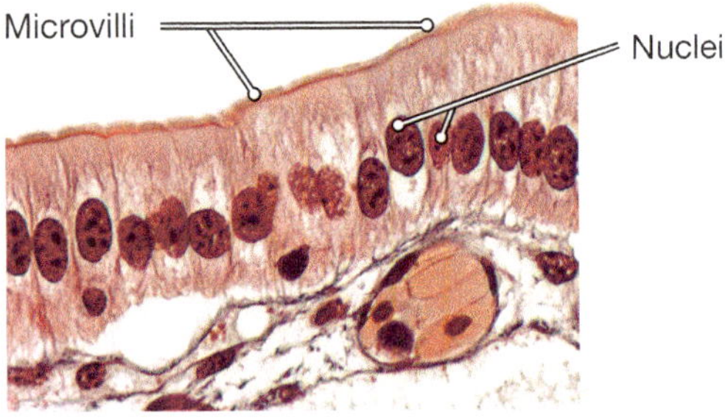

Figure 1.13: Simple columnar epithelium, mucosa of small intestine.
© David G. Ward.

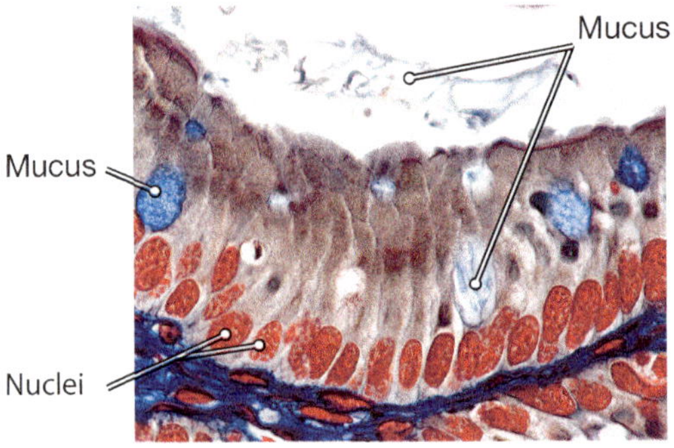

Figure 1.14: Simple columnar epithelium, mucosa of small intestine.
© David G. Ward.

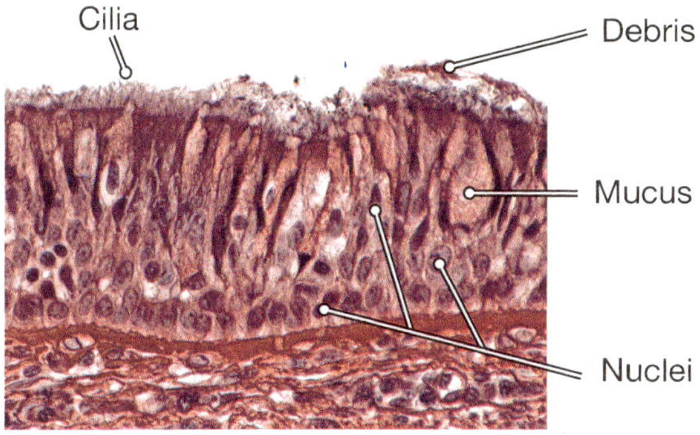

Figure 1.15: Pseudostratified ciliated columnar epitheium, mucosa of trachea.

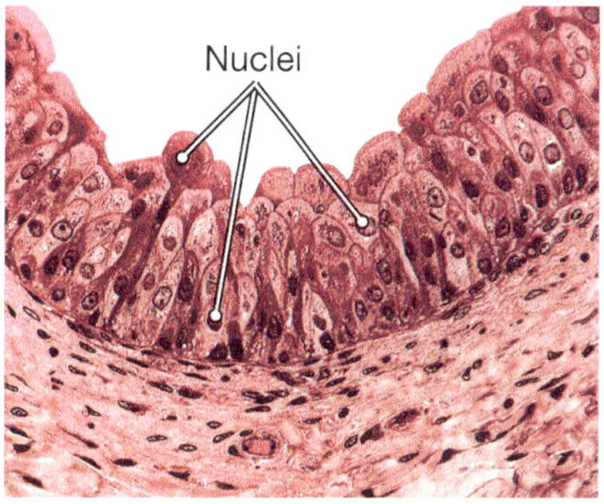

Figure 1.16: Transitional epithelium, mucosa of ureter.

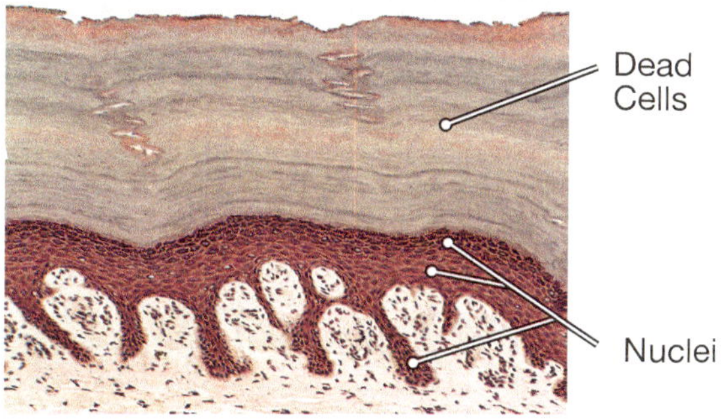

Figure 1.17: Stratified squamous epithelium (keratinized), epidermis of skin.

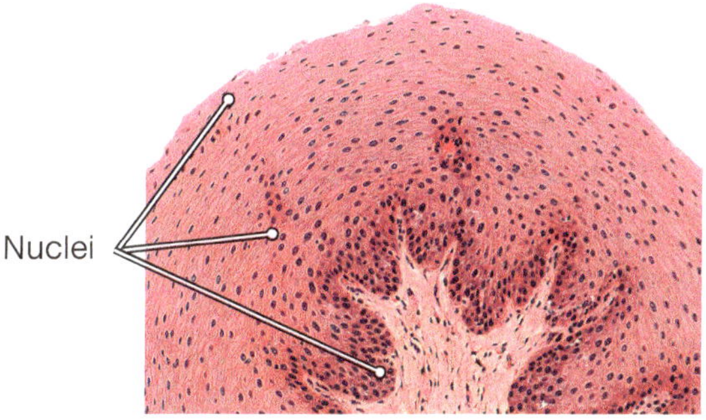

Figure 1.18: Stratified squamous epithelium (non-keratinized) mucosa of esophagus.

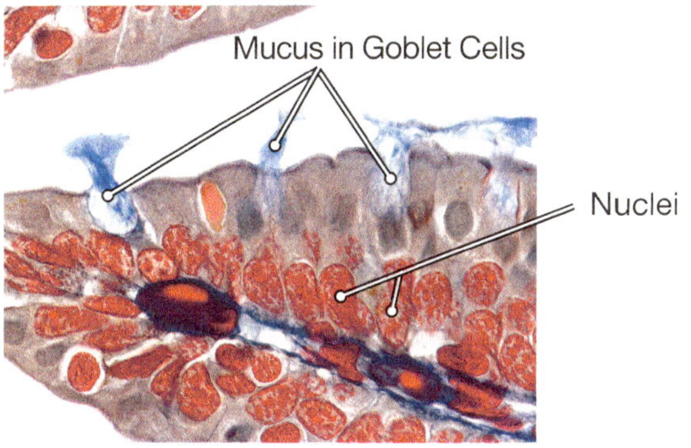

Figure 1.19: Gladular epithelium, goblet cells of small intestine.

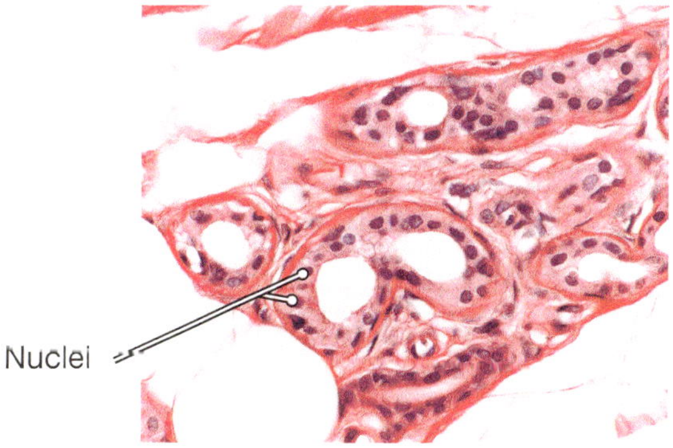

Figure 1.20: Glandular epithelium, sweat gland of skin.

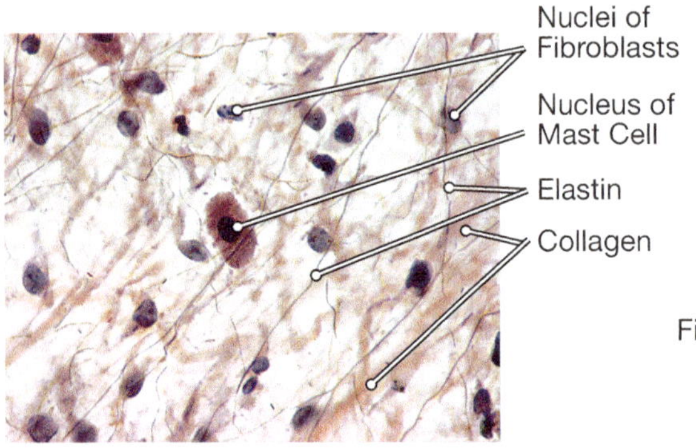

Figure 1.21: Areolar fibrous tissue, under epithelia.

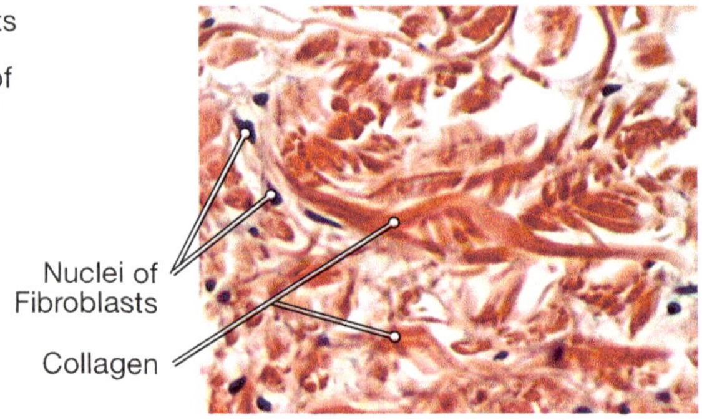

Figure 1.22: Dense irregular fibrous tissue, dermis of skin.

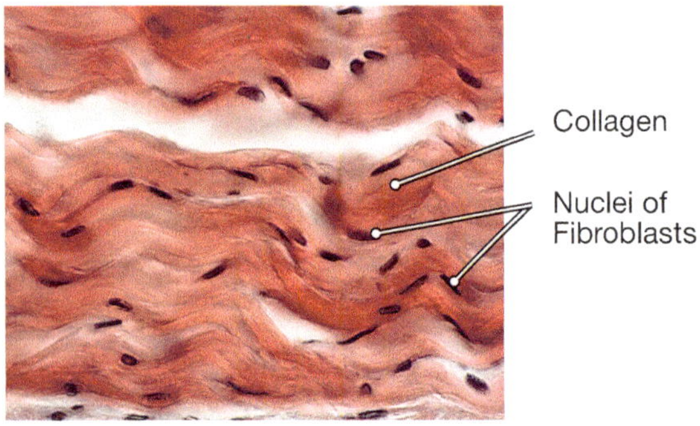

Figure 1.23: Dense regular fibrous tissue, tendon.

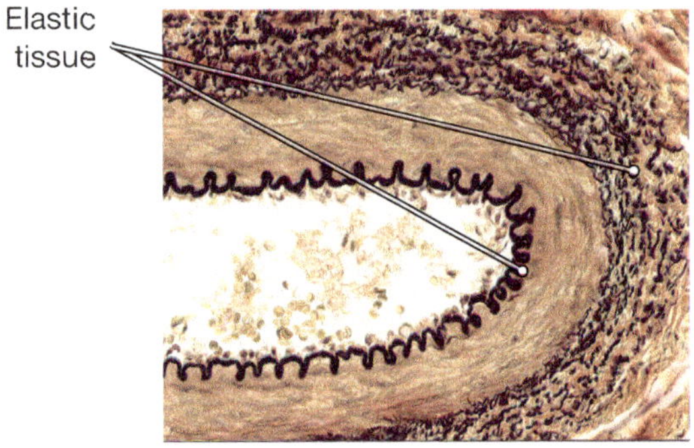

Figure 1.24: Elastic tissue

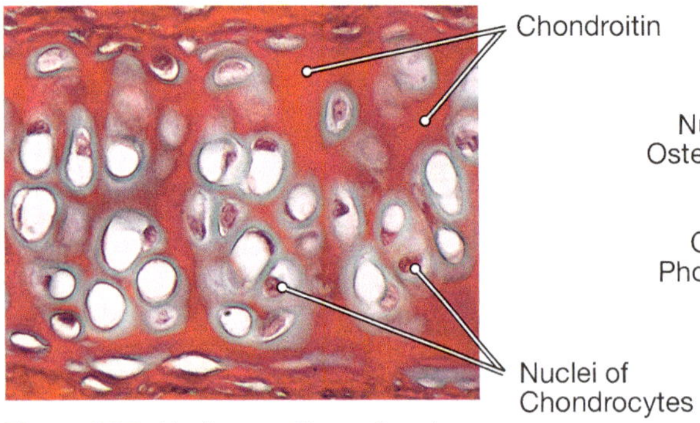

Figure 1.25: Hyaline cartilage of trachea.

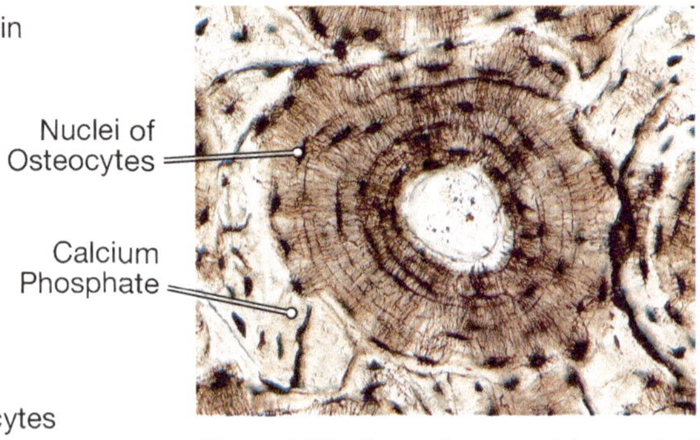

Figure 1.26: Dense (compact) bone, shaft of bone.

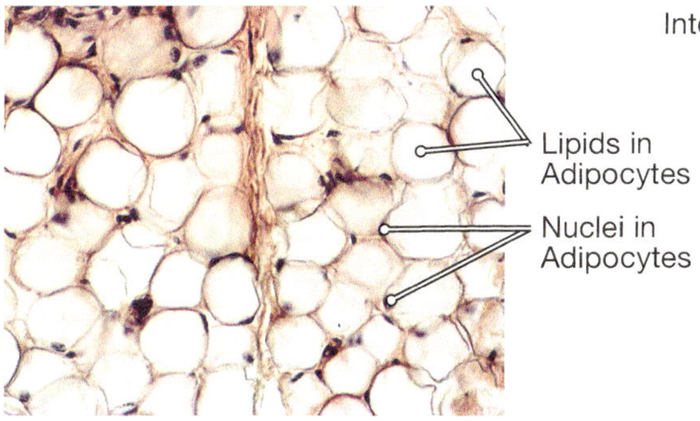

Figure 1.27: Adipose tissue, hypodermis of skin.
© David G. Ward.

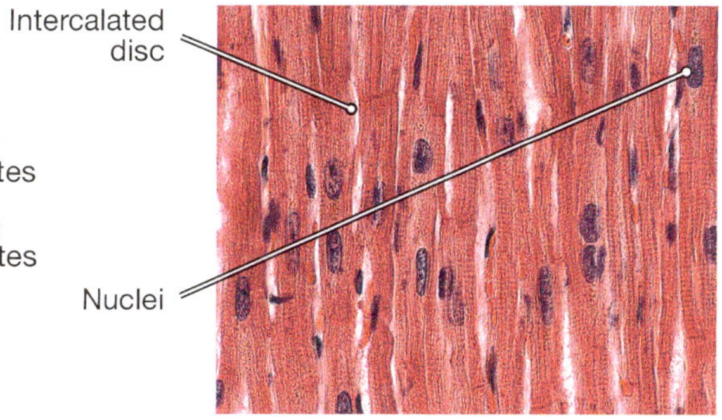

Figure 1.28: Cardiac muscle.
© Robert Sorenson

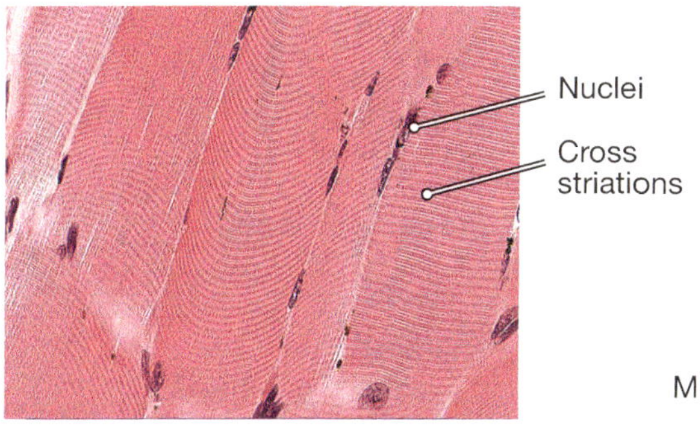

Figure 1.29: Skeletal muscle.
© Robert Sorenson

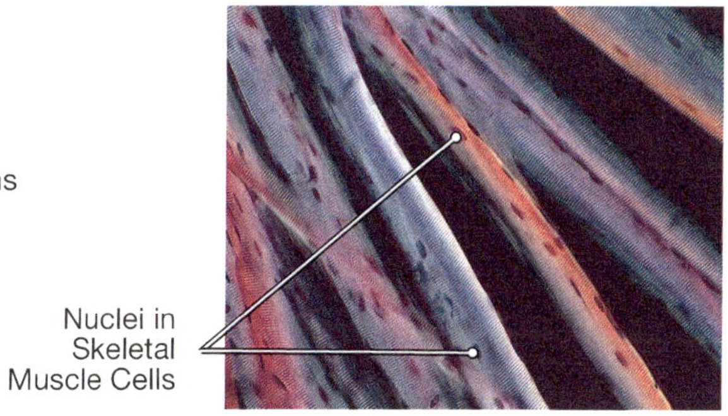

Figure 1.30: Skeletal muscle (poloarized light).
© bluedoor, LLC.

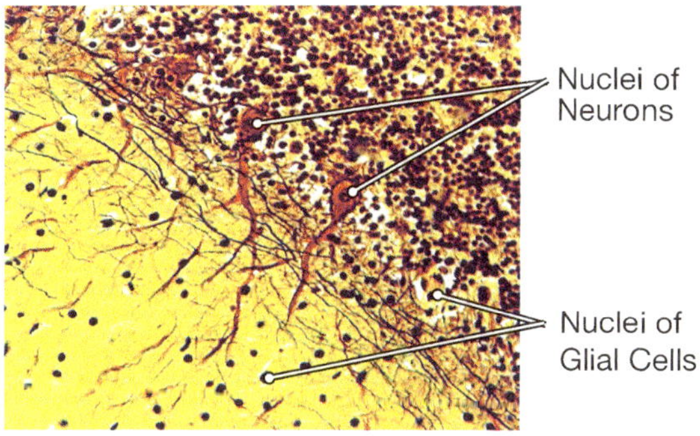

Figure 1.31: Nervous tissue, purkinje cells in cerebellum.
© David G. Ward.

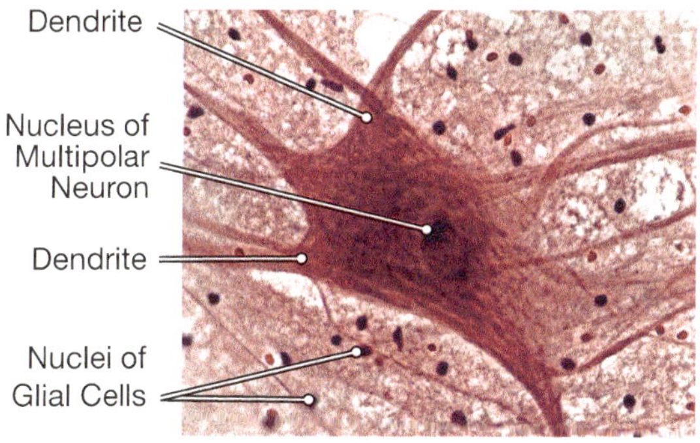

Figure 1.32: Nervous, motor neuron in spinal cord.
© David G. Ward.

Skin and Hair Follicles

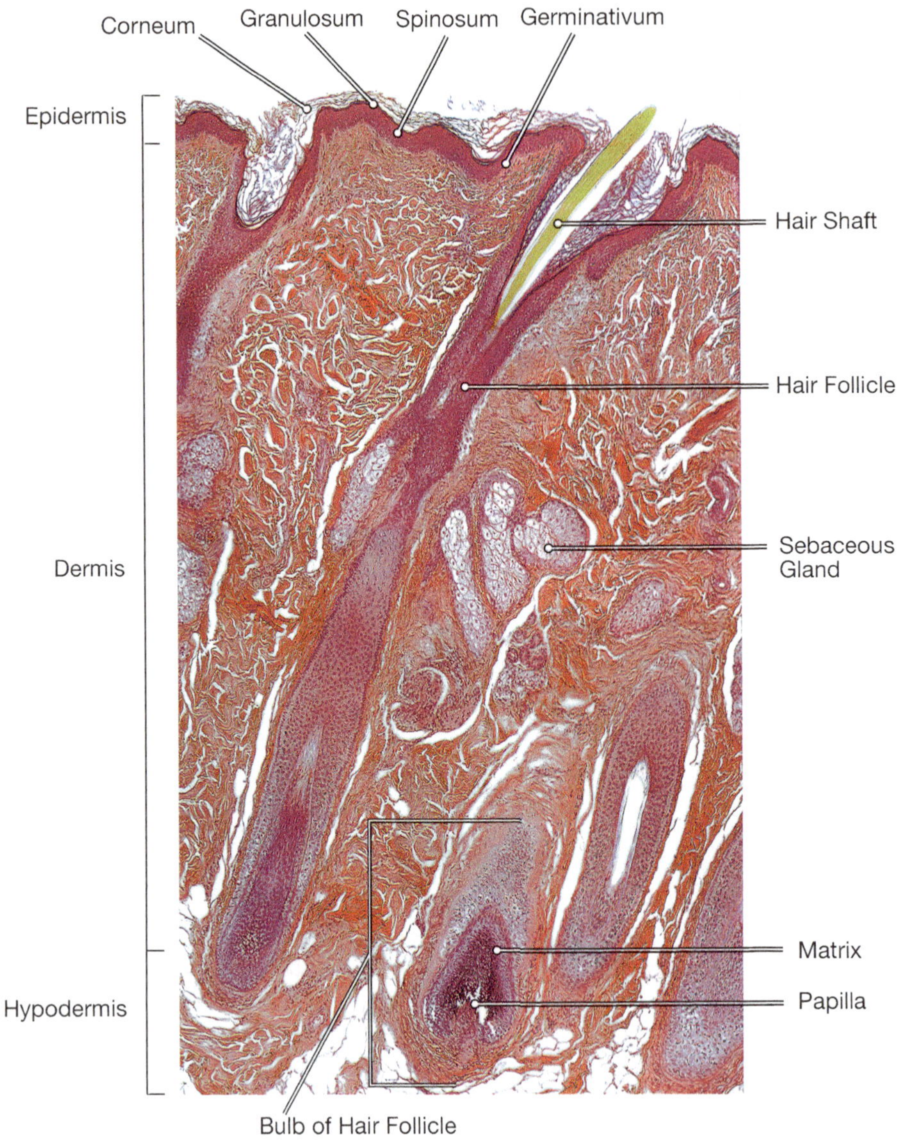

Figure 1.33: Skin.
© David G. Ward.

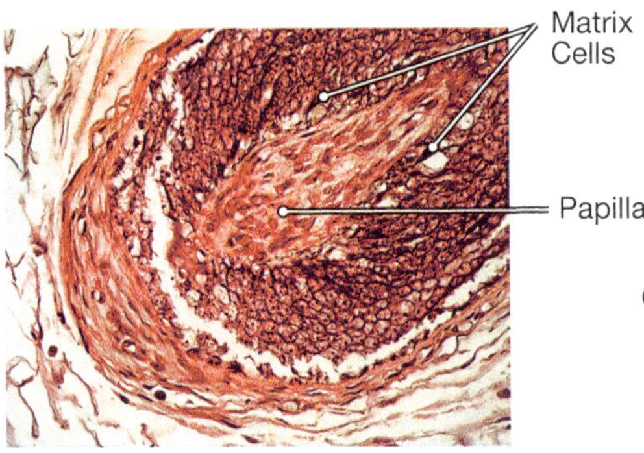

Figure 1.34: Detail of bulb of hair follicle.
© David G. Ward.

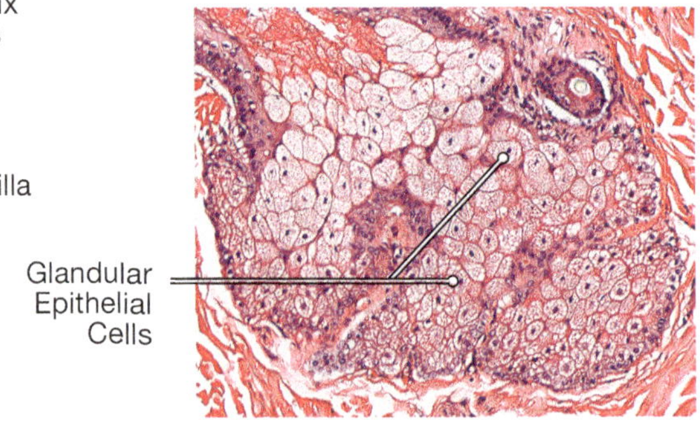

Figure 1.35: Sebaceous gland.
© David G. Ward.

Skin Full Thickness

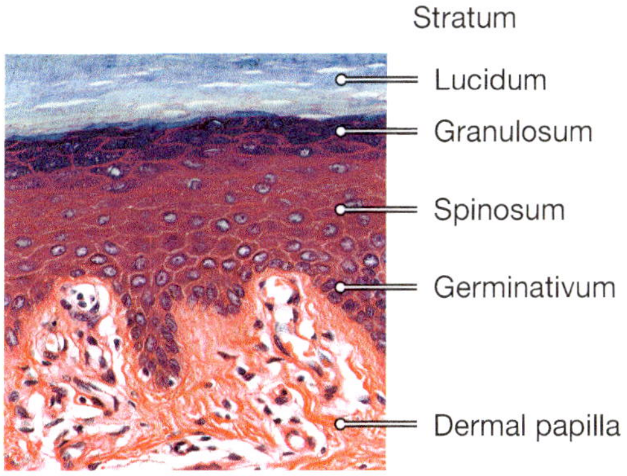

Figure 1.36: Epidermis and dermal papilla.
© David G. Ward.

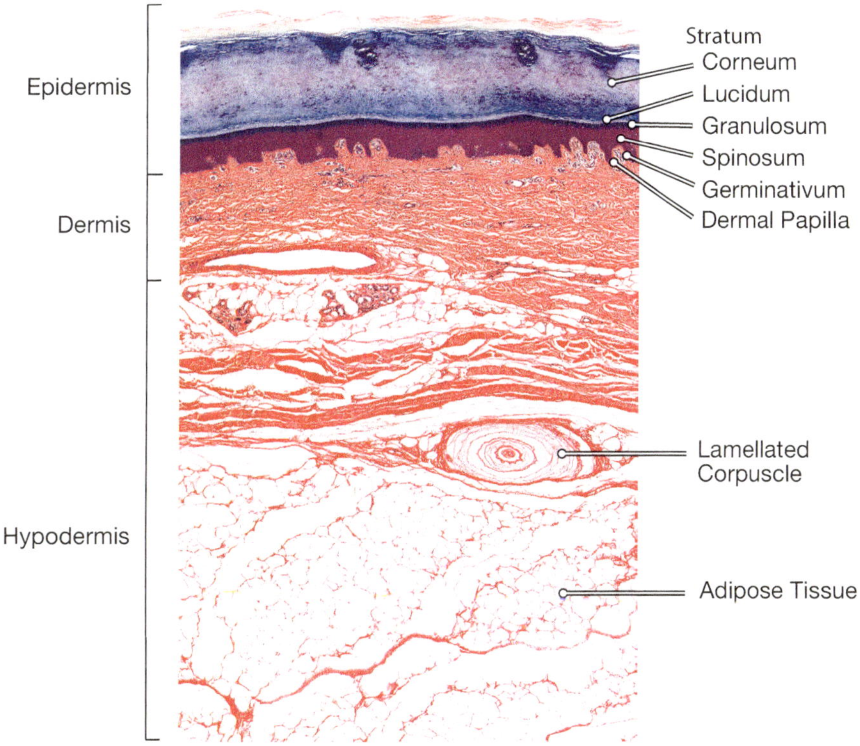

Figure 1.37: Thick skin.
© David G. Ward.

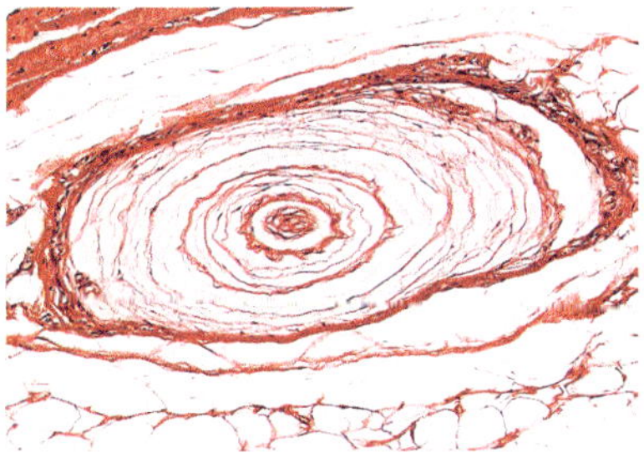

Figure 1.38: Lamellated corpuscle.
© David G. Ward.

Review: Labeling Activity

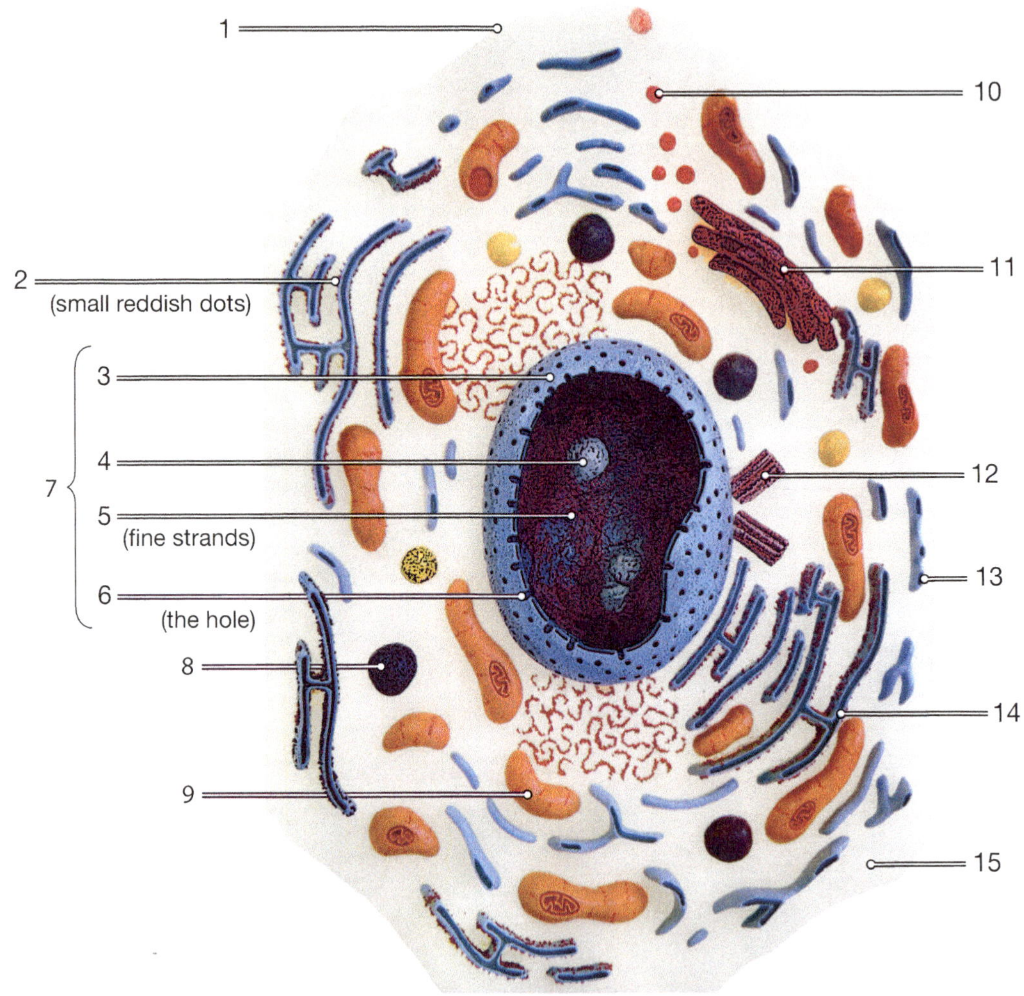

Figure 1.39: Mammalian cell model, labeling actitivy.

1. ______________________
2. ______________________
3. ______________________
4. ______________________
5. ______________________
6. ______________________
7. ______________________
8. ______________________
9. ______________________
10. ______________________
11. ______________________
12. ______________________
13. ______________________
14. ______________________
15. ______________________

Review: Key Term Matching

____	1.	interphase	A.	sister chromatids line up on equator
____	2.	nucleus	B.	connects two sister chromatids
____	3.	S phase	C.	site of ATP synthesis
____	4.	centromere	D.	structures within the cell
____	5.	metaphase	E.	sister chromatids split into chromosomes
____	6.	ribosome	F.	outer boundary of a cell
____	7.	nucleolus	G.	group of organs related by function
____	8.	cell	H.	ribosomes attach to its outer surface
____	9.	centriole	I.	period of cell division
____	10.	RER	J.	chromosomes reach cell poles
____	11.	prophase	K.	DNA replication takes place
____	12.	mitochondrion	L.	time of cell growth and function
____	13.	lysosome	M.	involved in lipid synthesis
____	14.	SER	N.	cell divides into two daughter cells
____	15.	tissue	O.	packages proteins for secretion
____	16.	cytokinesis	P.	organizes spindle fibers
____	17.	organ	Q.	houses DNA
____	18.	organelles	R.	nuclear envelope breaks down, spindle fibers form
____	19.	cytosol	S.	an area found inside the nucleus
____	20.	system	T.	smallest unit that is alive
____	21.	telophase	U.	group of identical cells
____	22.	Golgi apparatus	V.	organelles are suspended in this semi-solid fluid
____	23.	mitosis	W.	contains digestive enzymes
____	24.	plasma membrane	X.	composed of two or more tissues
____	25.	anaphase	Y.	site of mRNA translation

Review: Crossword Puzzle

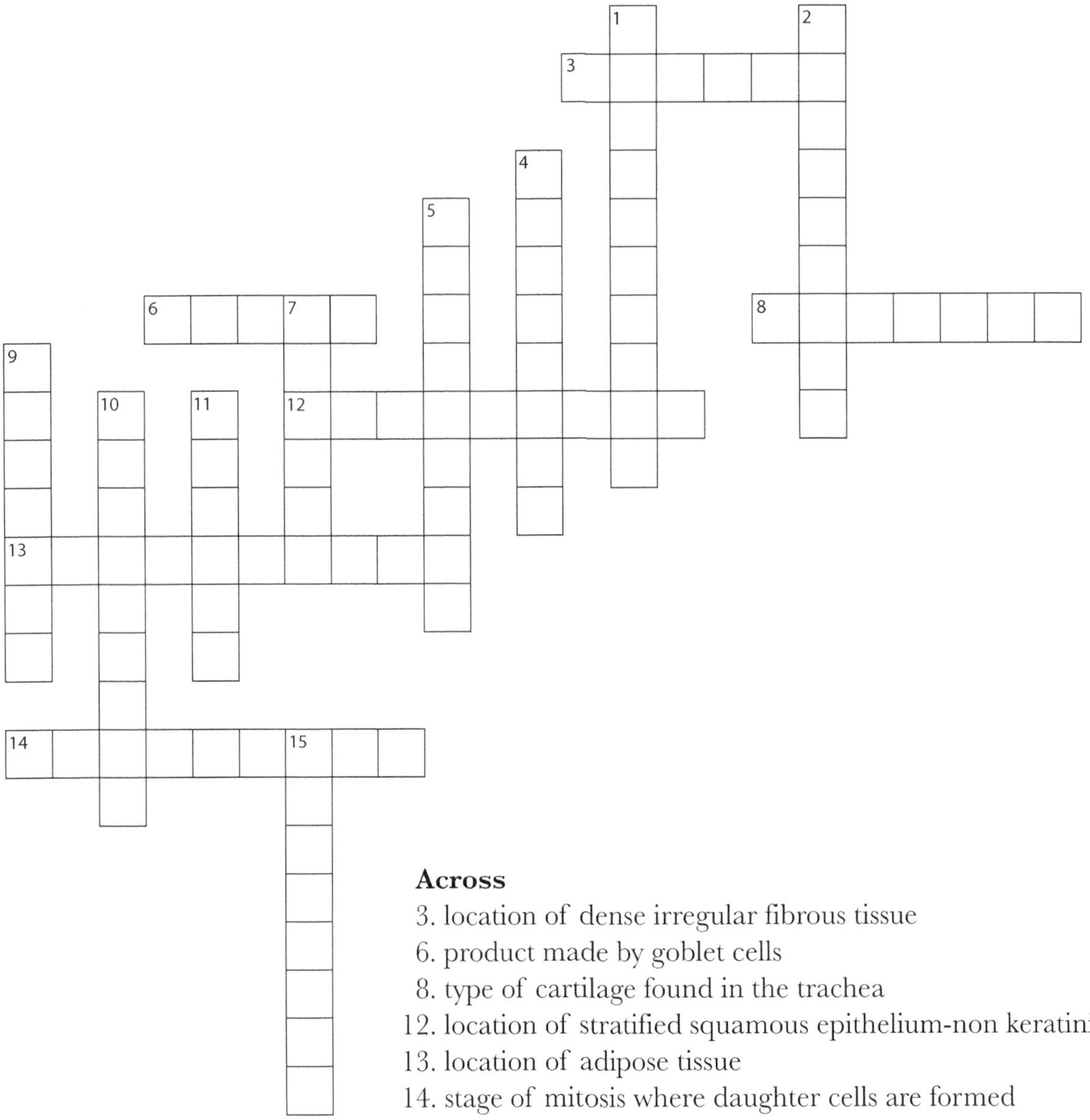

Across

3. location of dense irregular fibrous tissue
6. product made by goblet cells
8. type of cartilage found in the trachea
12. location of stratified squamous epithelium-non keratinized
13. location of adipose tissue
14. stage of mitosis where daughter cells are formed

Down

1. contains purkinje cells
2. type of cell found in the bone
4. stage of mitosis where the chromatin condenses
5. stage of mitosis where chromosomes align in the middle of the cell
7. contains transitional epithelial cells
9. contains pseudostratified ciliated columnar epithelial cells
10. lamellated ______
11. location of dense regular fibrous tissue
15. stage of mitosis where chromosomes separate

Chapter 2

Bones, Osseous Tissue, and the Skeleton

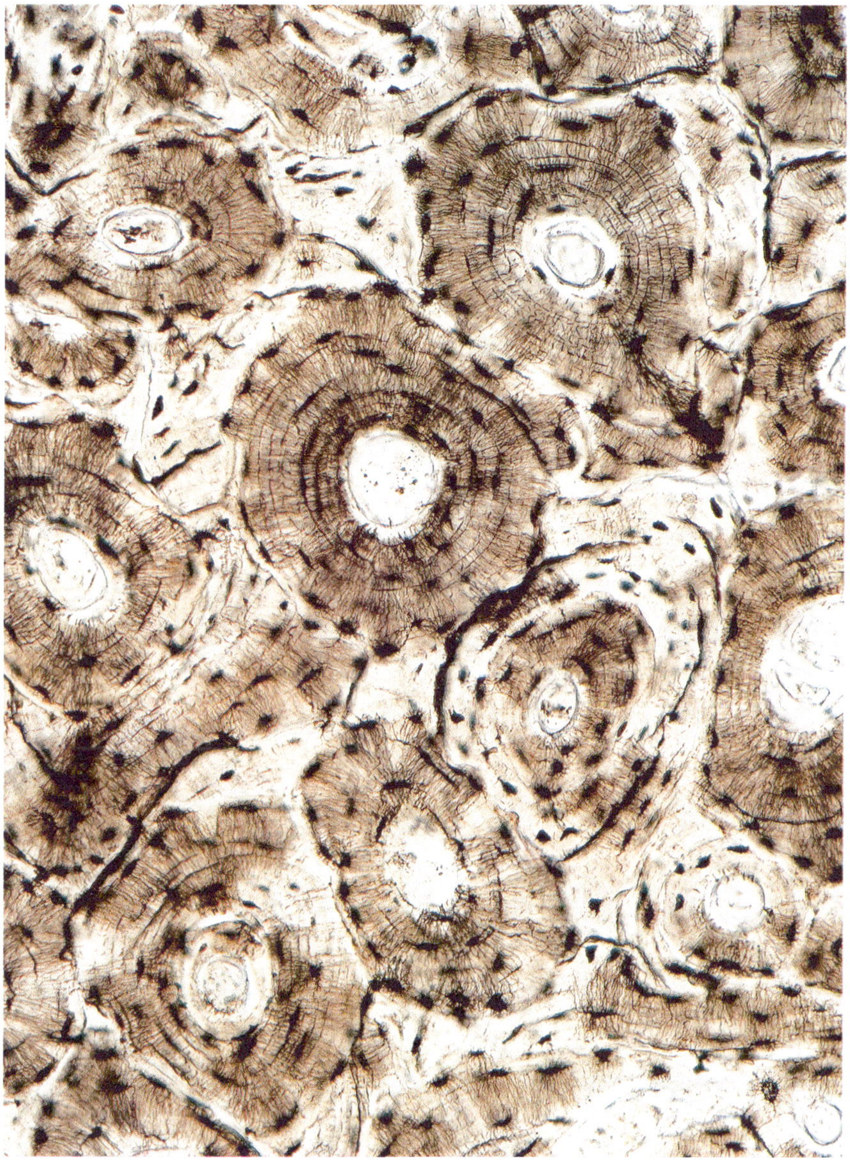

Compact Bone Showing Osteons in Shaft of Bone (Brightfield, x430)

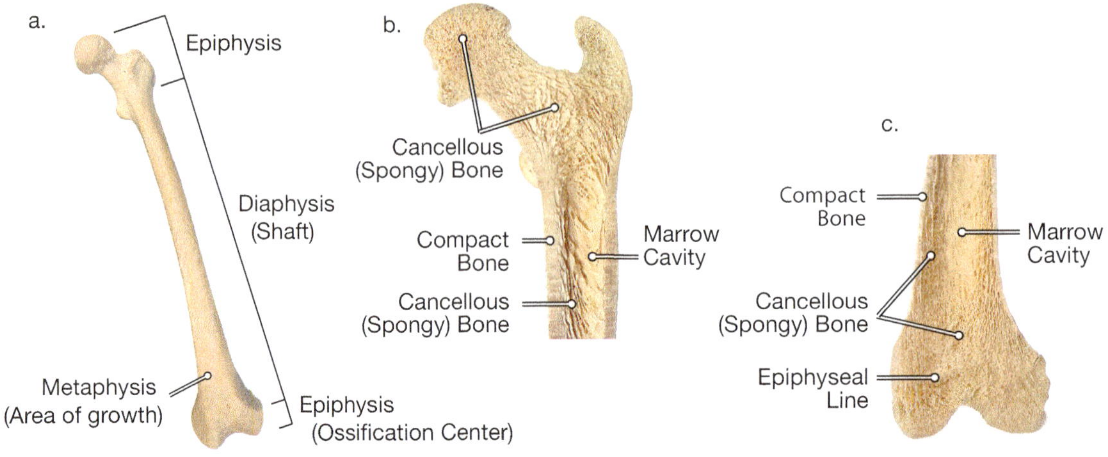

Figure 2.1: Long Bone. a. Femur. b. Interior, proximal view. c. Interior, distal view.
© David G. Ward.

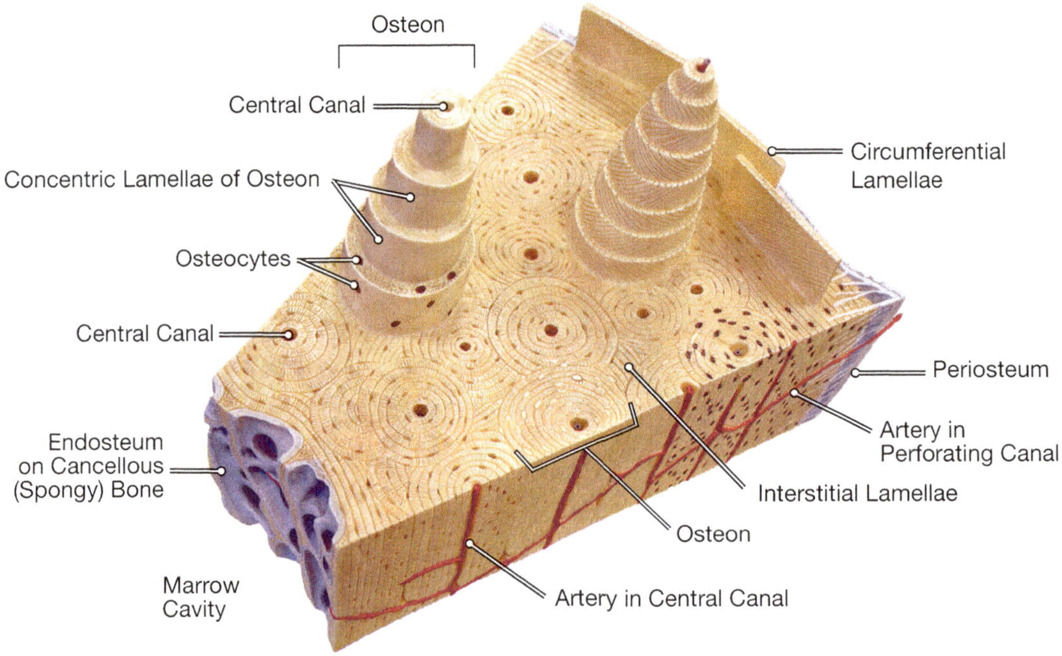

Figure 2.2: Compact bone model.
© David G. Ward.

Compact and Spongy Bone

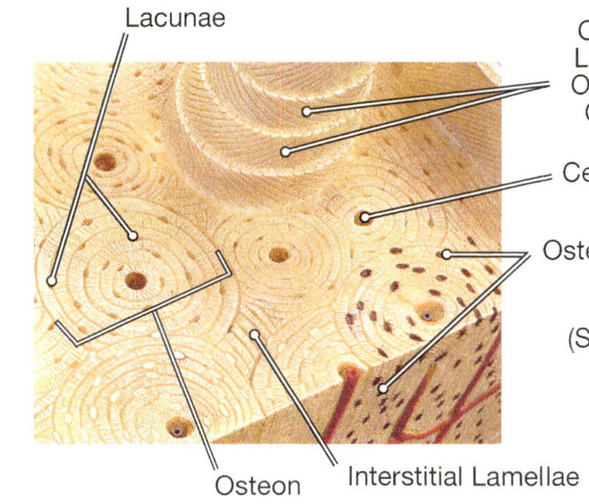

Figure 2.3: Compact bone.

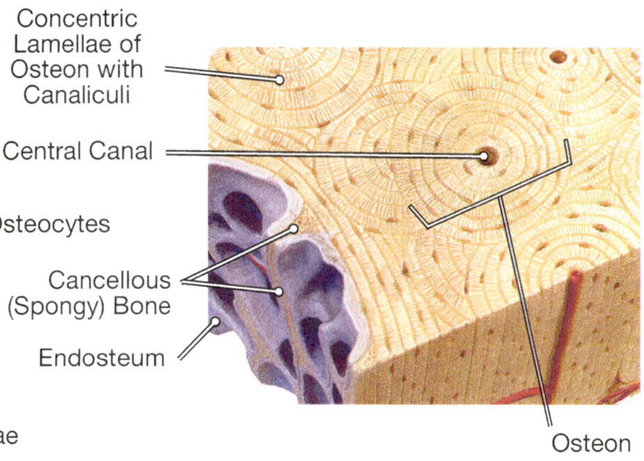

Figure 2.4: Compact and cancellous (spongy) bone.

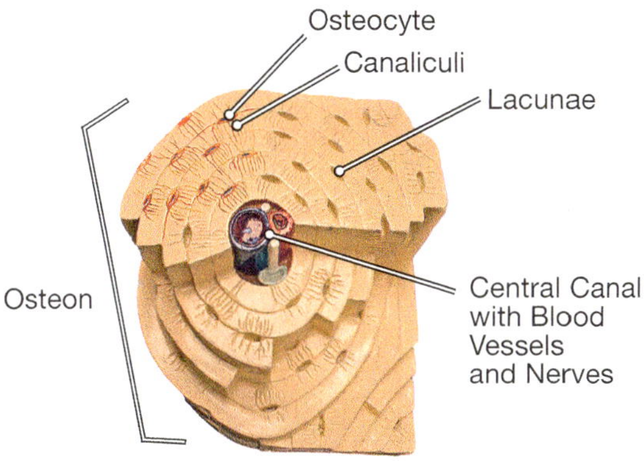

Figure 2.5: Compact bone.

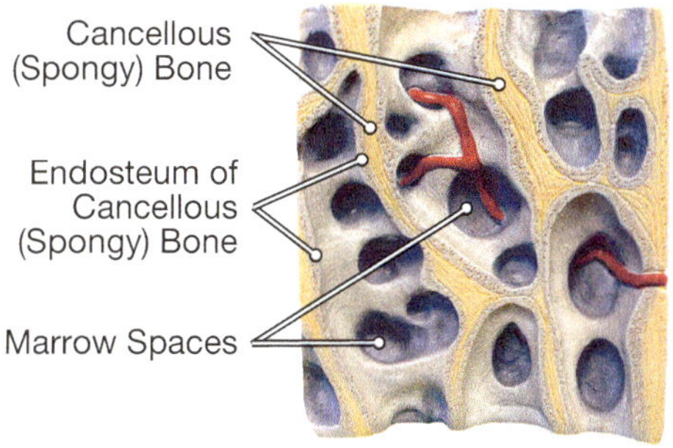

Figure 2.6: Cancellous (spongy) bone.

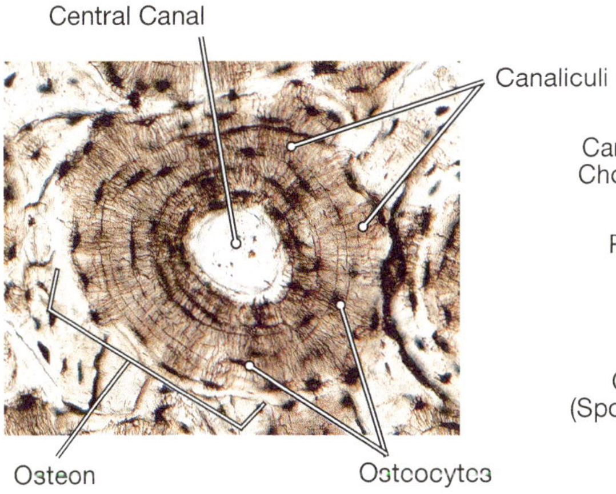

Figure 2.7: Compact bone.

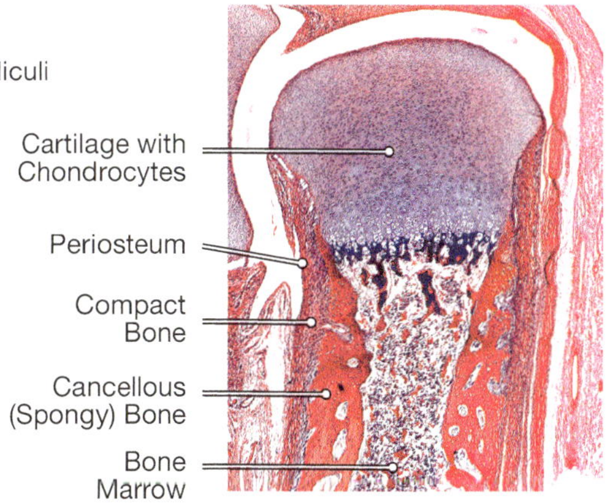

Figure 2.8: Developing bone.

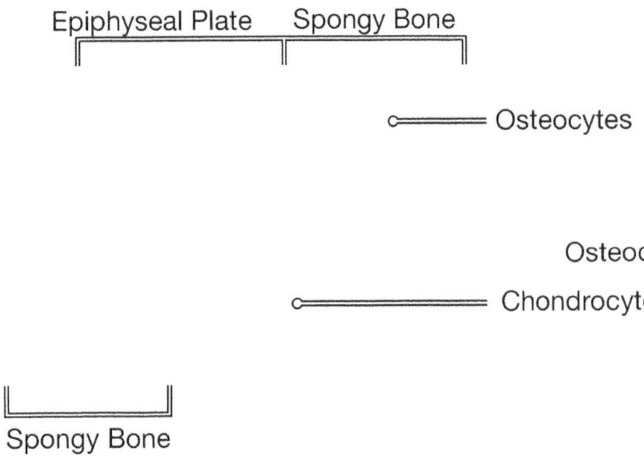

Epiphysis and spongy bone.
© David G. Ward.

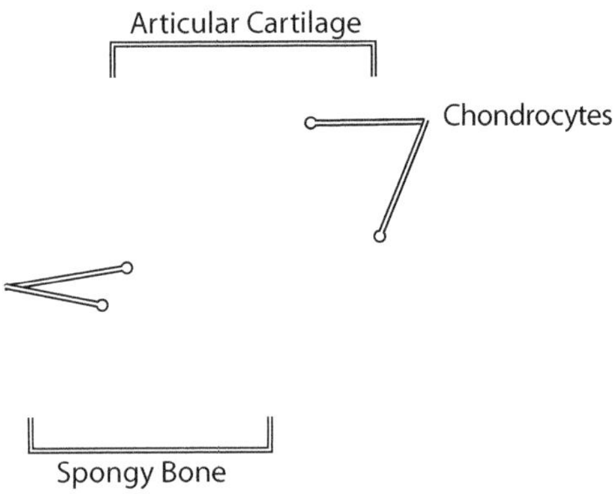

Epiphysis and spongy bone.
© David G. Ward.

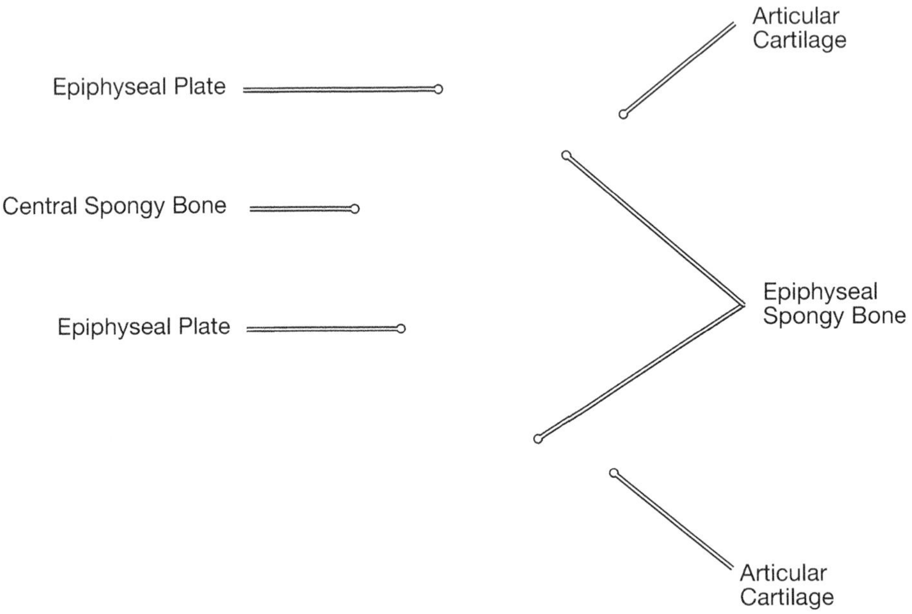

Epiphyseal plate.
© David G. Ward.

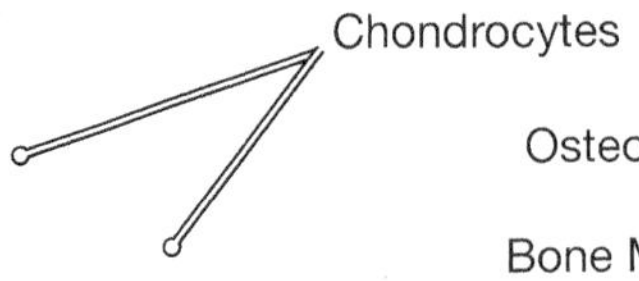

Epiphyseal plate.
© David G. Ward.

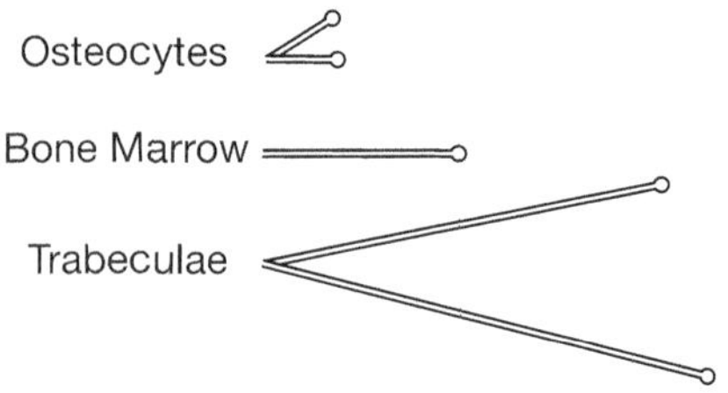

Epiphyseal spongy bone.
© David G. Ward.

Spongy Bone

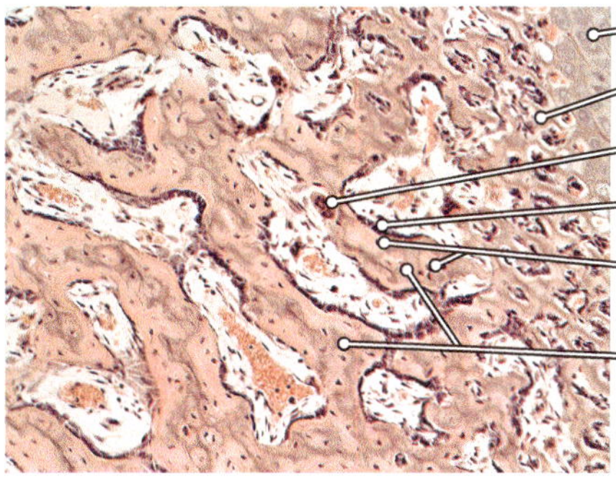

Figure 2.14: Central spongy bone.

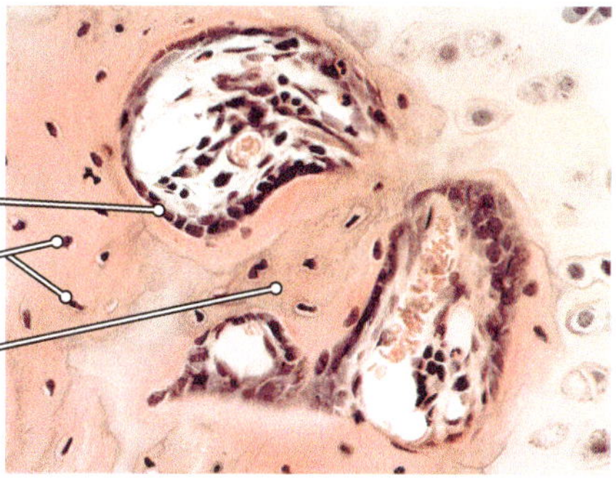

Figure 2.15: Epiphyseal spongy bone.

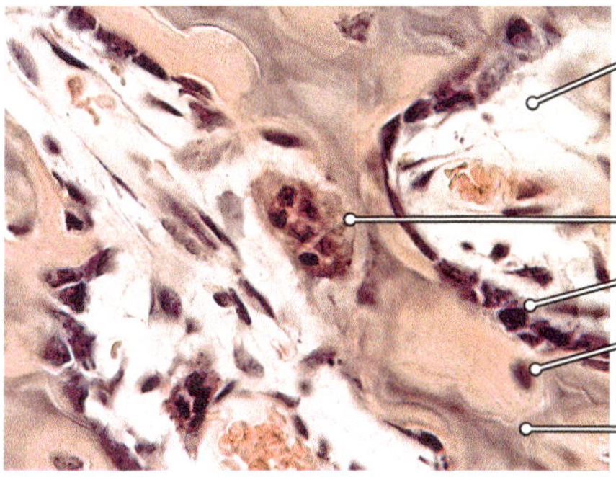

Figure 2.16: Central spongy bone.

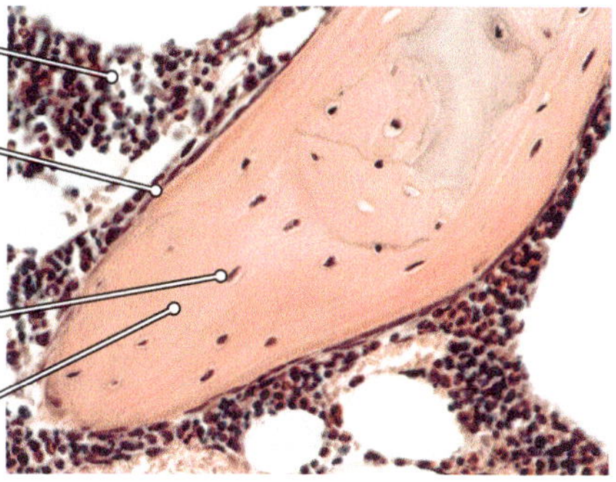

Figure 2.17: Epiphyseal spongy bone.

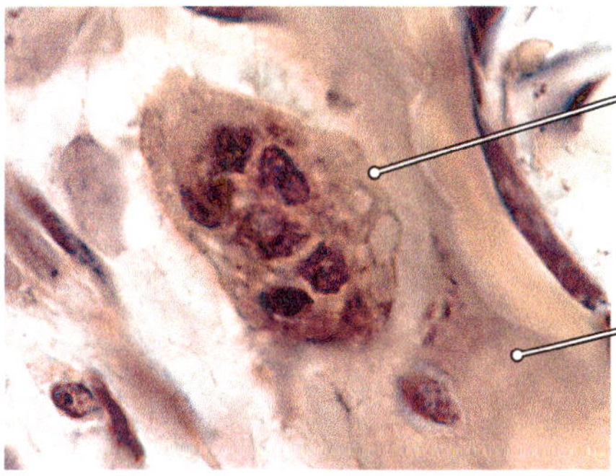

Figure 2.18: Central spongy bone.

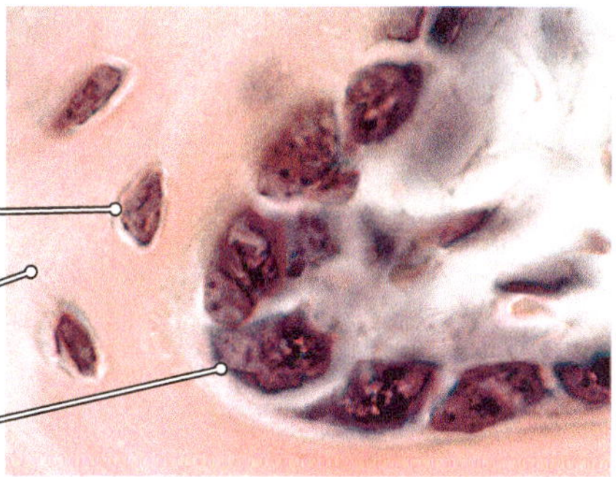

Figure 2.19: Epiphyseal spongy bone.

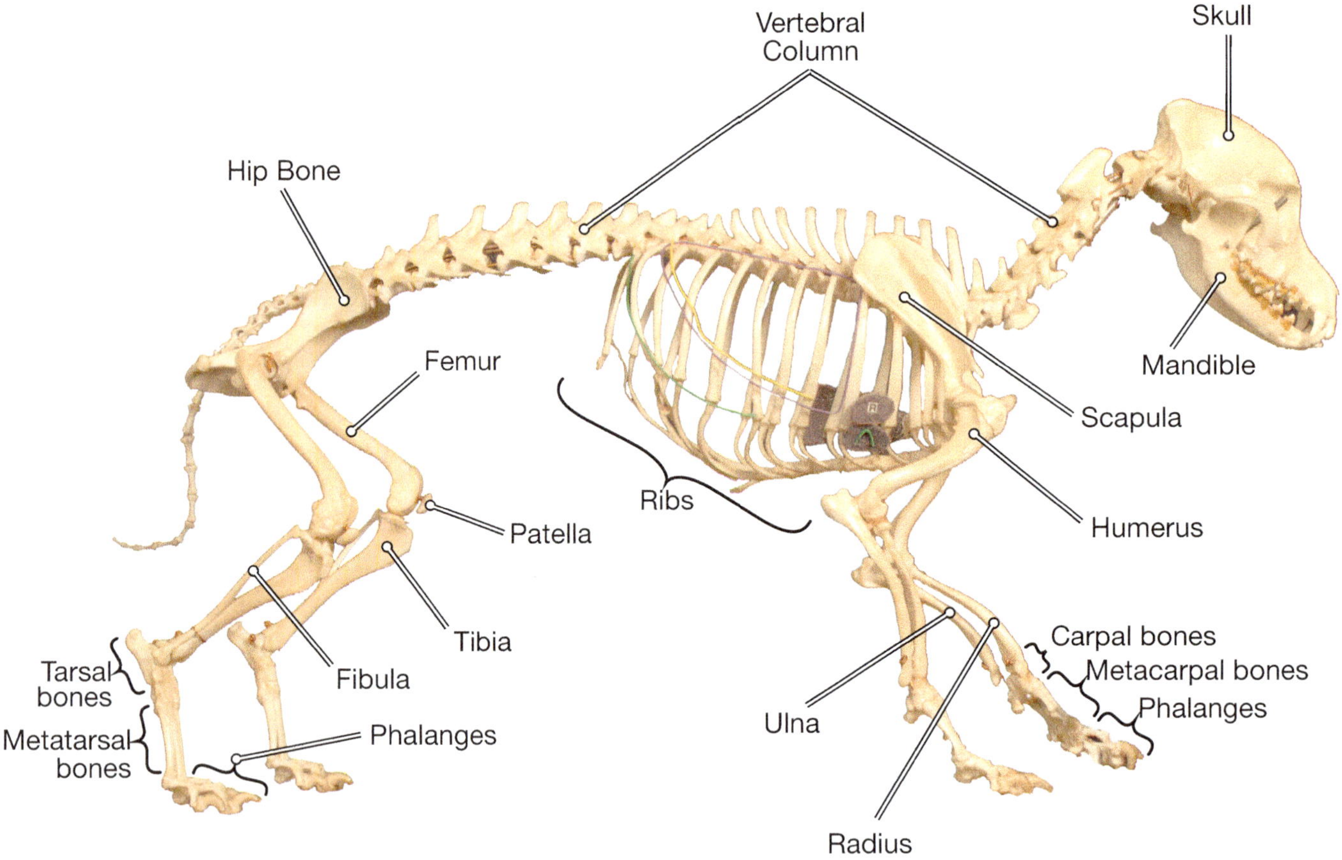

Figure 2.20: Skeleton, overview.
© NCSU.

Axial Skeleton

Figure 2.21: Canine skeleton, axial skeleton highlighted.

Axial Skeleton
Skull

Figure 2.22: Canine skeleton, skull highlighted.
© NCSU.

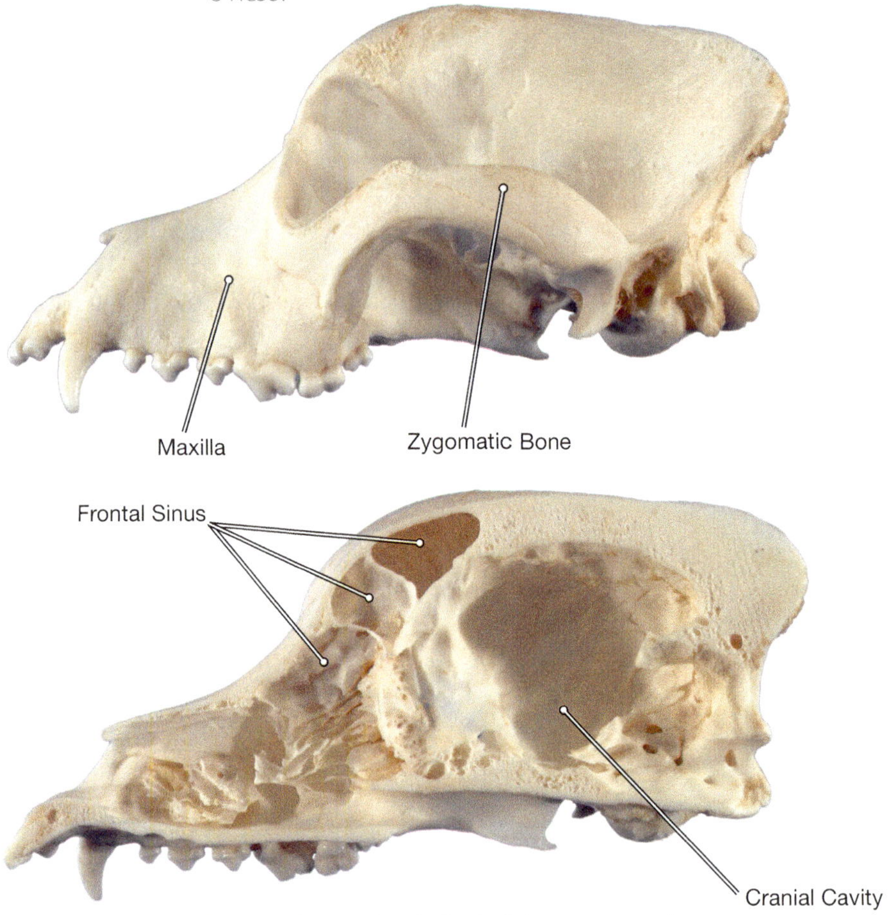

Figure 2.23: Mesaticephalic skull (most common).
© NCSU.

a.

b.

Figure 2.24: Skull. a. Dolichocephalic skull (Greyhound, Collie), ventral view with mandible (lower jaw) in place. b. Brachycephalic skull (English Bulldog, Pekingnese, Pug, Boston Terrier), dorsal view with skull removed at frontal bone.

Axial Skeleton

Skull

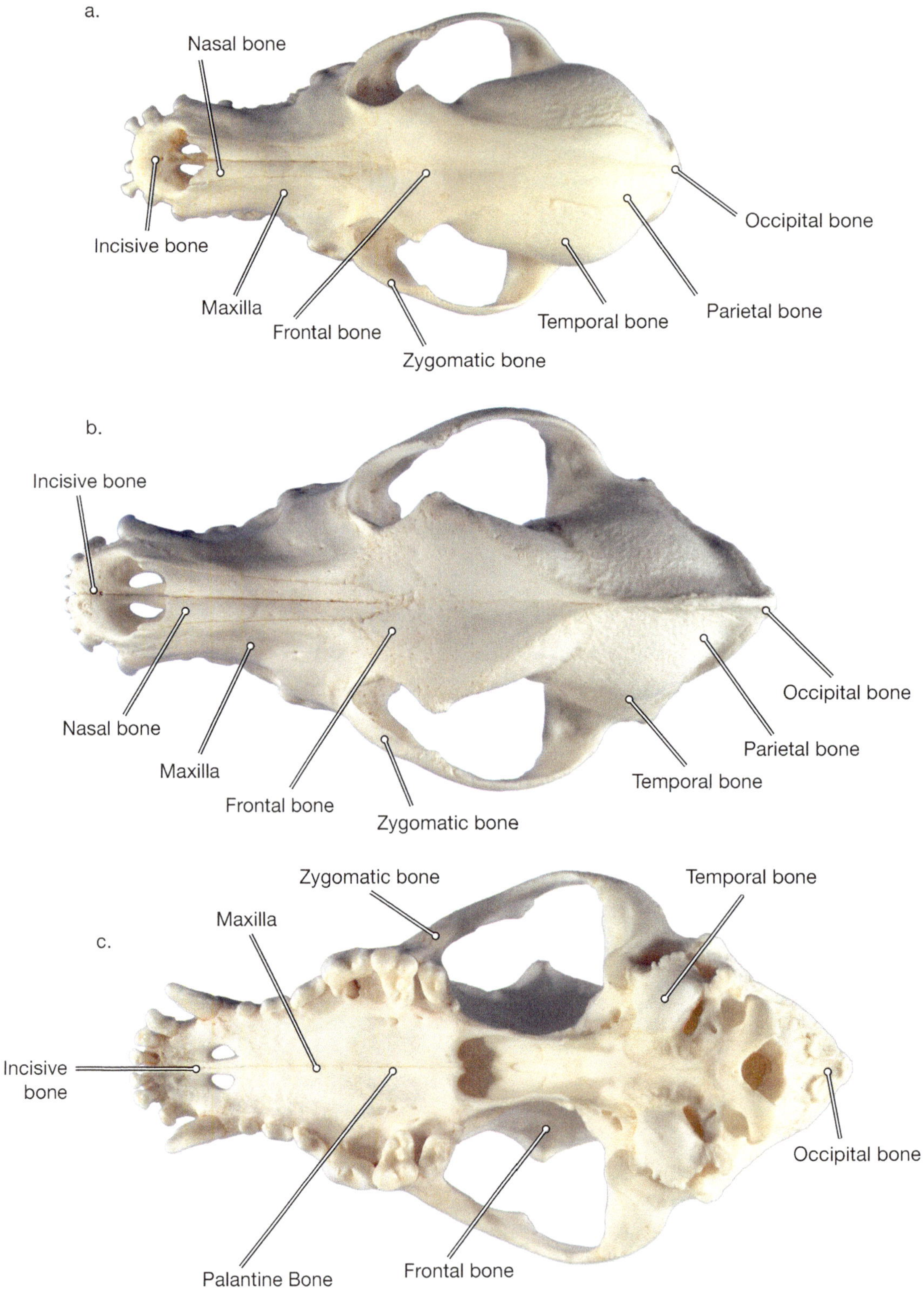

Figure 2.25: Skull. a. Small breed (Terrier), dorsal view. b. Large breed (Golden Retriever), dorsal view. c. Large breed (Golden Retriever), ventral view.

Axial Skeleton

Skull: Exploded

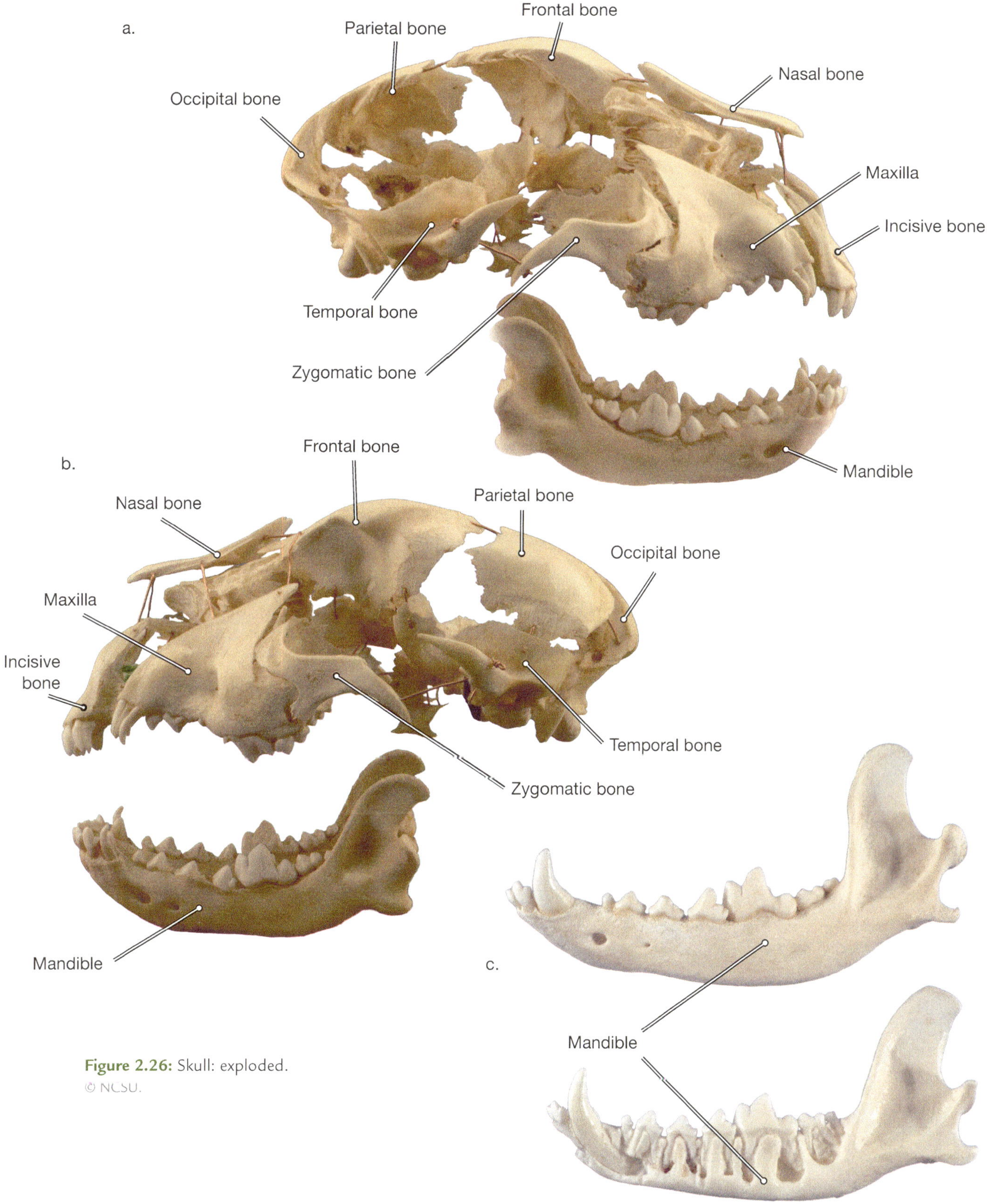

Figure 2.26: Skull: exploded.

Axial Skeleton
Vertebrae

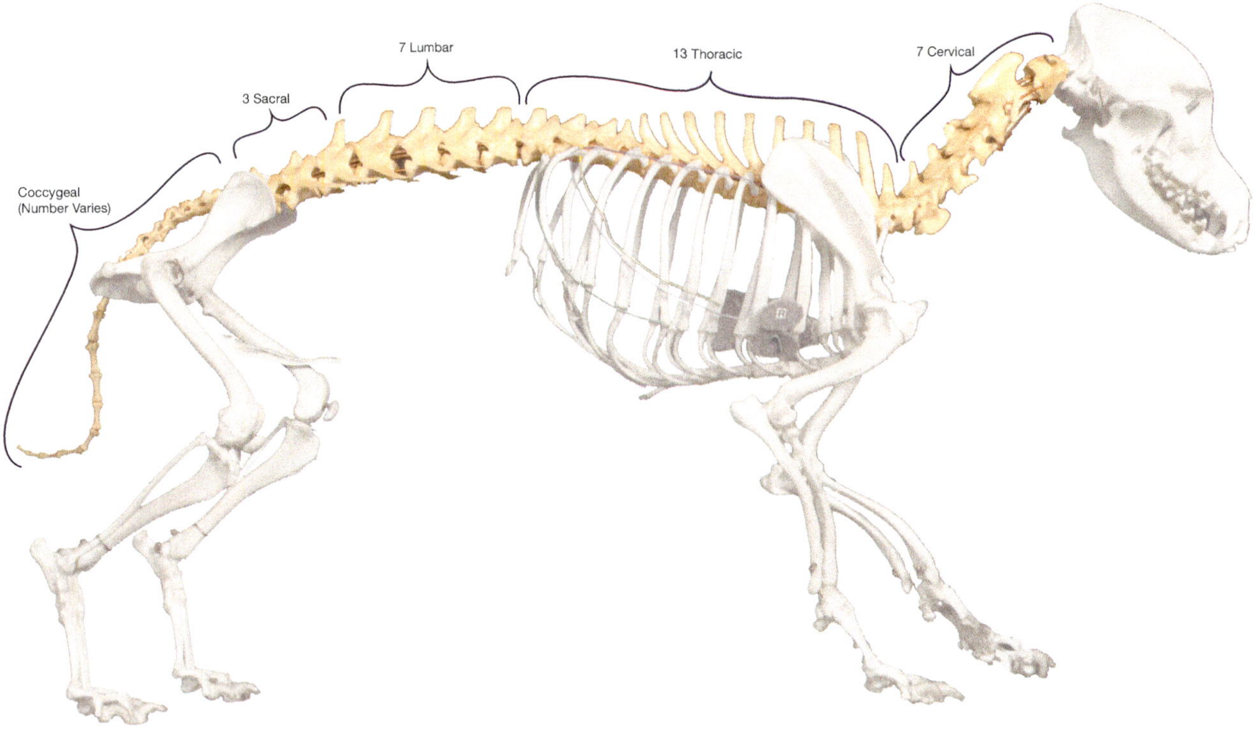

Figure 2.27: Canine skeleton, vertebrae highlighted.
© NCSU.

Axial Skeleton

Vertebrae: Cervical

Figure 2.28: Canine skeleton, cervical highlighted.

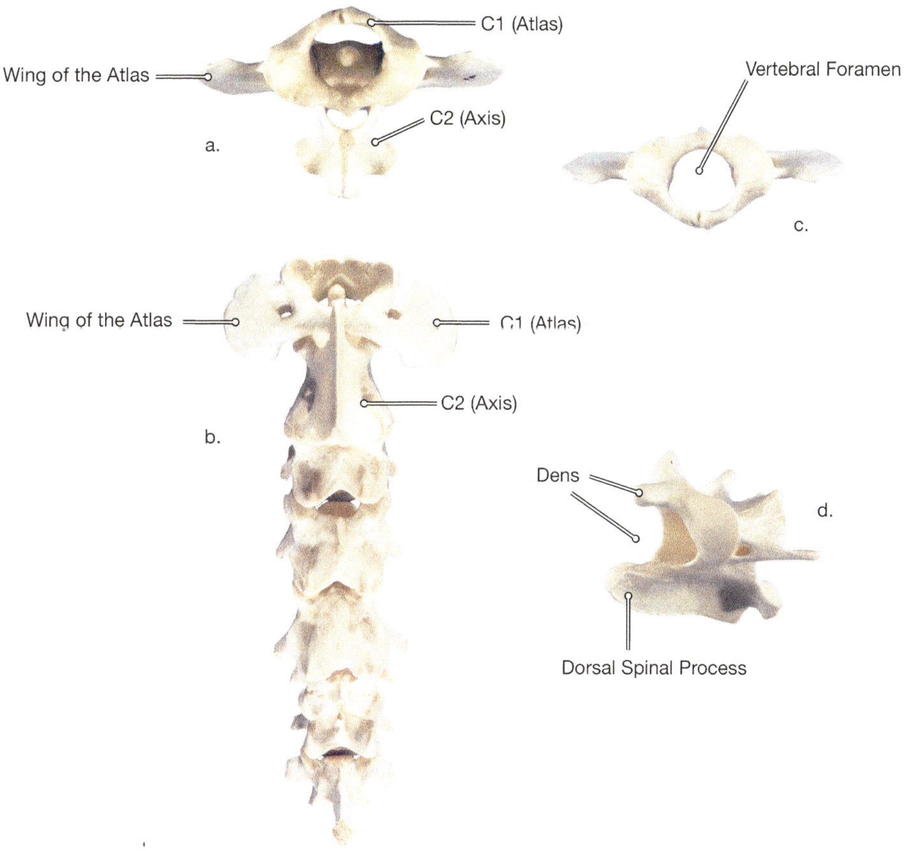

Figure 2.29: Vertebrae: cervical. a. Vertebrae, ventral view. b. Cervical, dorsal view, c. Atlas, cutaway view, d. Axis, lateral view.

Axial Skeleton

Vertebrae: Thoracic

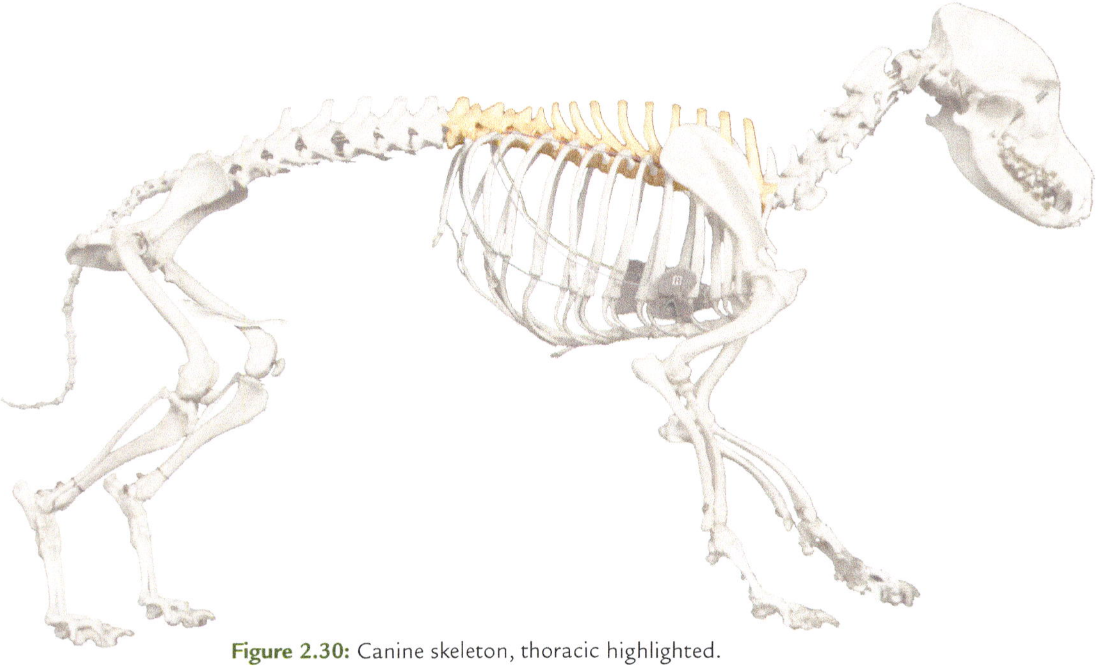

Figure 2.30: Canine skeleton, thoracic highlighted.

a.

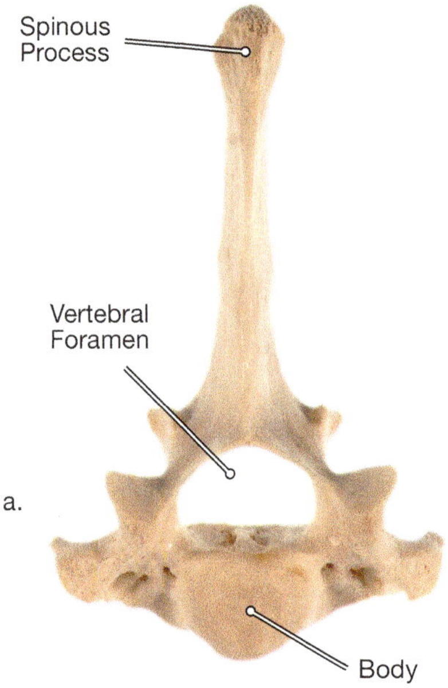

(Front)

b.

Figure 2.31: Vertebrae: thoracic.

Axial Skeleton
Vertebrae: Lumbar

Figure 2.32: Canine skeleton, lumbar highlighted.

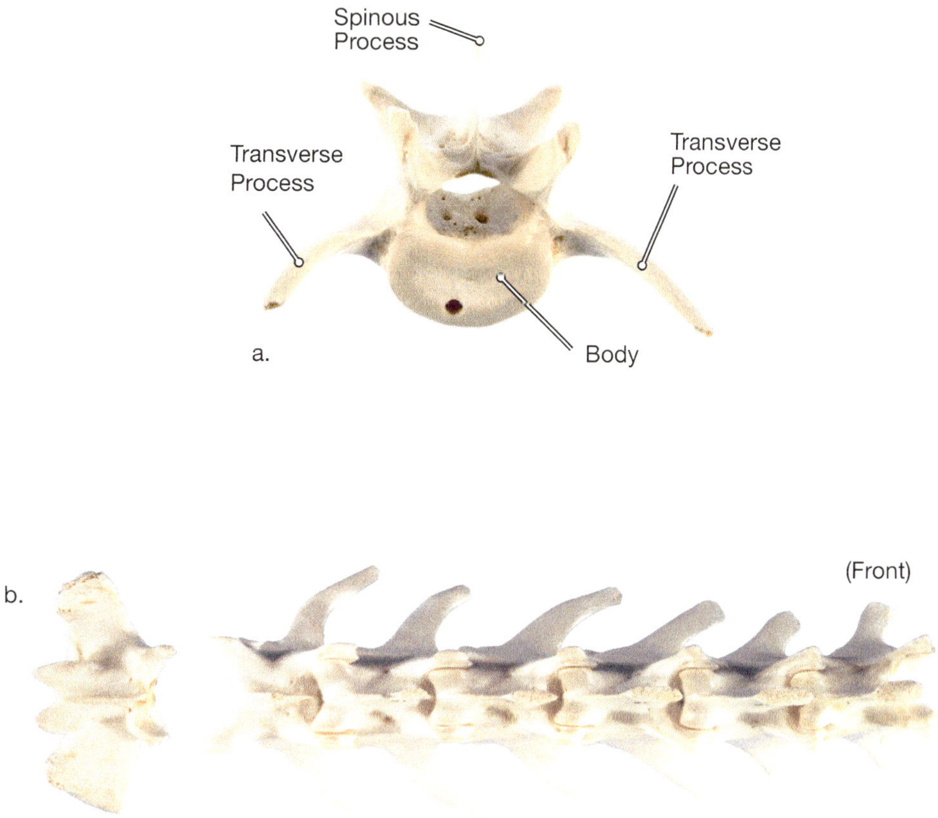

Figure 2.33: Vertebrae: lumbar.

Figure 2.34: Canine skeleton, coccygeal highlighted.
© NCSU.

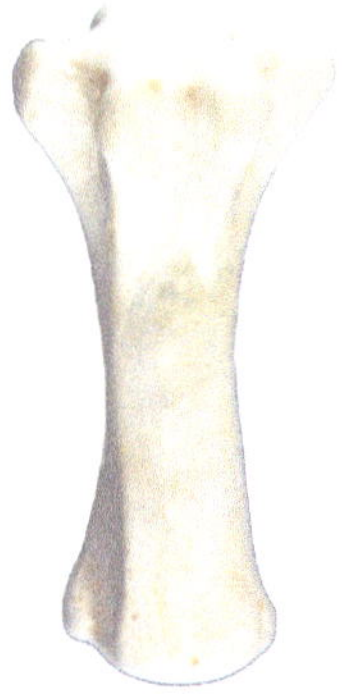

Figure 2.35: Coccygeal vertebra #8.
© NCSU.

Axial Skeleton
Sternum

Figure 2.36: Canine skeleton, sternum highlighted.
© NCSU.

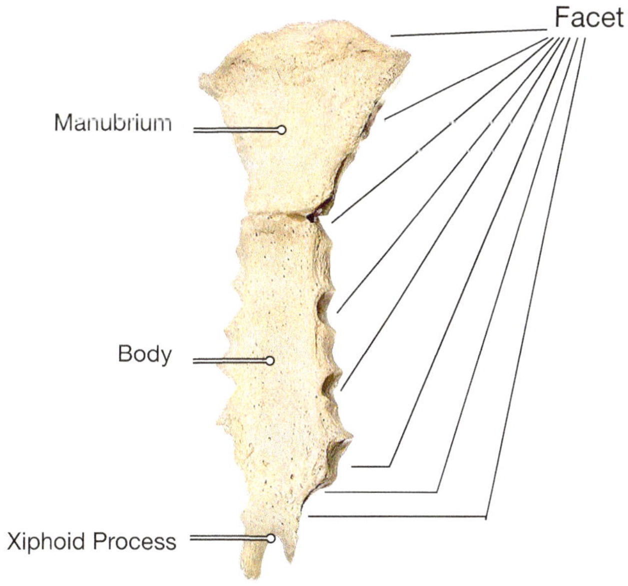

Figure 2.37: Sternum.
© NCSU.

Axial Skeleton
Ribs

Figure 2.38: Canine skeleton, ribs highlighted.
© NCSU.

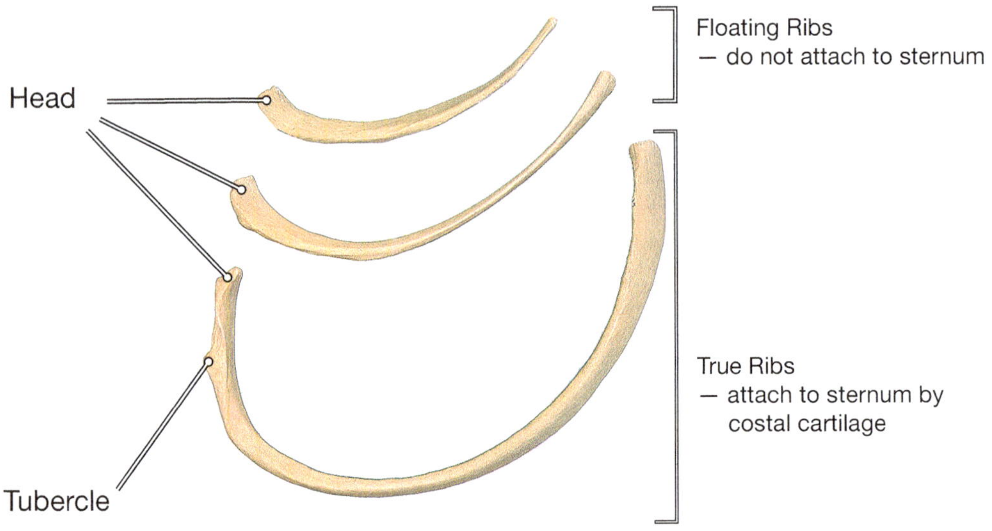

Figure 2.39: Ribs.
© NCSU.

Appendicular Skeleton

Figure 2.40: Canine skeleton, appendicular skeleton highlighted.
© NCSU.

Thoracic Limb: Scapula

Figure 2.41: Canine skeleton, scapula highlighted.

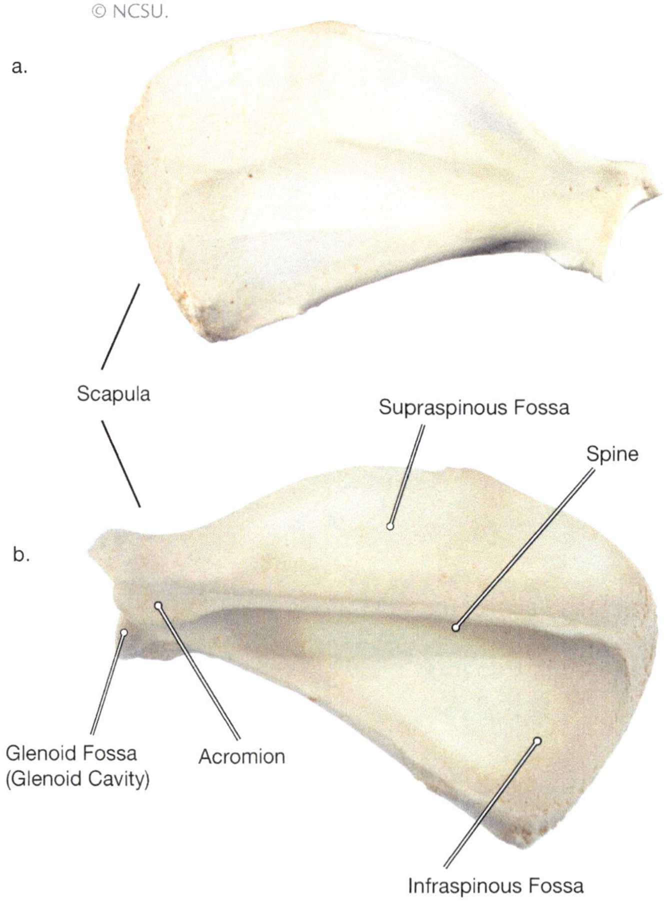

Figure 2.42: Scapula. a. Medial view. b. Lateral view.

Appendicular Skeleton

Thoracic Limb: Humerus

Figure 2.43: Canine skeleton, humerus highlighted.
© NCSU.

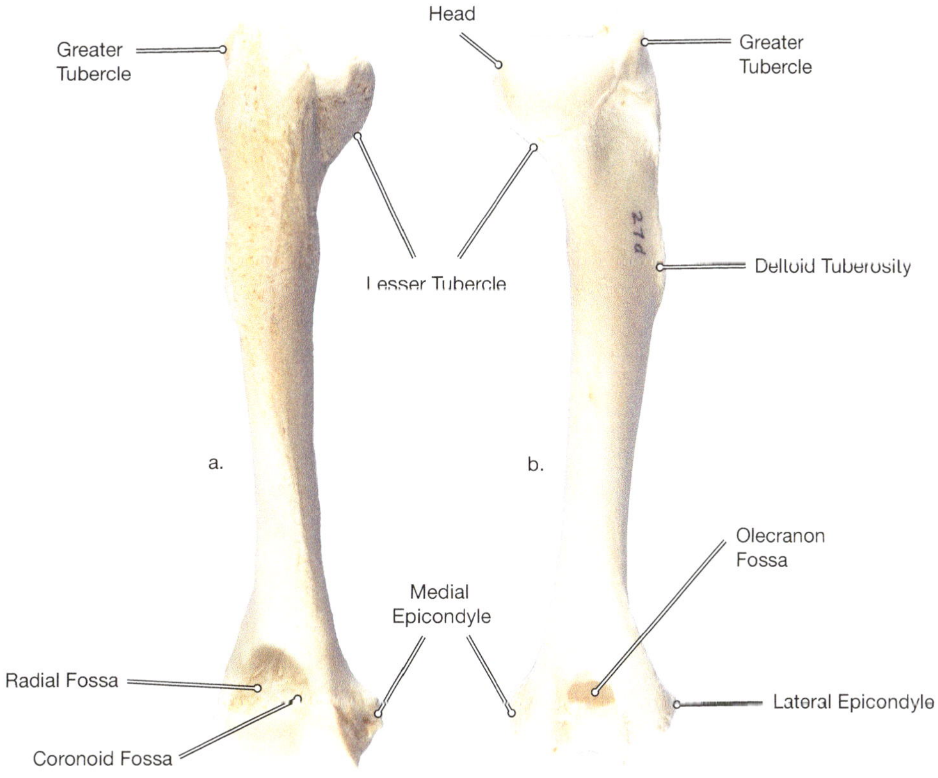

Figure 2.44: Humerus. a. Anterior view (cranial). a. Posterior view (caudal).
© NCSU.

Appendicular Skeleton
Thoracic Limb: Ulna

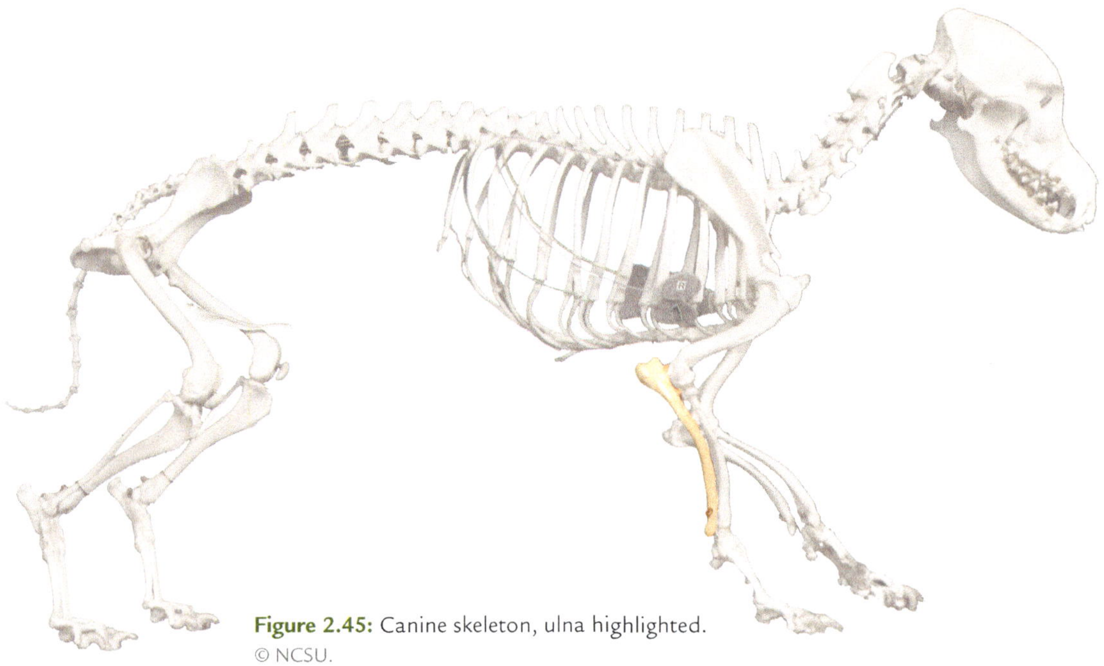

Figure 2.45: Canine skeleton, ulna highlighted.
© NCSU.

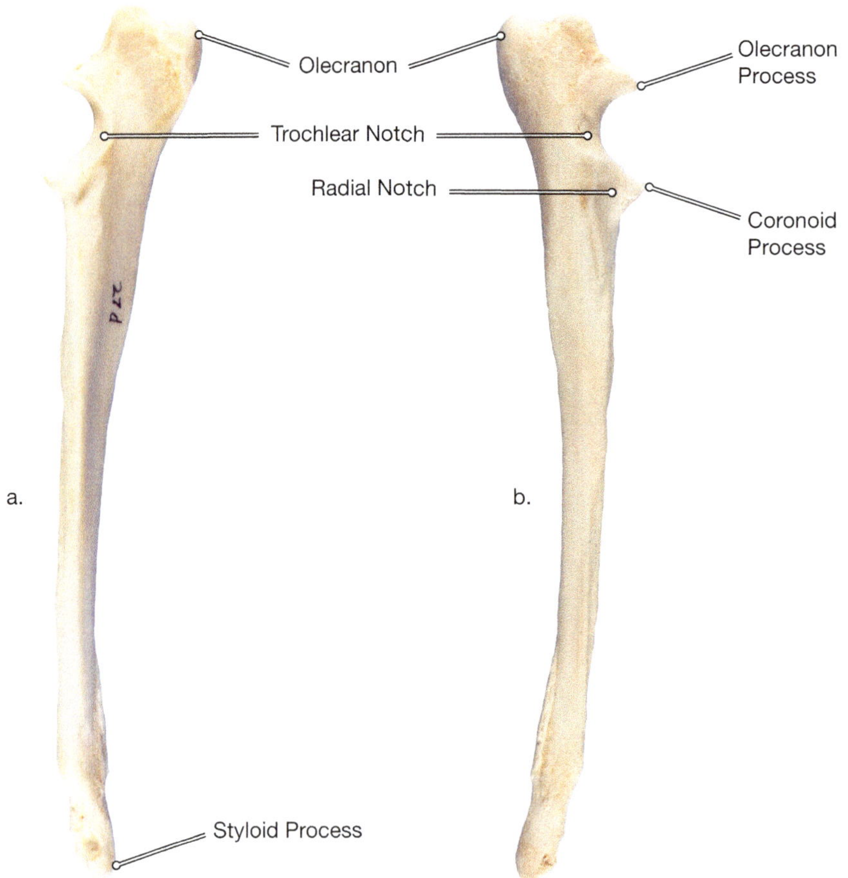

Figure 2.46: Ulna. a. Medial view. b. Lateral view.
© NCSU.

Appendicular Skeleton
Thoracic Limb: Radius

Figure 2.47: Canine skeleton, radius highlighted.

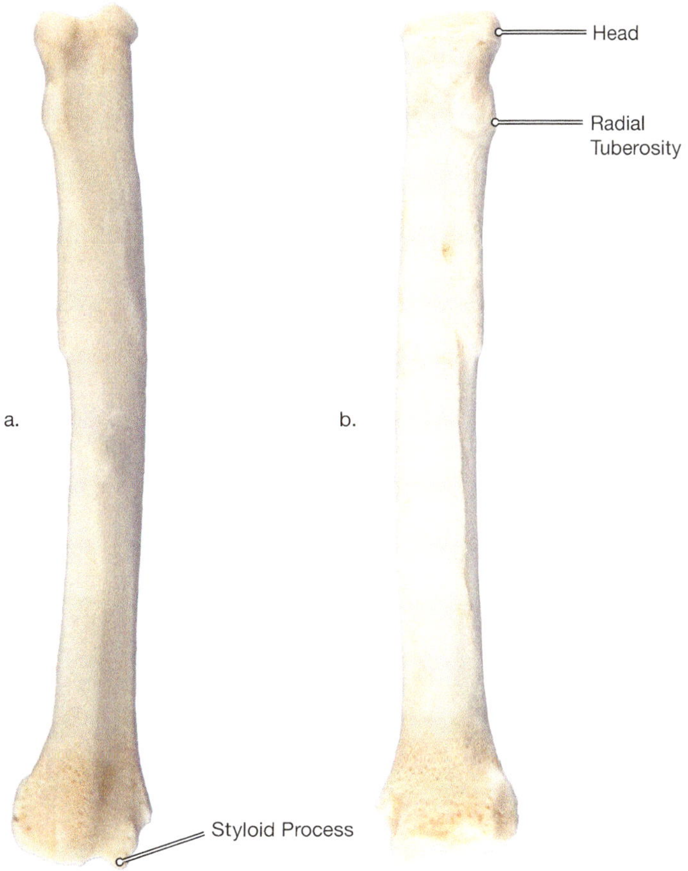

Figure 2.48: Radius. A. Posterior view. B. Anterior view.

Distal Limbs

Figure 2.49: Canine skeleton, distal limbs highlighted.
© NCSU.

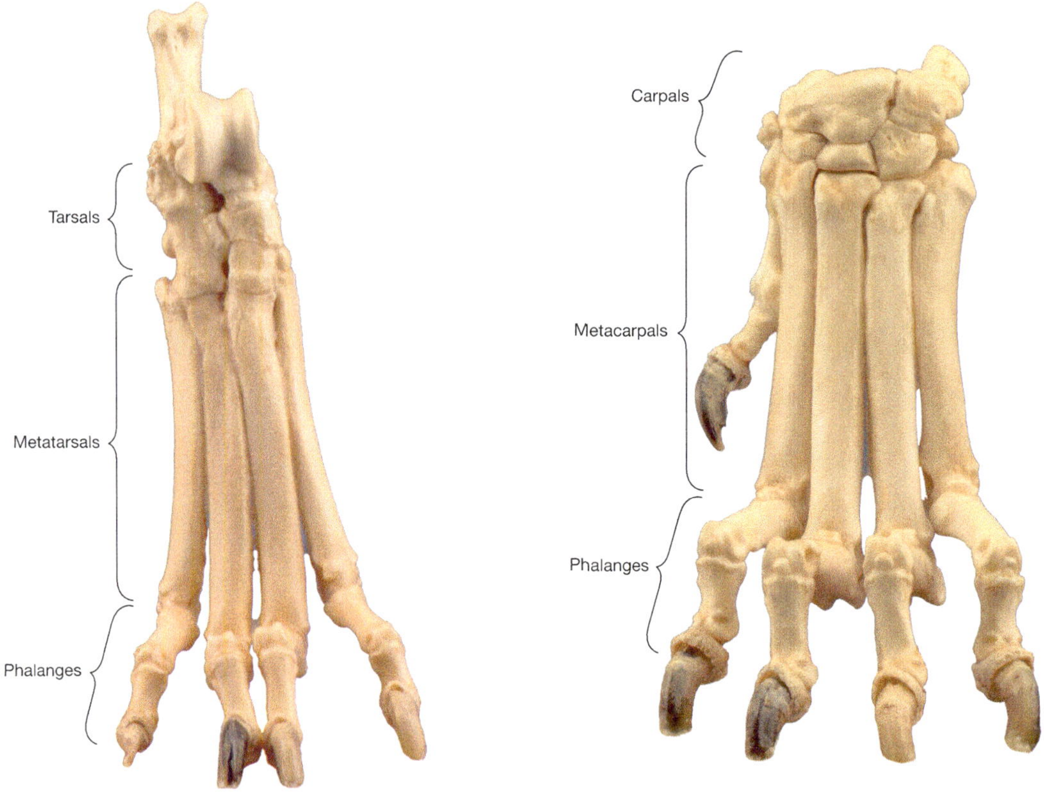

Figure 2.50: Distal limbs. a. Rear paw (Pes). b. Front paw (Manus).
© NCSU.

Appendicular Skeleton
Pelvis

Figure 2.51: Canine skeleton, pelvis highlighted.

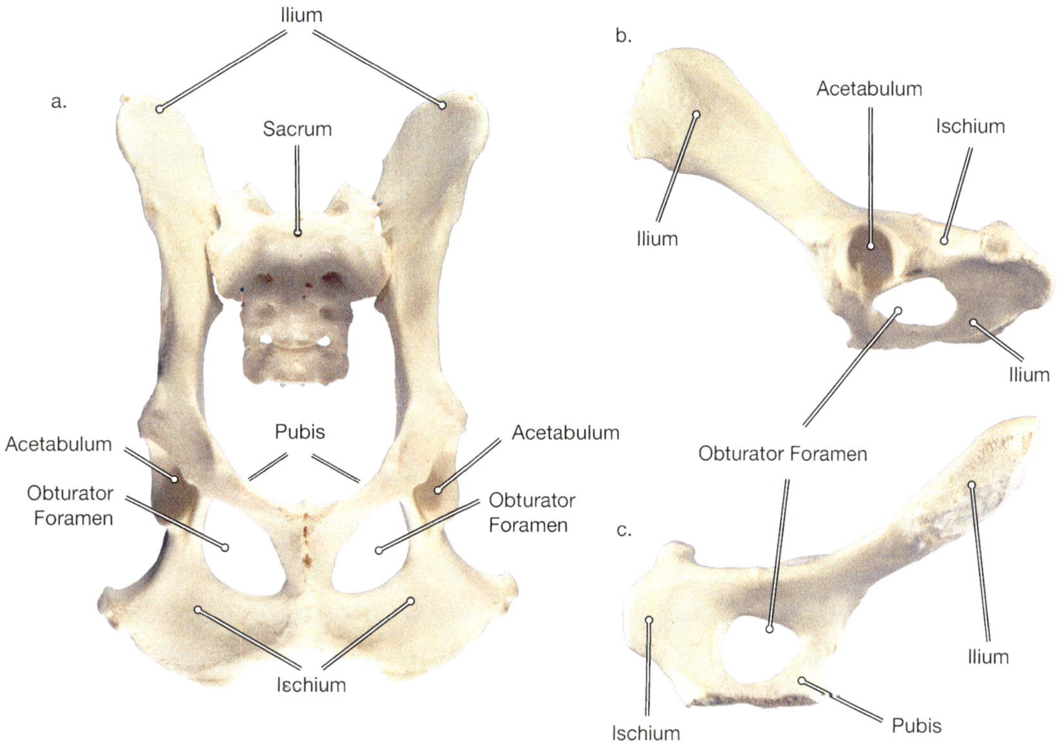

Figure 2.52: Pelvis, a. Ventral view, b. Pelvis (cutaway), lateral view, c. Pelvis (cutaway), medial view.

Appendicular Skeleton
Pelvic Limb: Femur

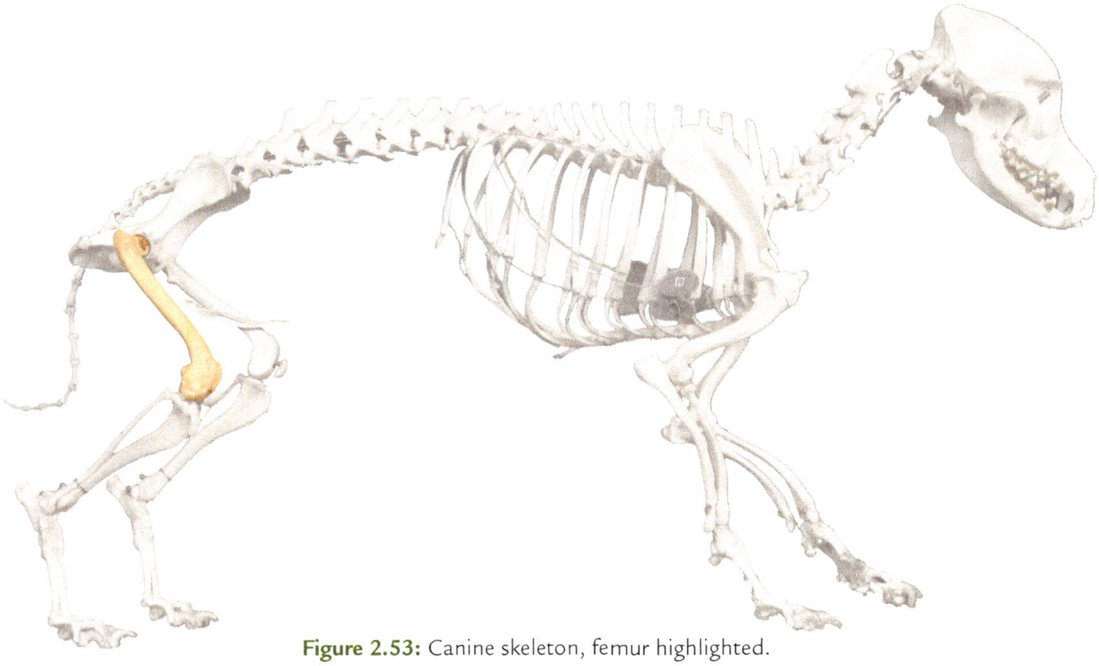

Figure 2.53: Canine skeleton, femur highlighted.

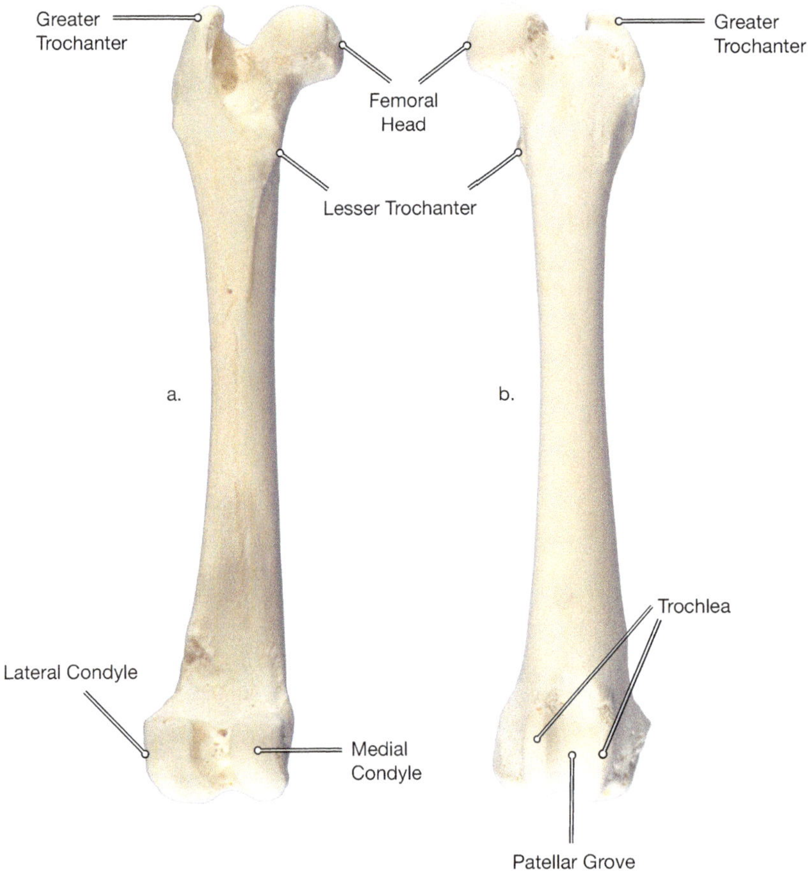

Figure 2.54: Femur. a. Posterior view. b. Anterior view.

Appendicular Skeleton
Pelvic Limb: Patella

Figure 2.55: Canine skeleton, patella highlighted.

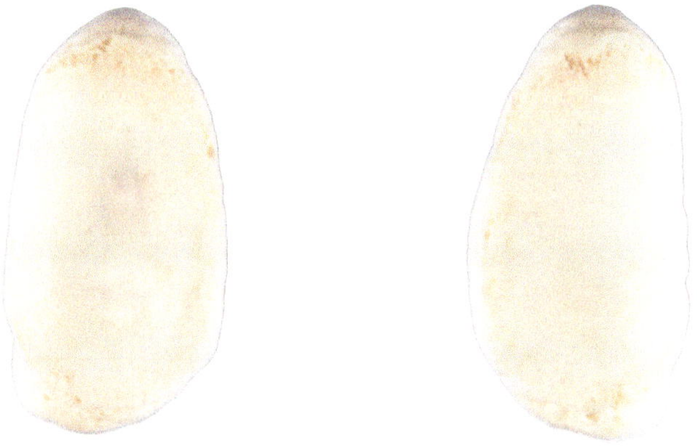

Figure 2.56: Patella.

Pelvic Limb: Tibia

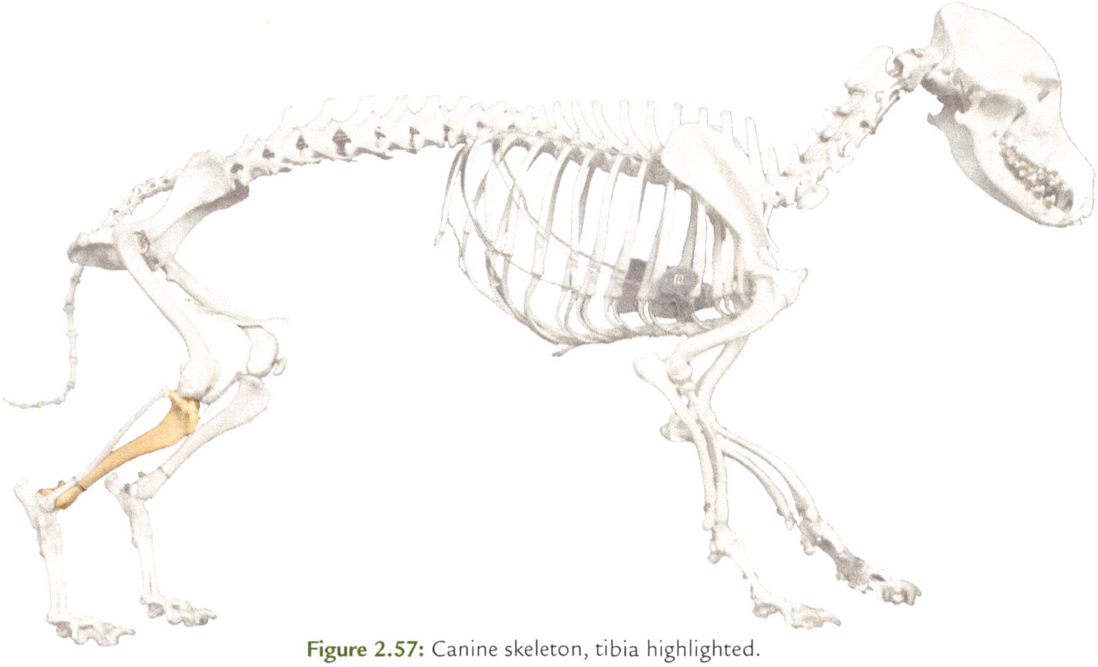

Figure 2.57: Canine skeleton, tibia highlighted.

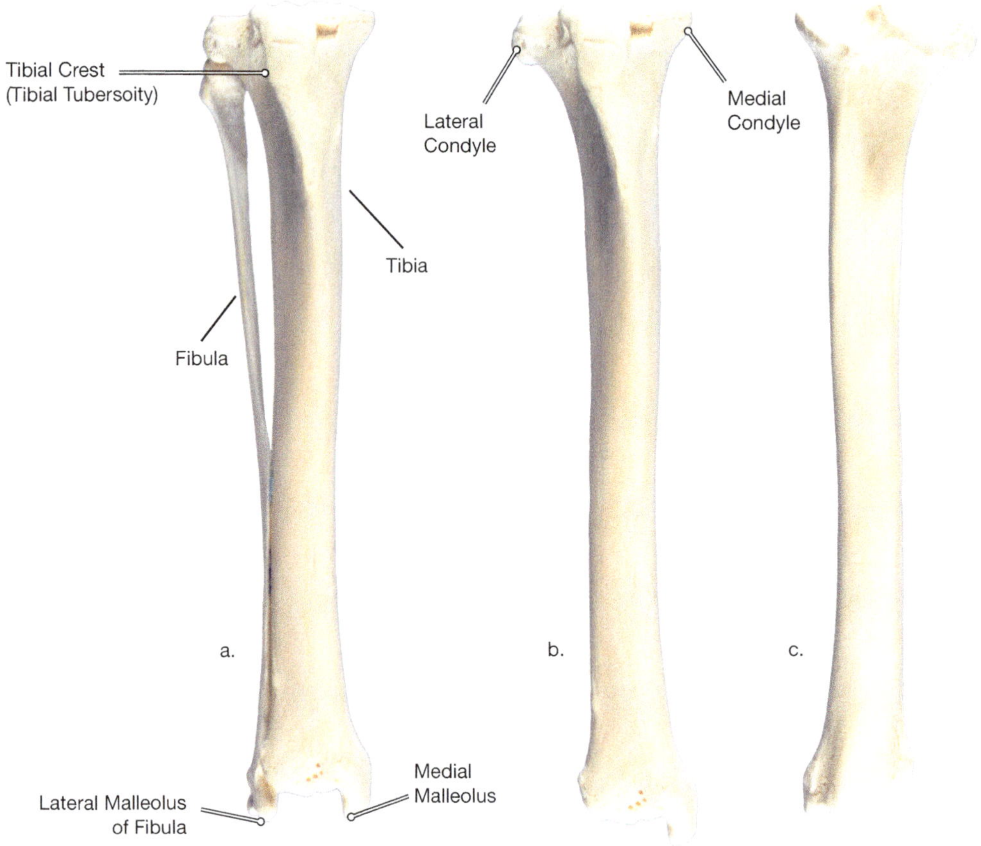

Figure 2.58: Tibia. a. Tibia with fibula, anterior view b. Tibia (fibula removed), anterior view. c. Tibia (fibula removed), posterior view.

Appendicular Skeleton
Pelvic Limb: Fibula

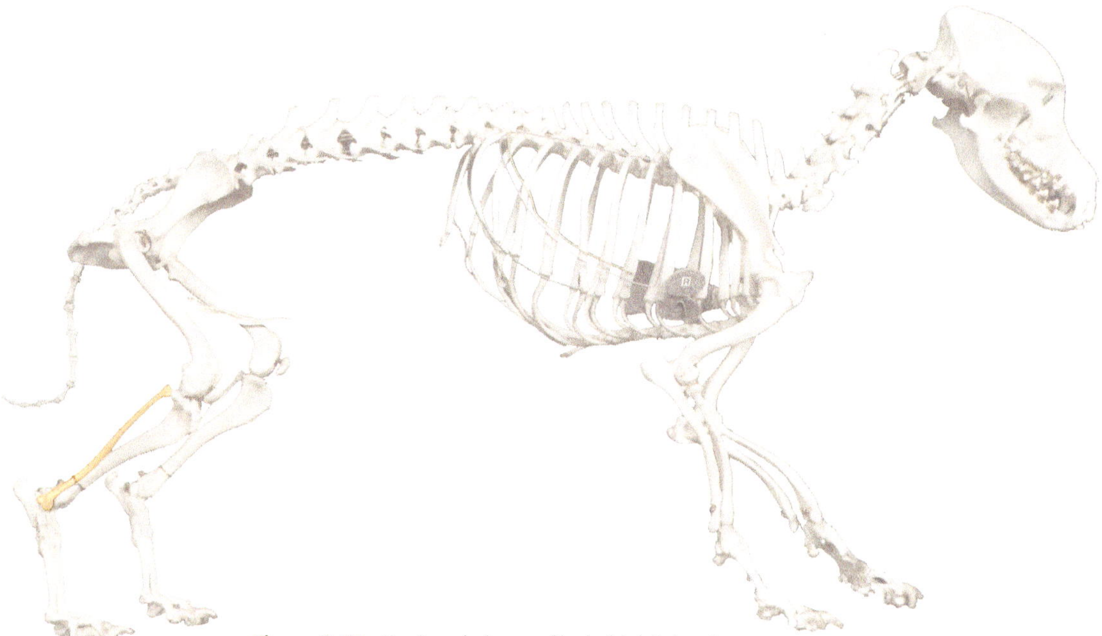

Figure 2.59: Canine skeleton, fibula highlighted.

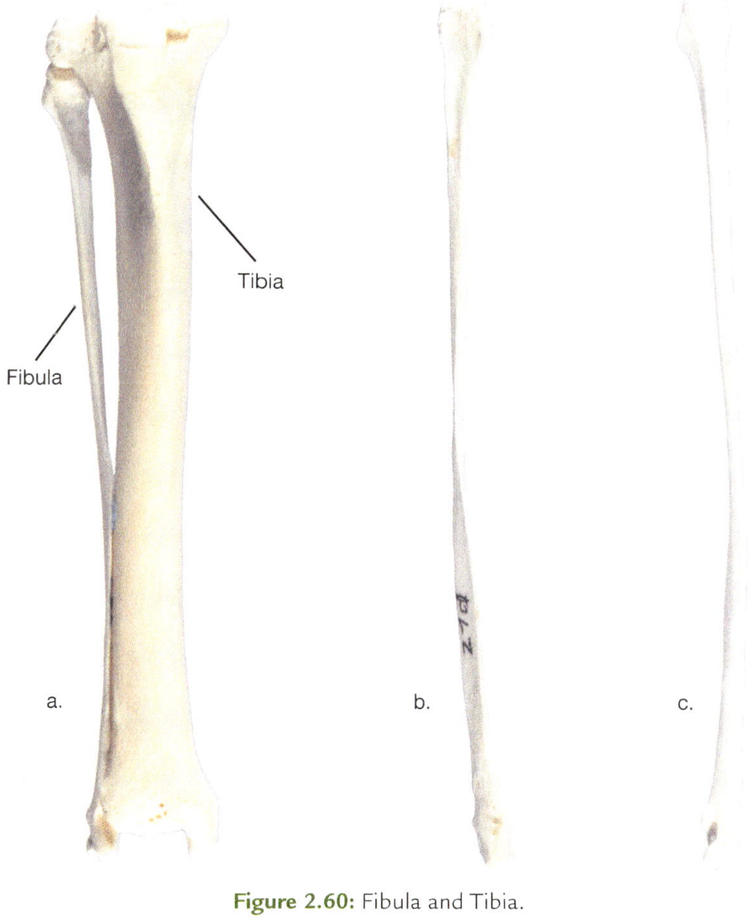

Figure 2.60: Fibula and Tibia.

Appendicular Skeleton
Os Penis

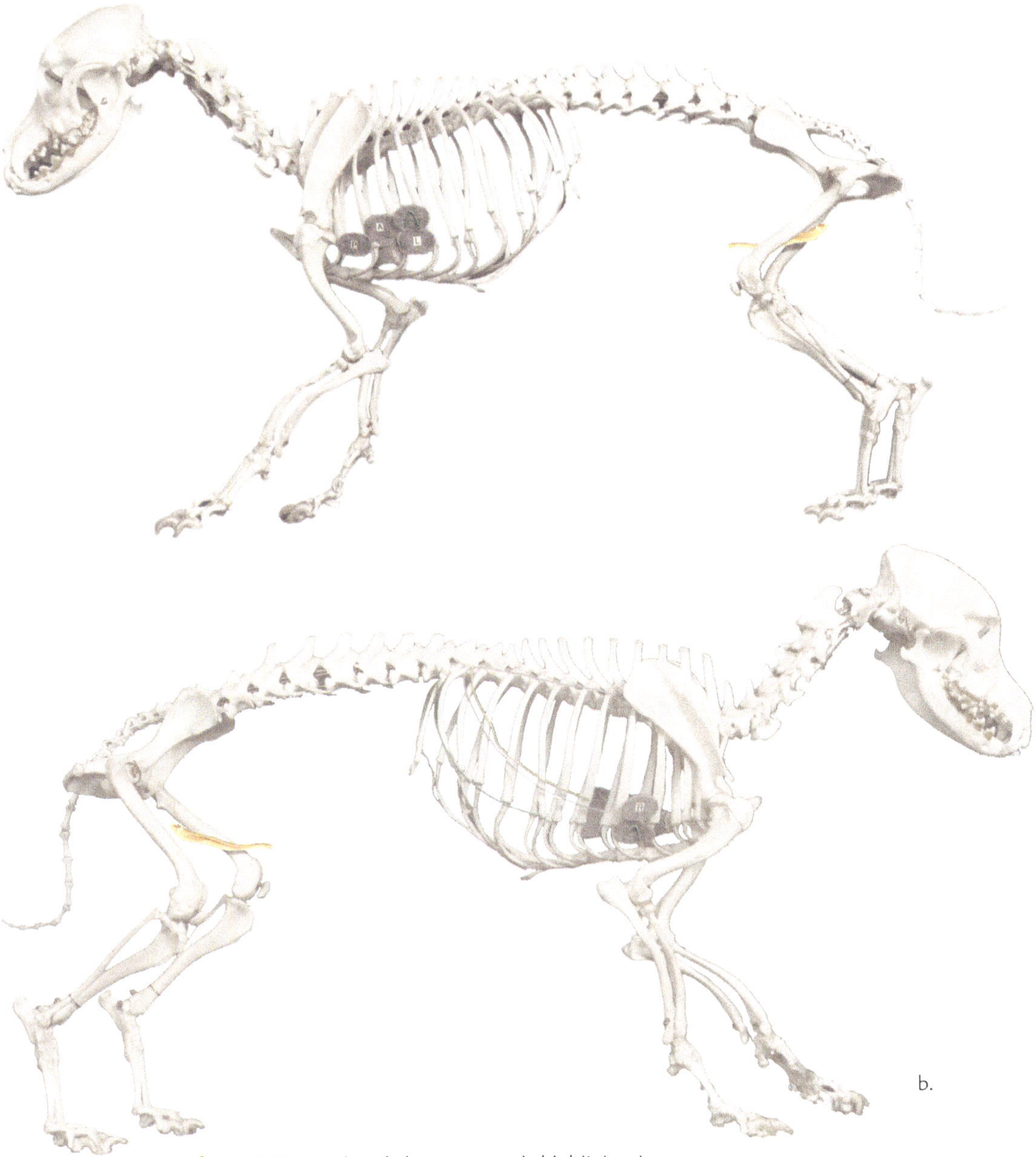

Figure 2.61: Canine skeleton, os penis highlighted.

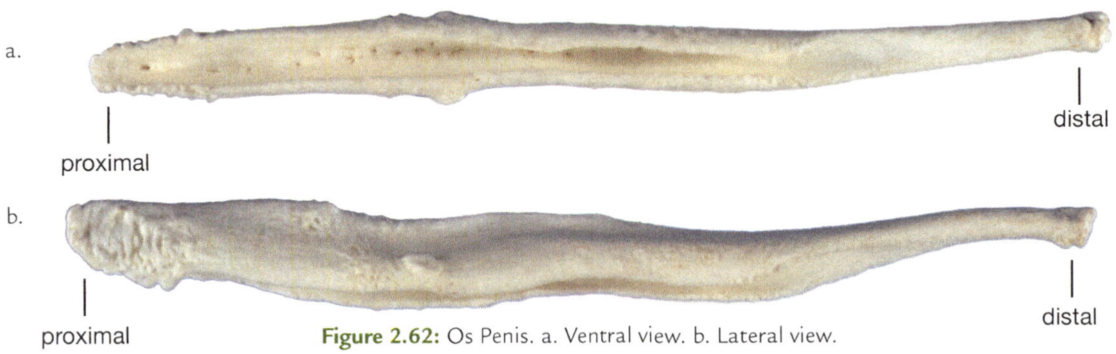

Figure 2.62: Os Penis. a. Ventral view. b. Lateral view.

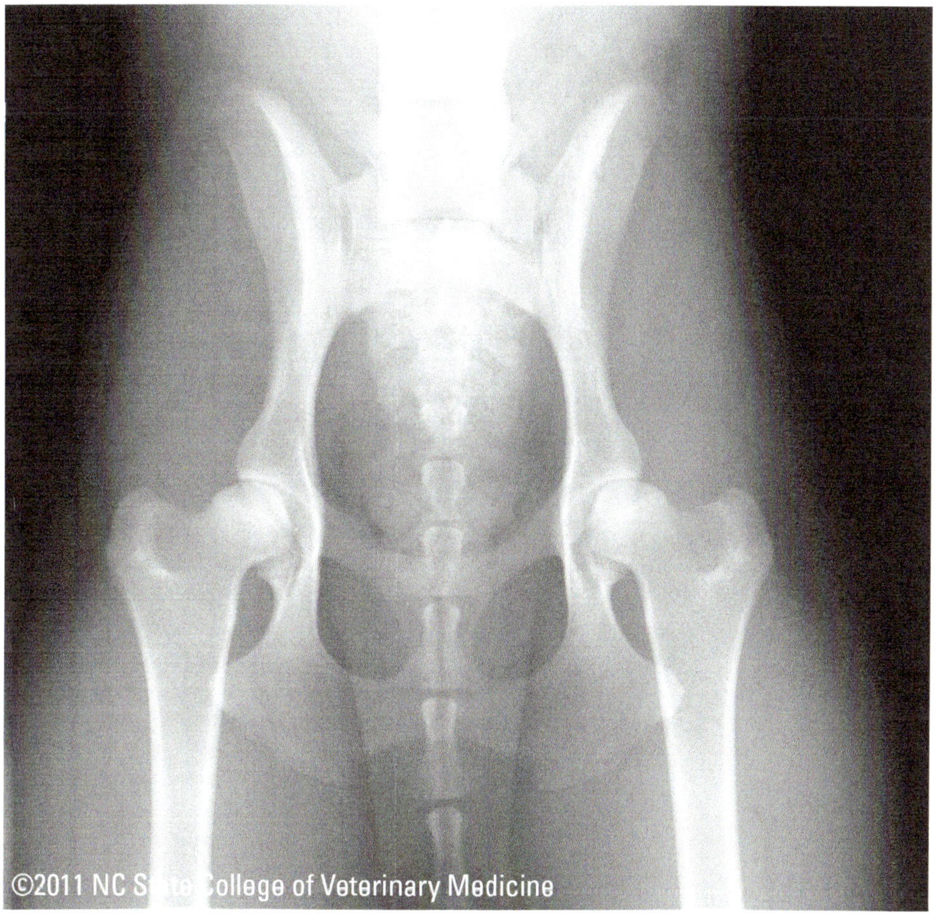

Figure 2.63: Radiograph, pelvis and hip joints.

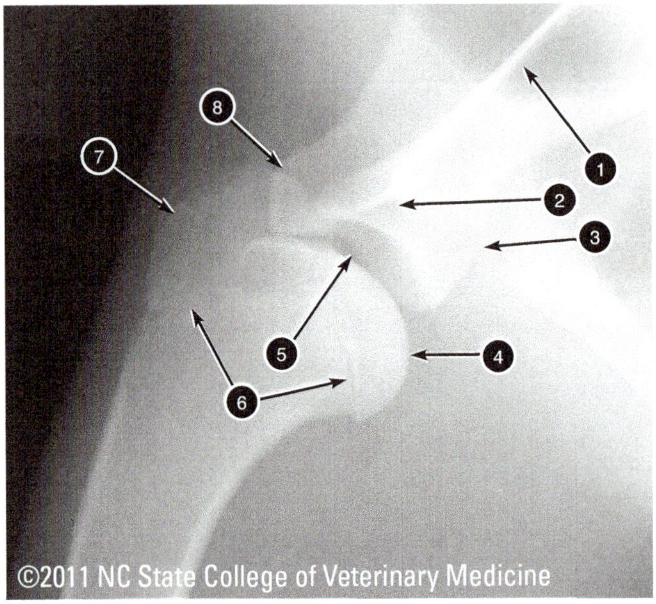

Figure 2.64: Medial lateral (ML) view.

1. Spine of scapula
2. Acromion of scapula
3. Infraglenoid tubercle of scapula
4. Head of humerus
5. Glenoid cavity of scapula
6. Proximal physis of humerus
7. Greater tubercle of humerus
8. Supraglenoid tubercle of scapula

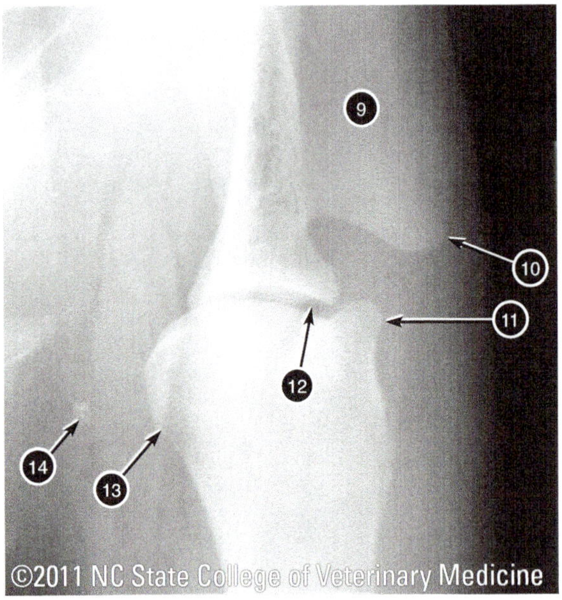

Figure 2.65: Caudocranial (Cd-Cr) view.

9. Spine of scapula
10. Acromion of scapula
11. Greater tubercle of humerus
12. Humeral (scapulohumeral) joint
13. Lesser tubercle of humerus
14. Clavicle

Joints: Shoulder with OCD (Osteochondritis Dissecans)

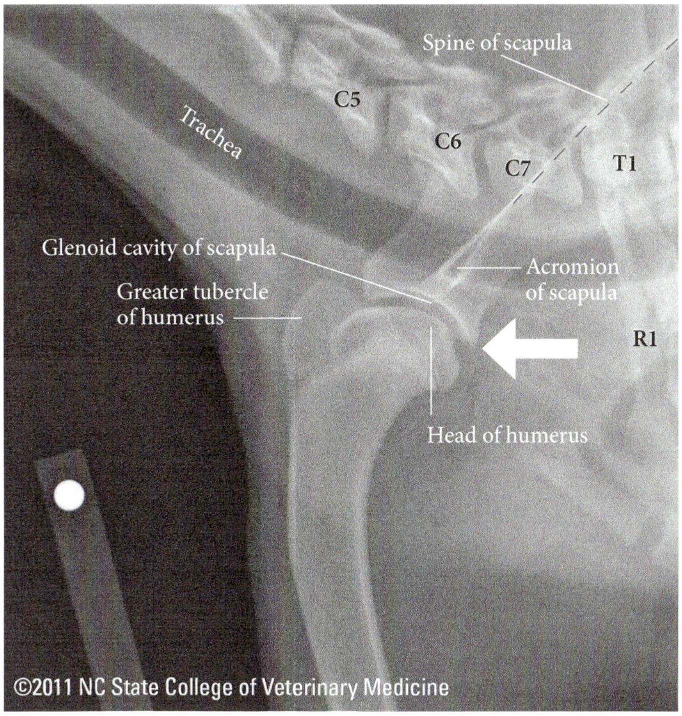

Figure 2.66: Medial lateral (ML) view.

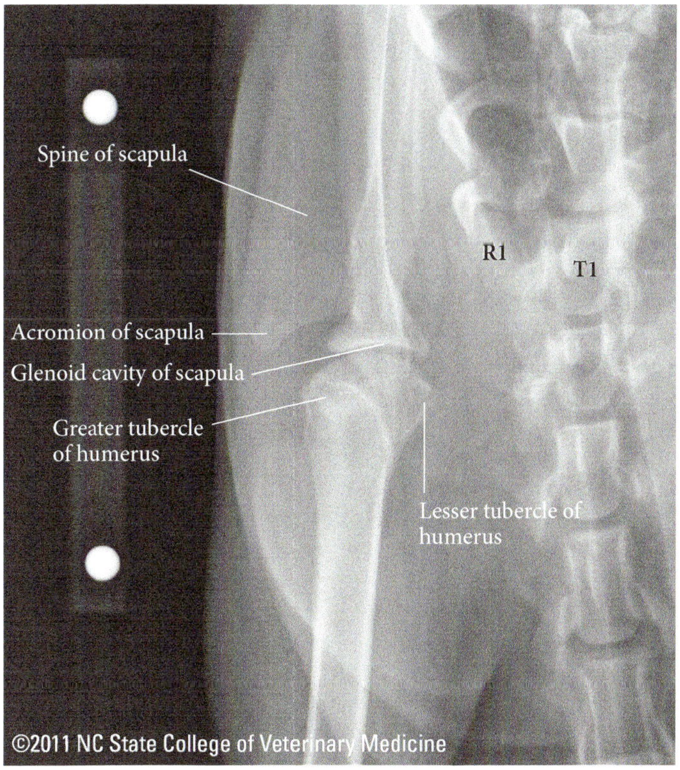

Figure 2.67: Caudocranial (Cd-Cr) view.

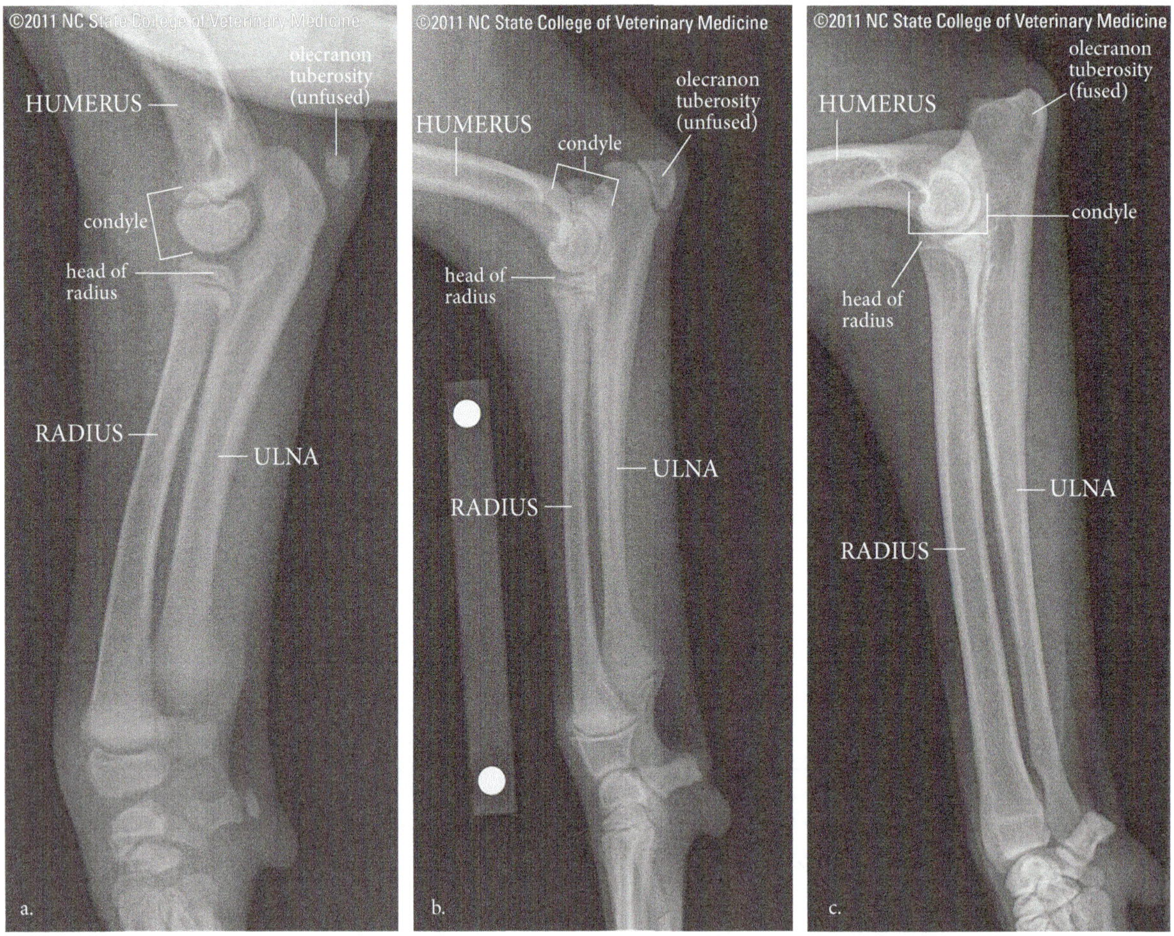

Figure 2.68: Various stages of development, mediolateral view. Epiphyseal (cartilage) plates unfused. a. Immature approximately three months. b. Immature less than six months. c. Adult.

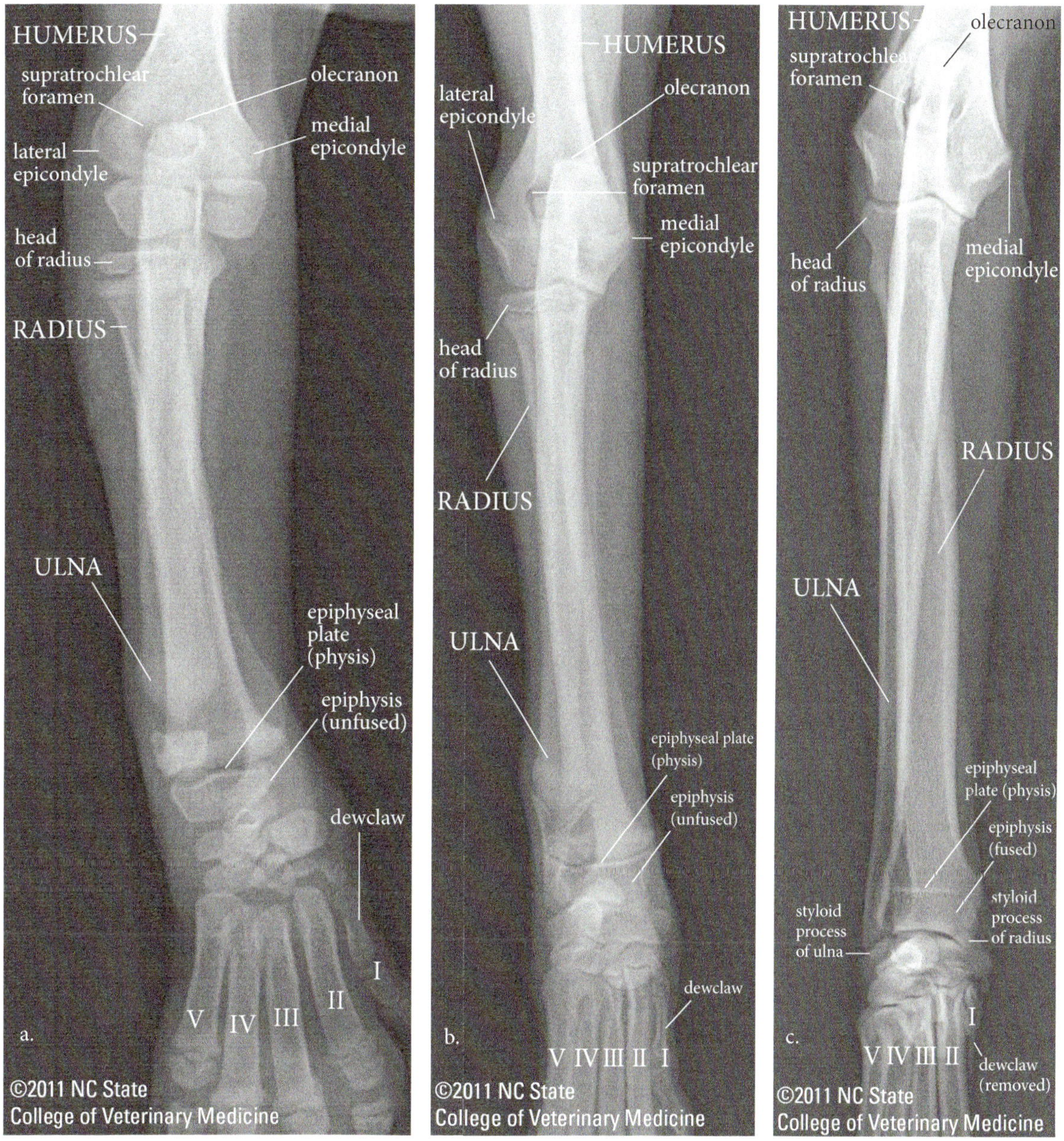

Figure 2.69: Various stages of development, craniocaudal view. Epiphyseal (cartilage) plates unfused. a. Immature approximately three months. b. Immature less than six months. c. Adult.

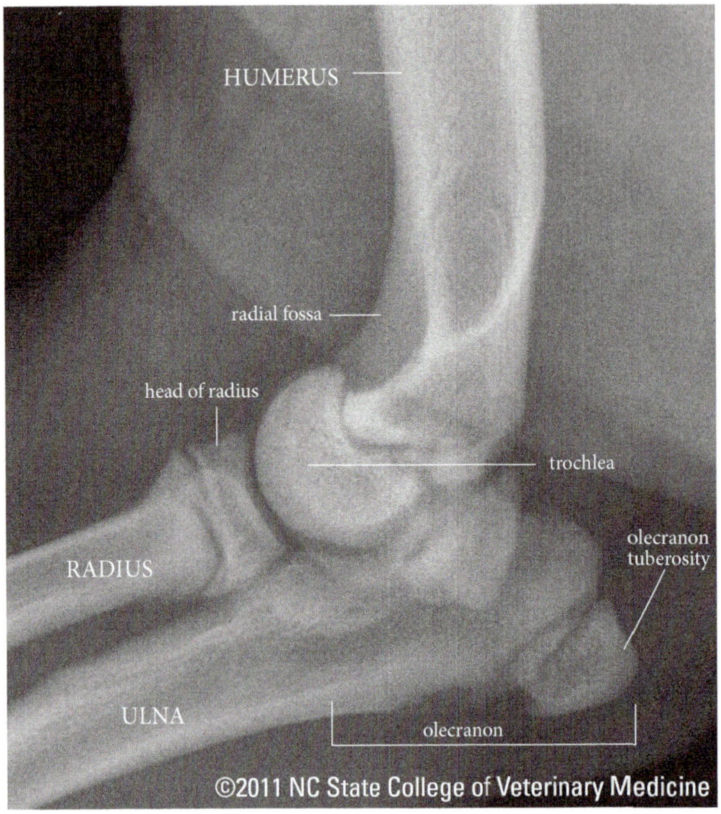

Figure 2.70: Immature appoximately three months, mediolateral view.

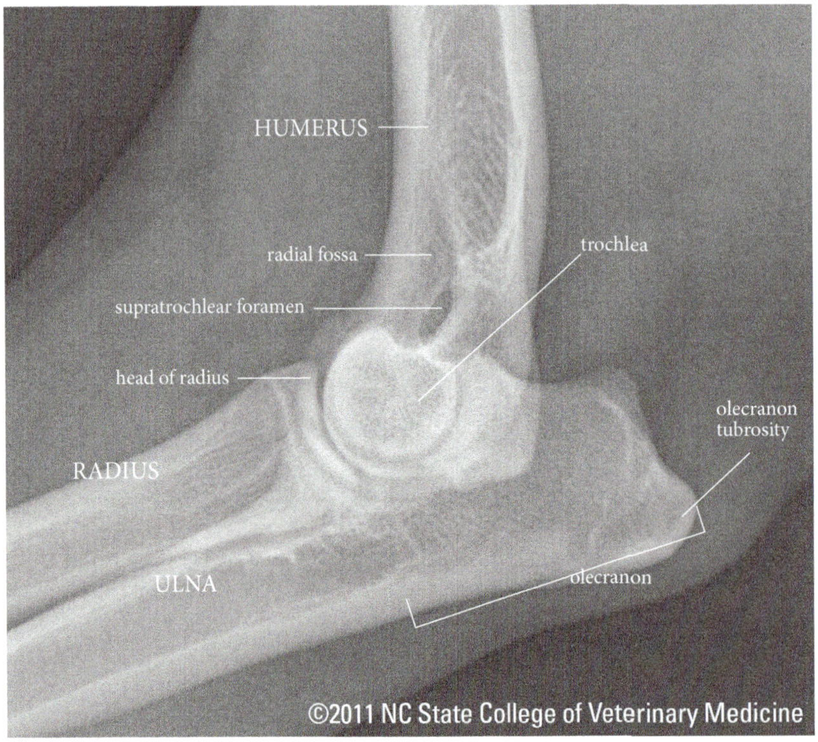

Figure 2.71: Adult, mediolateral view.

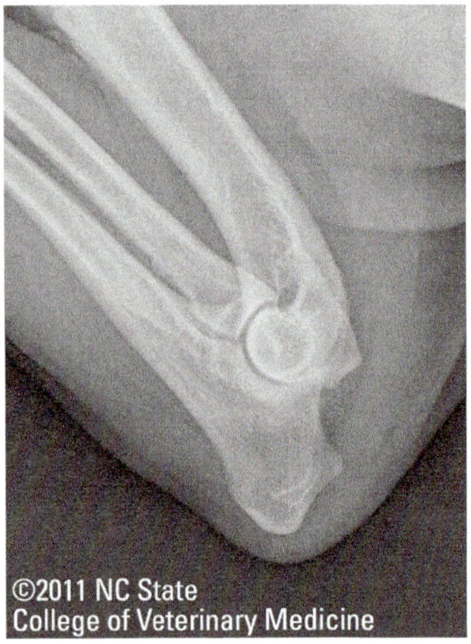

Figure 2.72: Adult, flexed elbow, mediolateral view.

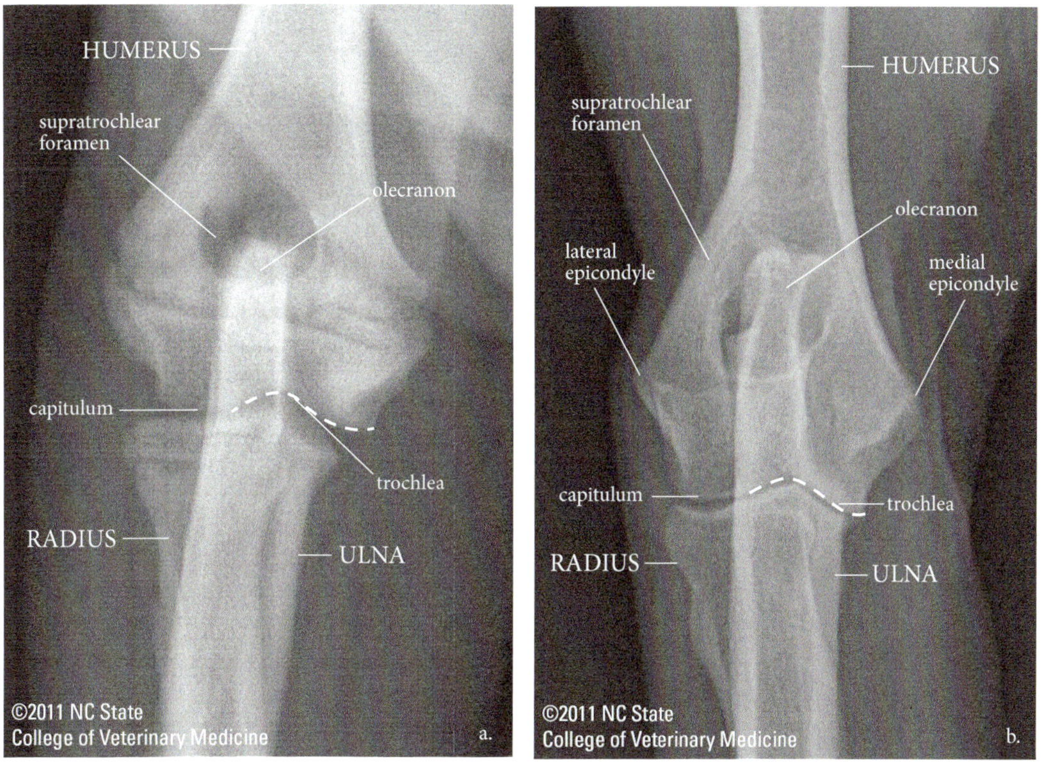

Figure 2.73: Elbow, caudocranial view. a. Immature approximately three months. b. Adult.

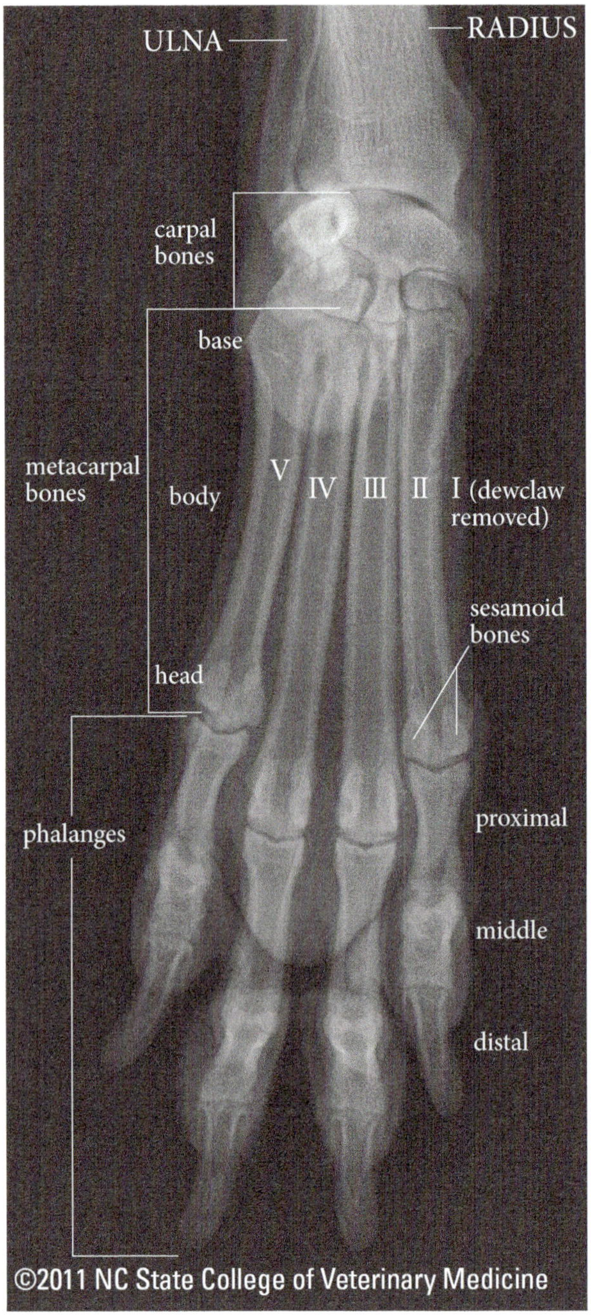

Figure 2.74: Dorsopalmar view.

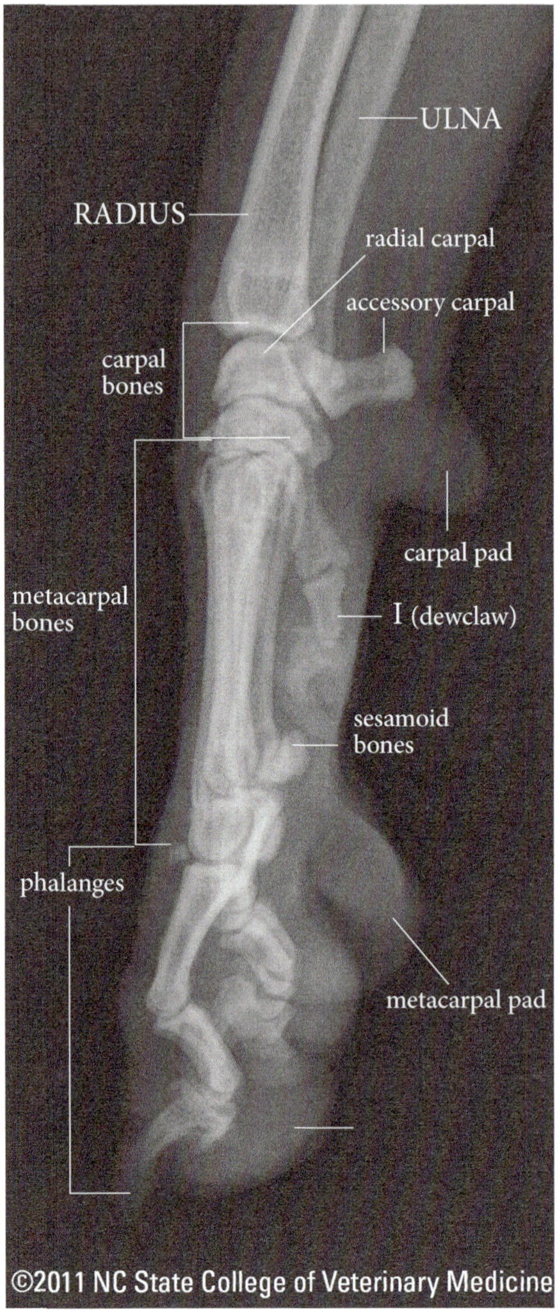

Figure 2.75: Mediolateral view.

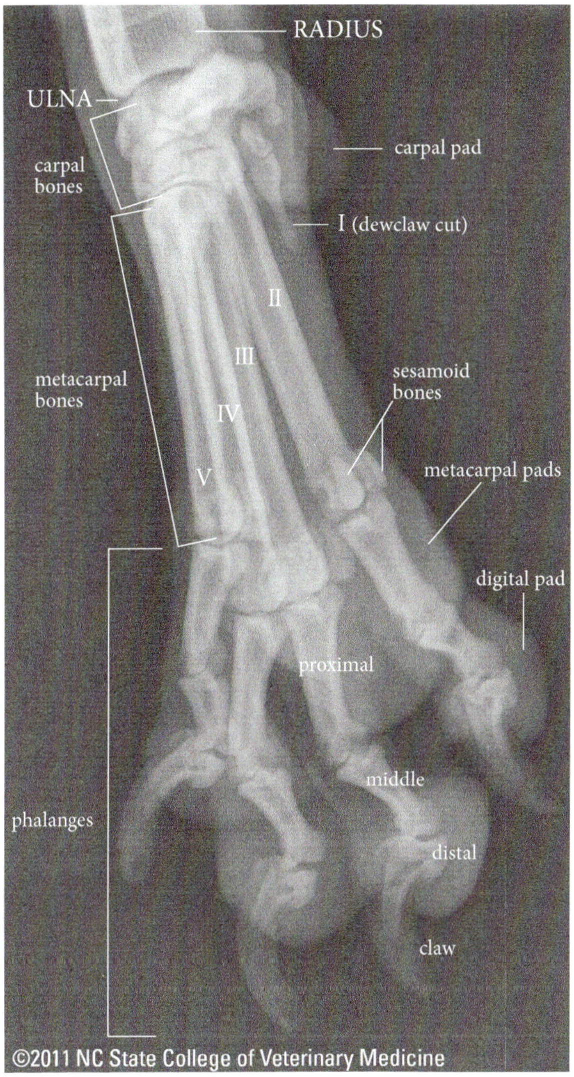

Figure 2.76: Pads down (splayed), dorsomedial-palmarolateral oblique (DPMLO) view.

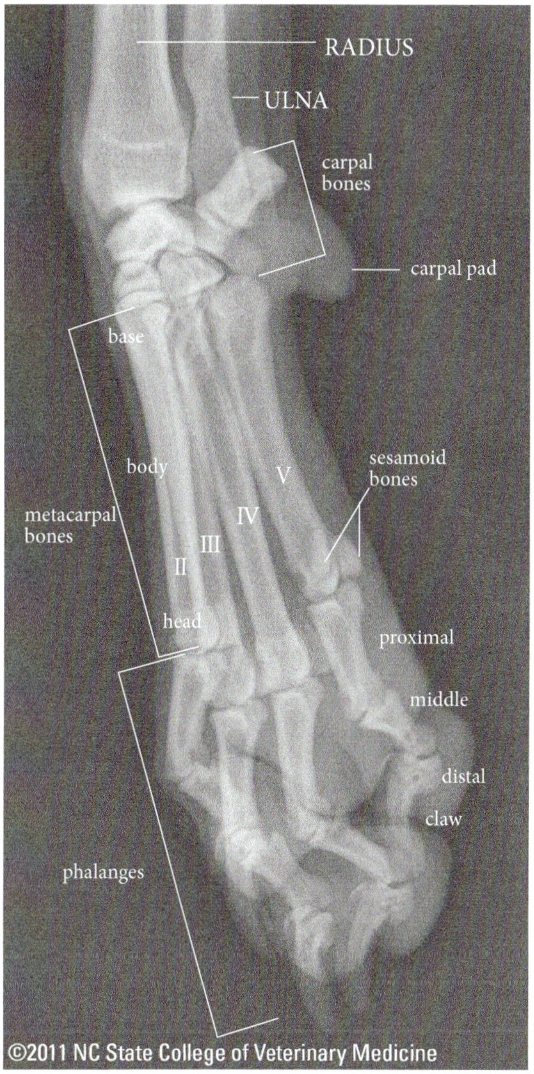

Figure 2.77: Rotated 180° from previous figure, pads up, dorsomedial-palmarolateral oblique (DPMLO) view.

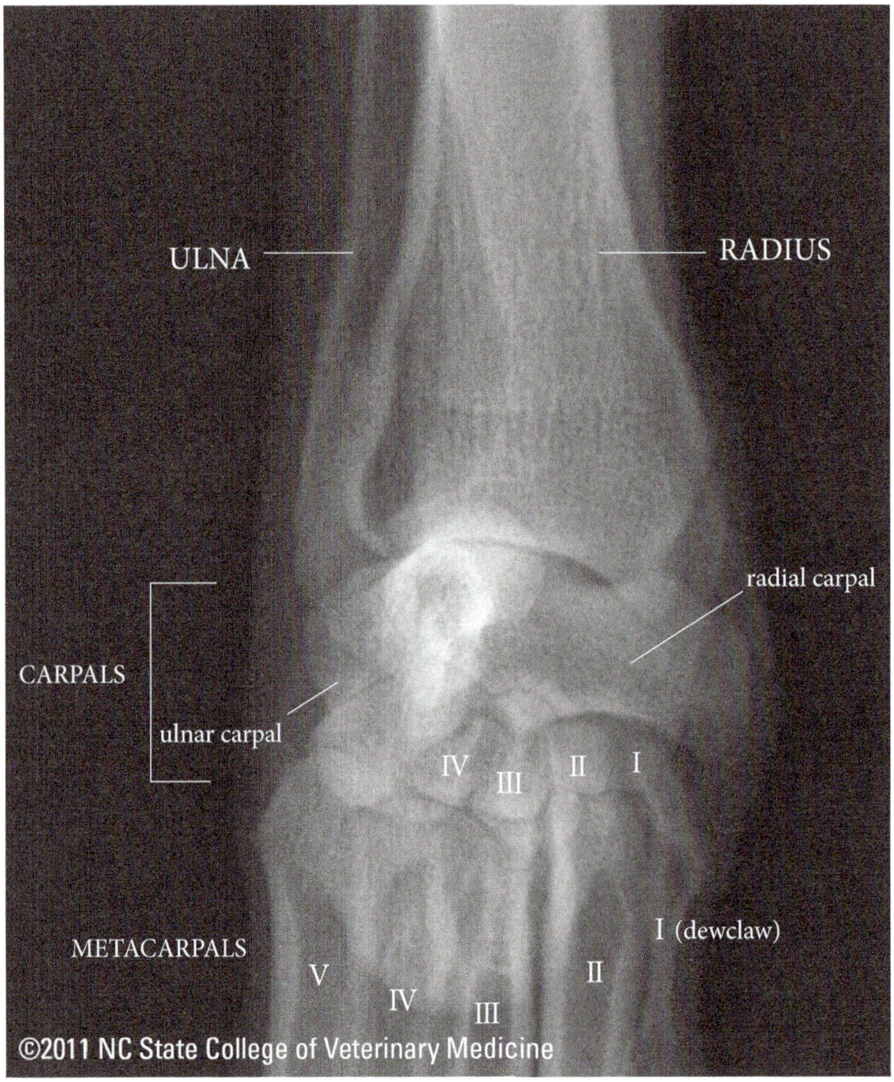

Figure 2.78: Carpus, dorsopalmar view.

Joints: Hip

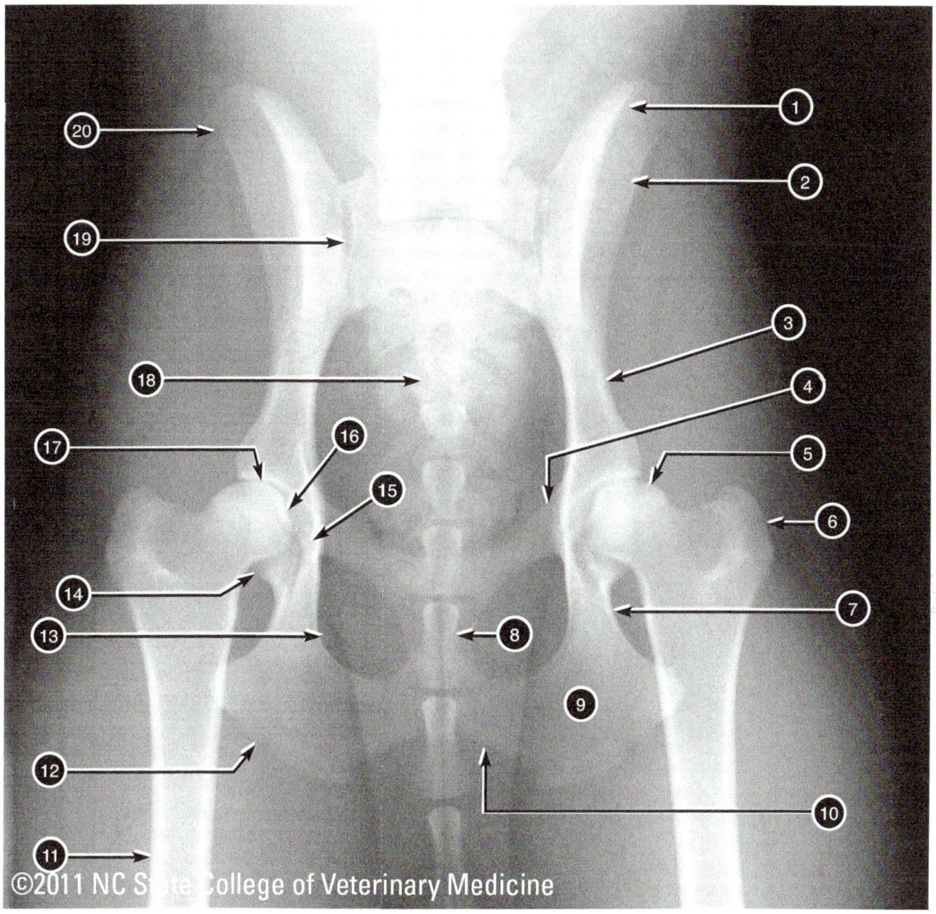

Figure 2.79: Hip, pelvis overview, ventrodorsal view.

1. Crest of left ilium
2. Wing of left ilium
3. Body of left ilium
4. Body of left pubis
5. Head of left femur
6. Greater trochanter of left femur
7. Body of left ischium
8. Caudal vertebra 4
9. Table of left ischium
10. Ischial arch
11. Body of right femur
12. Right tuber ischiadicum (ischiatic tuberosity)
13. Right obturator foramen
14. Caudodorsal aspect of lunate surface of right acetabulum
15. Acetabular fossa
16. Fovea capitis of right femur
17. Cranioventral aspect of lunate surface of right acetabulum
18. Sacral vertebra 3
19. Right sacroiliac joint
20. Right tuber coxae (cranioventral iliac spine)

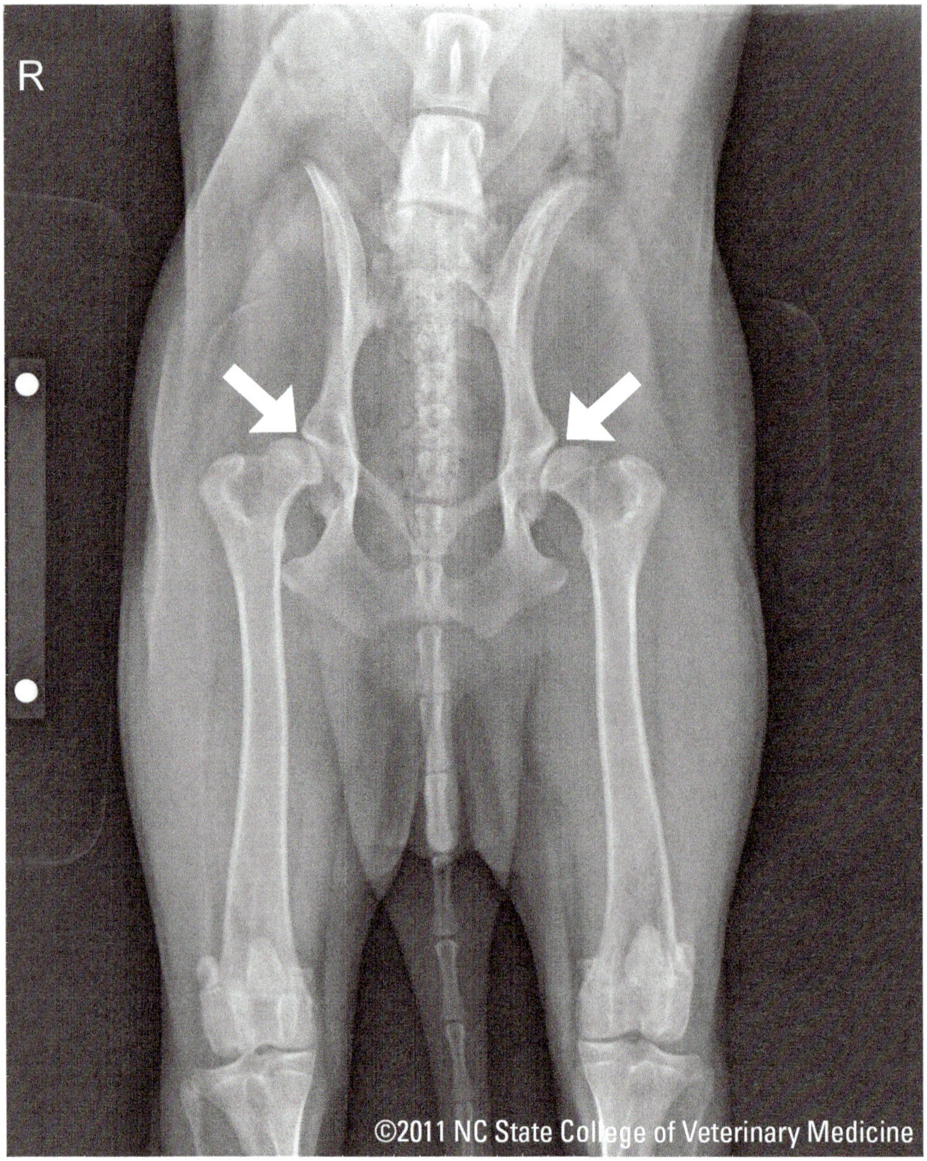

Figure 2.80: Hip dysplasia, vertrodorsal view.

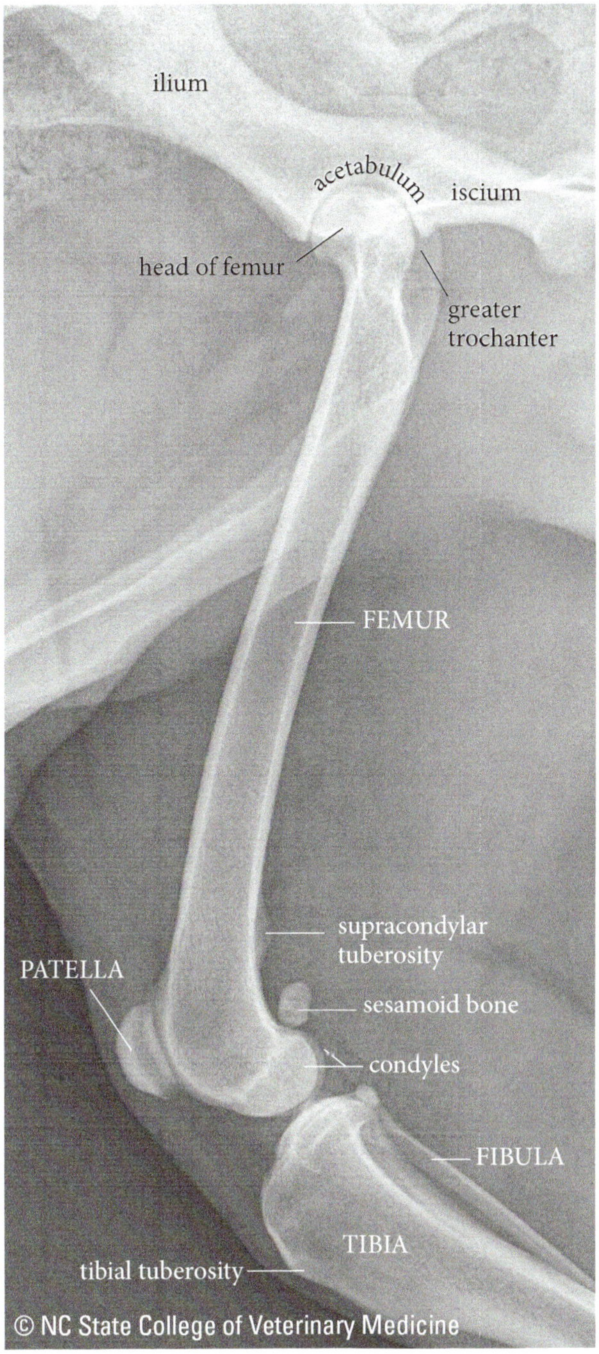

Figure 2.81: Femur, mediolateral view.

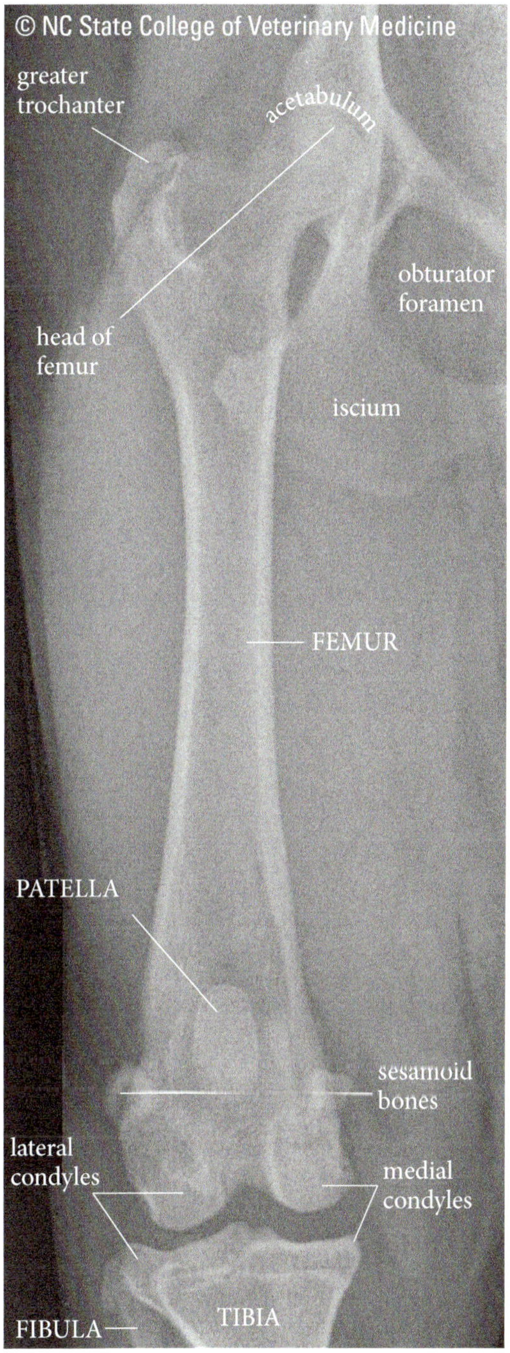

Figure 2.82: Femur, ventral view.

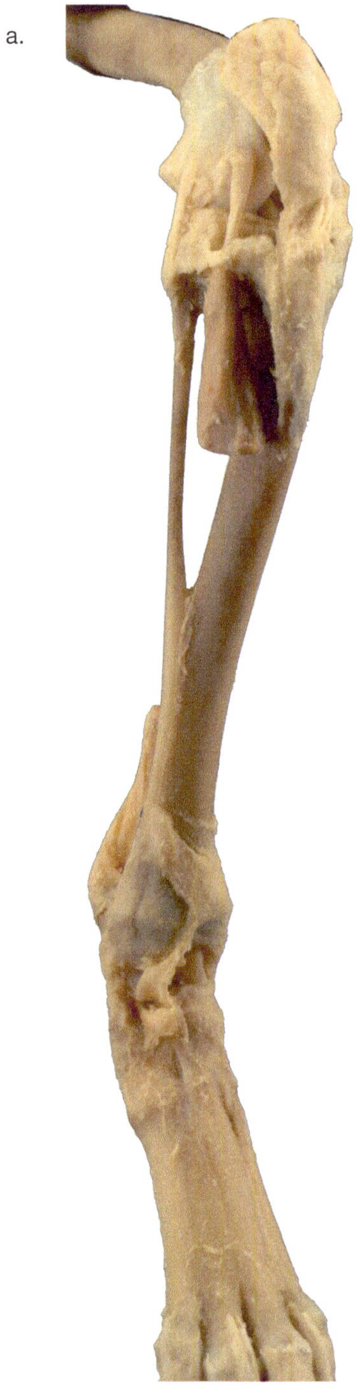

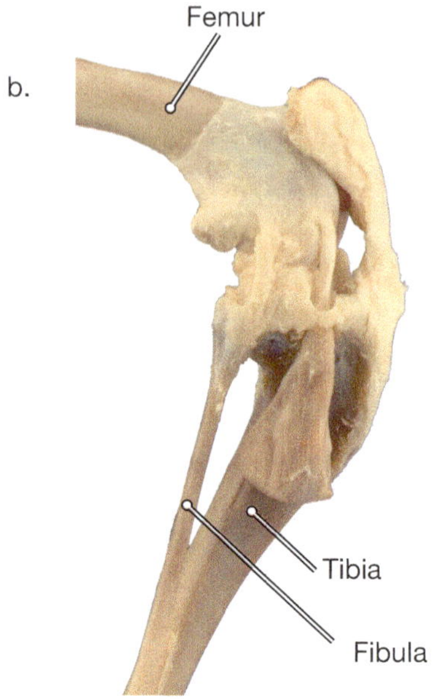

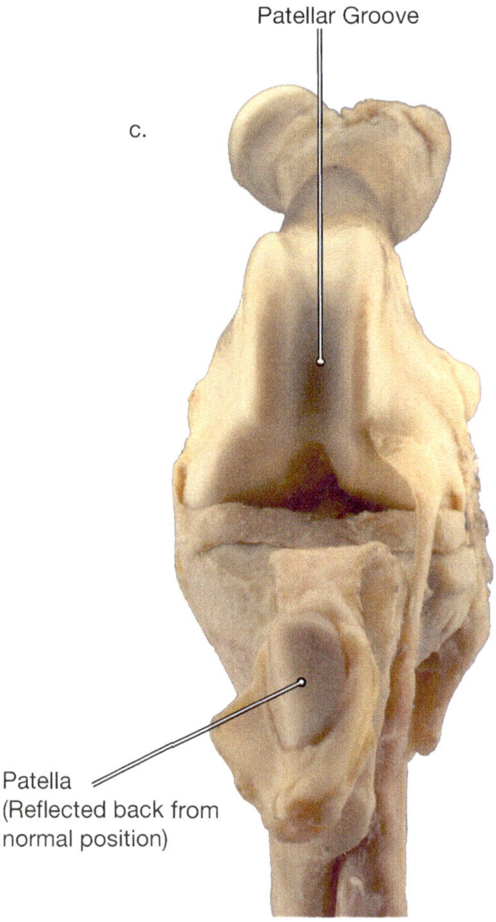

Figure 2.83: Stifle Joint with lower leg.
© NCSU.

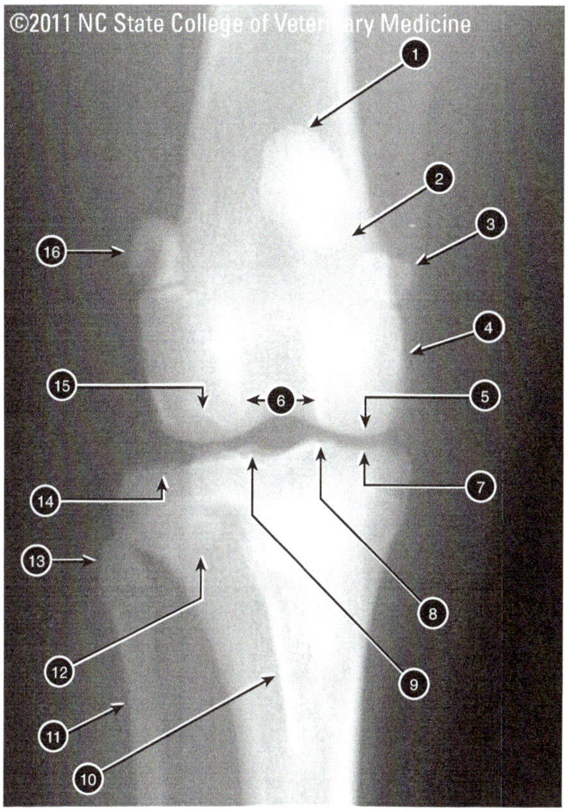

Figure 2.84: Stifle, craniocaudal view.

A. Craniocaudal (CrCd) View
1. Base of patella (anatomic)
2. Apex of patella
3. Medial sesamoid of gastrocnemius muscle (shorter than lateral sesamoid bone)
4. Medial epicondyle of femur
5. Medial condyle of femur
6. Intercondylar fossa of femur
7. Medial condyle of tibia
8. Medial tubercle of intercondylar eminence of tibia
9. Lateral tubercle of intercondylar eminence of tibia
10. Cranial border of tibia
11. Body of fibula
12. Sesamoid bone of popliteus muscle
13. Head of fibula
14. Lateral condyle of tibia
15. Extensor fossa of femur (origin of long digital extensor muscle)
16. Lateral sesamoid of gastrocnemius muscle (taller than medial sesamoid bone)

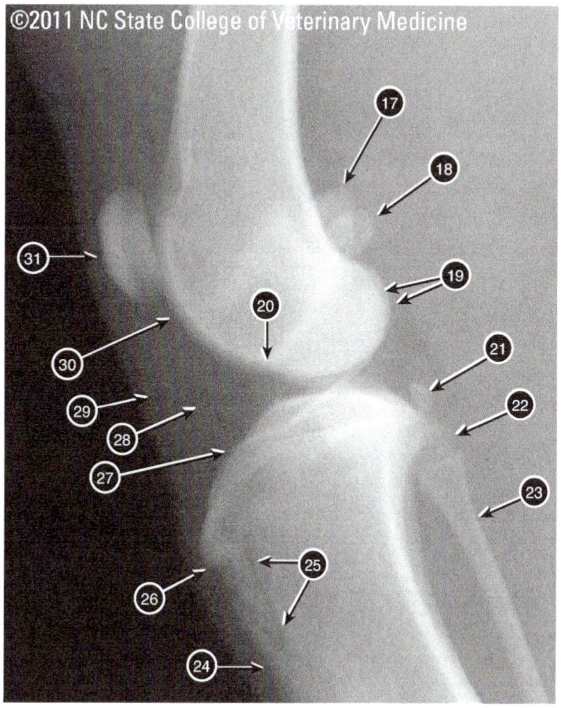

Figure 2.85: Stifle, mediolateral view.

17. Lateral sesamoid of gastrocnemius muscle (taller than medial sesamoid bone)
18. Medial sesamoid of gastrocnemius muscle (shorter than lateral sesamoid bone)
19. Superimposed medial and lateral condyles of femur
20. Extensor fossa of femur (origin of long digital extensor muscle)
21. Sesamoid bone of popliteus muscle (embedded in tendon of muscle)
22. Head of fibula
23. Body of fibula
24. Cranial border of tibia
25. Hyaline cartilage between tibial tuberosity and body of tibia (incomplete fusion)
26. Tibial tuberosity (termination of patellar ligament; quadriceps femoris muscle)
27. Cranial intercondylar area of tibia (attachment of cranial meniscal ligaments
28. Infrapatellar fat body (between fibrous and synovial layers of joint capsule)
29. Patellar ligament (functional continuation of quadriceps femoris muscle)
30. Superimposed ridges of femoral trochlea
31. Patella (functions as sesamoid bone for quadriceps femoris muscle)

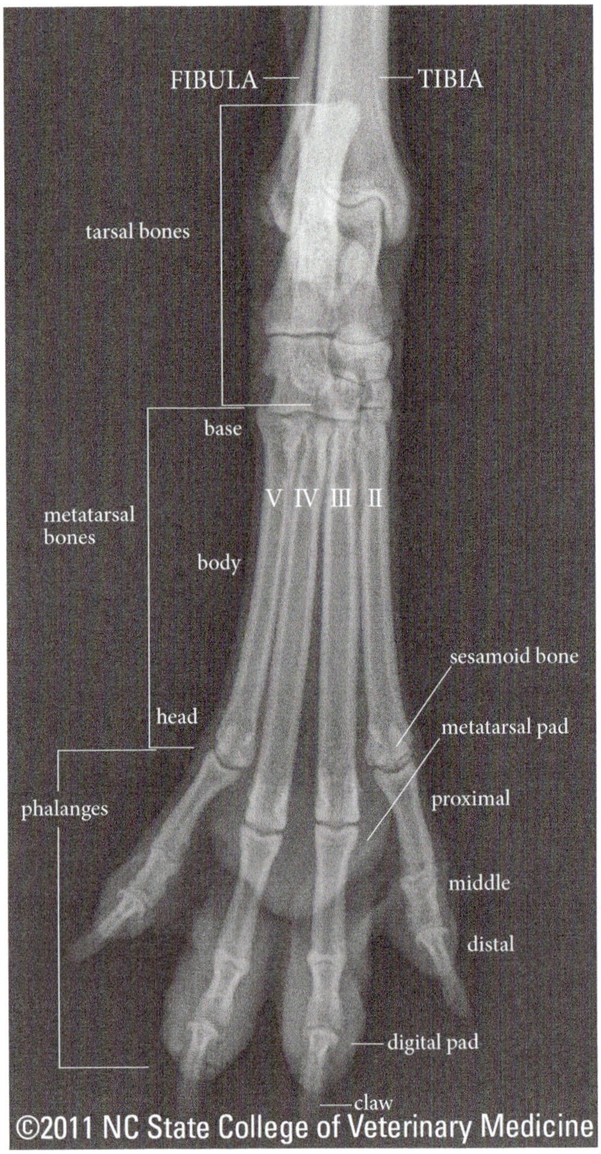

Figure 2.86: Tarsus/Pes, dorsopalmer view.

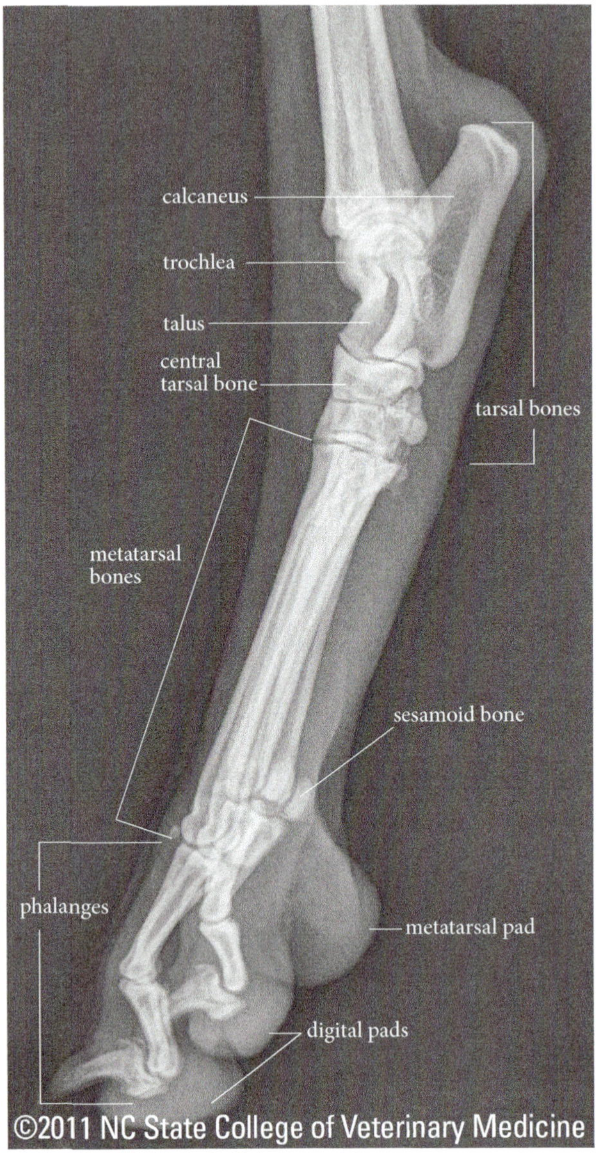

Figure 2.87: Tarsus/Pes, mediolateral view.

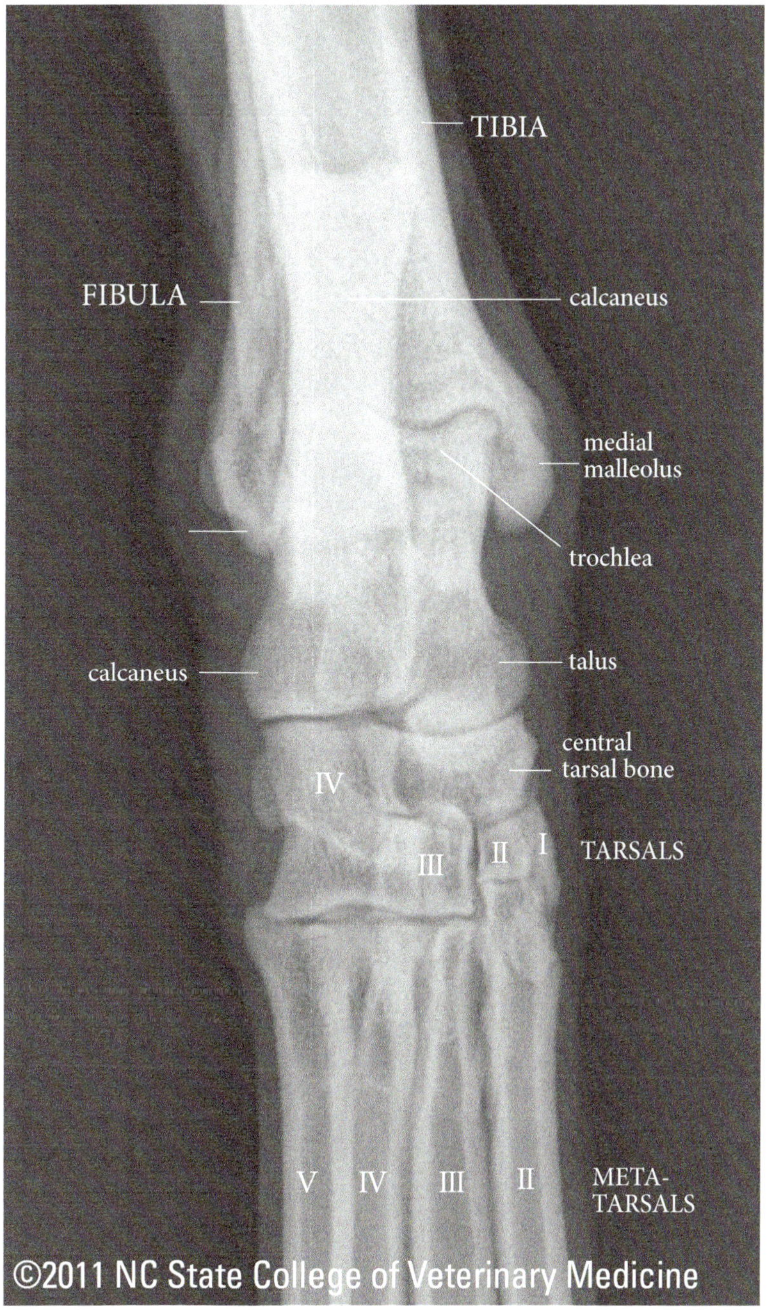

Figure 2.88: Tarsus, dorsopalmar view.

Articular cartilage

Compact bone of diaphysis

Distal epiphysis

Endosteum

Epiphyseal line

Medullary cavity

Periosteum

Red bone marrow

Spongy bone of diaphysis

Spongy bone of proximal epiphysis

Yellow bone marrow

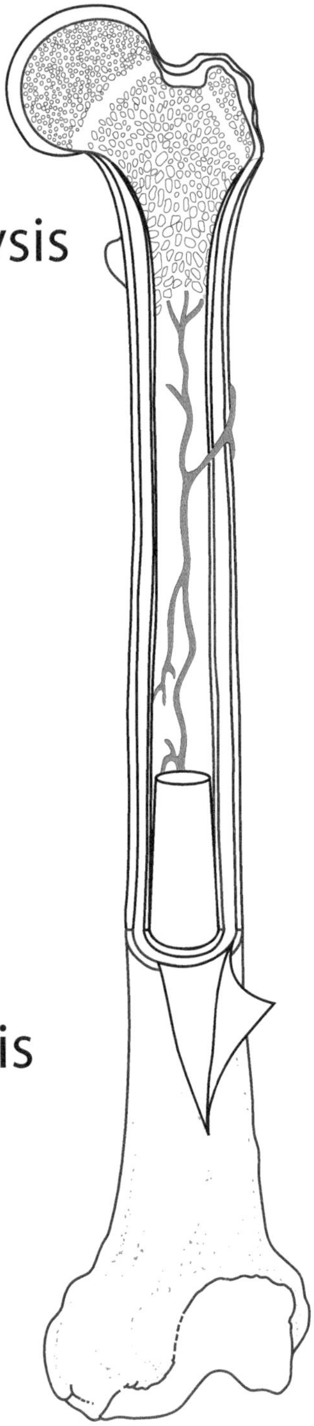

Figure 2.89: Bone coloring activity.

Review: Key Term Matching

____ 1. Epiphysis
____ 2. Diaphysis
____ 3. Periosteum
____ 4. Cancellous
____ 5. Central canal
____ 6. Epiphyseal plate
____ 7. Mesaticephalic skull
____ 8. Axis
____ 9. Appendicular
____ 10. Thoracic
____ 11. Metacarpals
____ 12. Mandible
____ 13. Ulna
____ 14. Axial
____ 15. Dolichocephalic skull
____ 16. Atlas
____ 17. Body
____ 18. Caudal
____ 19. Scapula
____ 20. Ribs
____ 21. Hip bone
____ 22. Brachycephalic skull
____ 23. Lumbar
____ 24. Vertebral foramen
____ 25. Occipital bone

A. first cervical vertebrae
B. lower jaw bone
C. outer membrane of compact bone
D. Boston Terrier skull
E. region where growth occurs in long bones
F. the end of a long bone
G. most posteriorly located cranial bone
H. ilium, ischium, pubis
I. the ribs are part of this skeletal region
J. the shaft of a long bone
K. most common canine skull
L. bones found in the front foot
M. opening in vertebrae where spinal cord is located
N. this bone contains a tubercle and head
O. another name for spongy bone
P. Greyhound skull
Q. center region of the spine
R. located in the center of the osteon
S. this bone contains the supraspinous fossa
T. most caudal region of the spine
U. tailbone
V. second cervical vertebrae
W. this bone contains a trochlear notch
X. center area of the sternum
Y. the humerus and tibia are parts of this skeletal region

Review: Labeling Activity

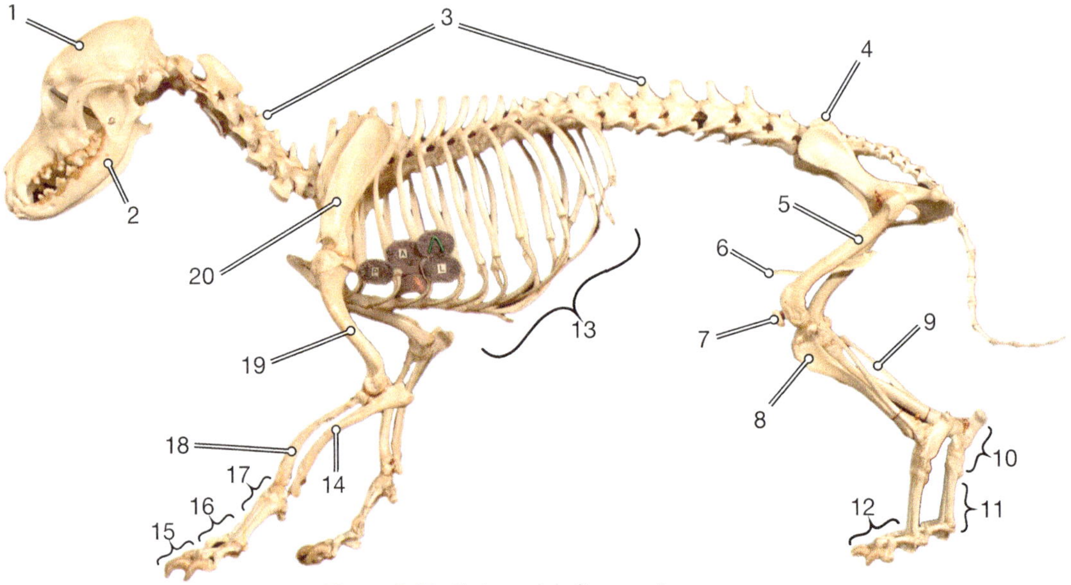

Figure 2.90: Skeleton labeling actvity.

1. ______	11. ______
2. ______	12. ______
3. ______	13. ______
4. ______	14. ______
5. ______	15. ______
6. ______	16. ______
7. ______	17. ______
8. ______	18. ______
9. ______	19. ______
10. ______	20. ______

Chapter 3

Muscle Tissues and Skeletal Muscles

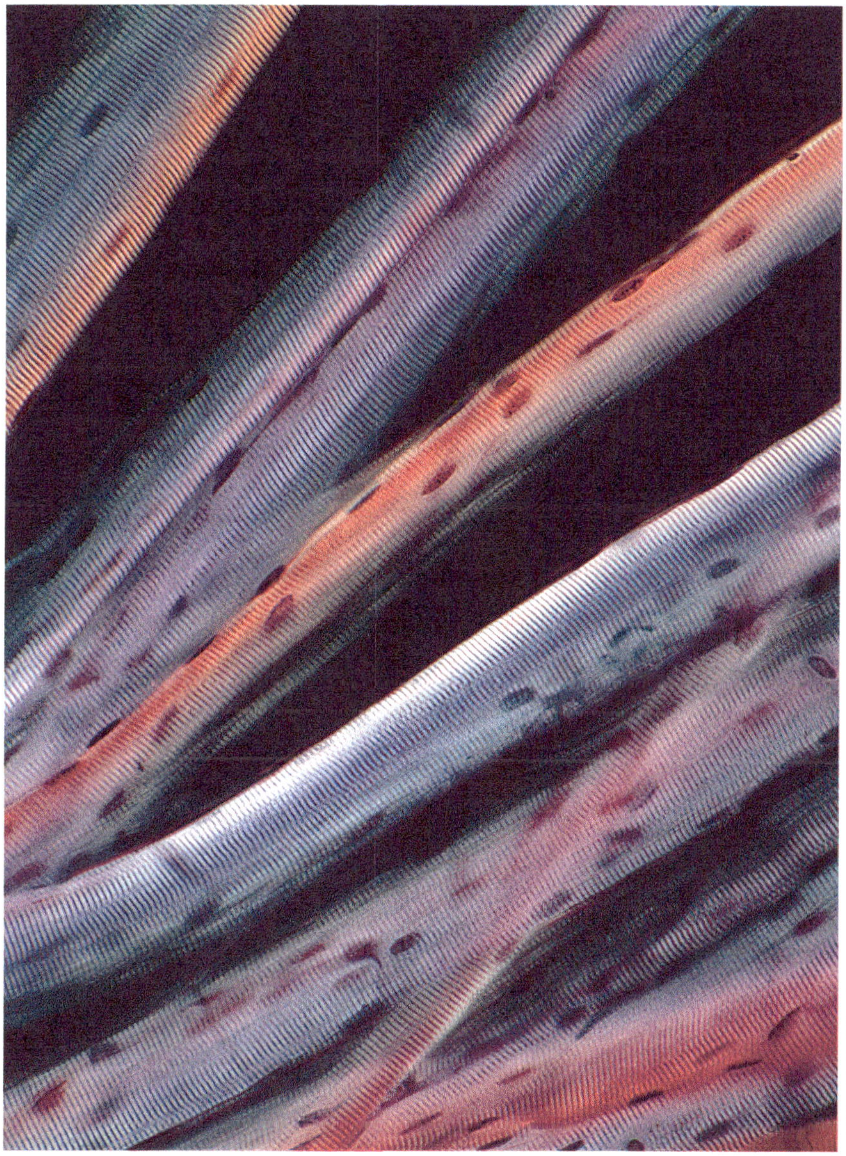

© David G. Ward

Skeletal Muscle Cells Showing Banding (Polarized Light, Lambda; x300)

Skeletal Muscle Tissue

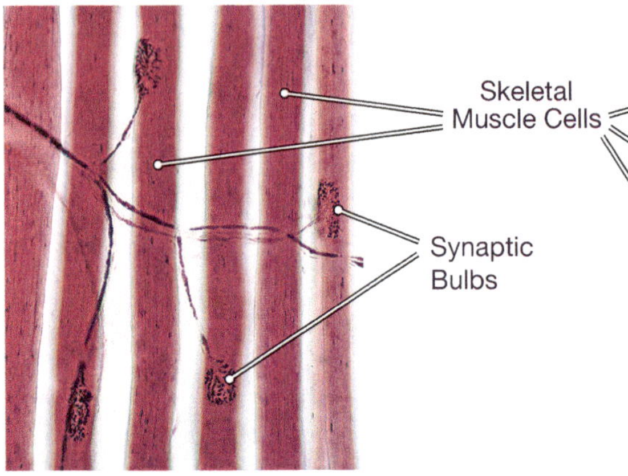

Figure 3.1: Nerve and muscle, longitudinal section (vertical).

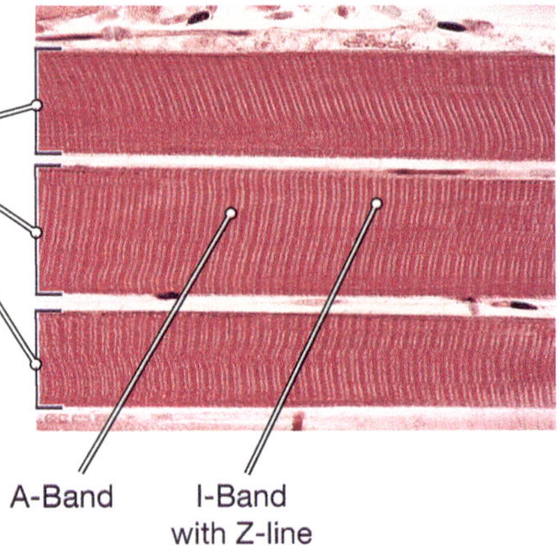

Figure 3.2: Skeletal muscle cells, longitudinal section (horizontal).

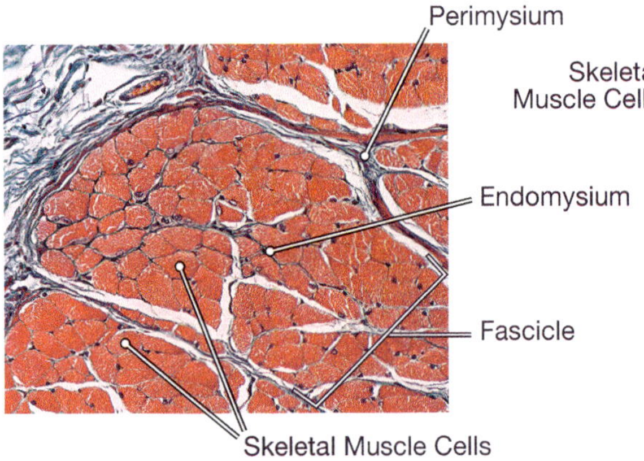

Figure 3.3: Skeletal muscle fascicles and cells, cross section.

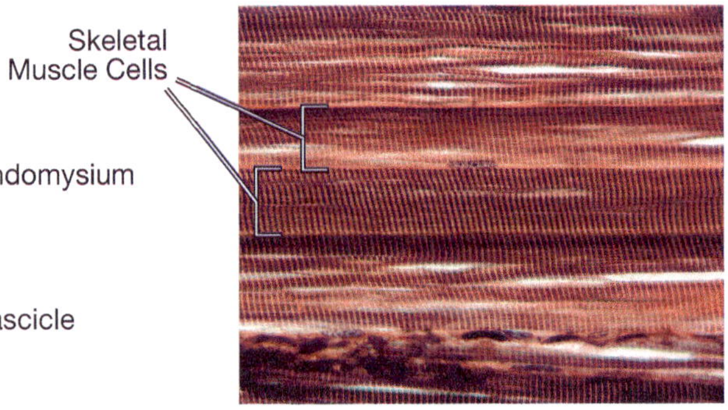

Figure 3.4: Skeletal muscle cells, longitudinal section.

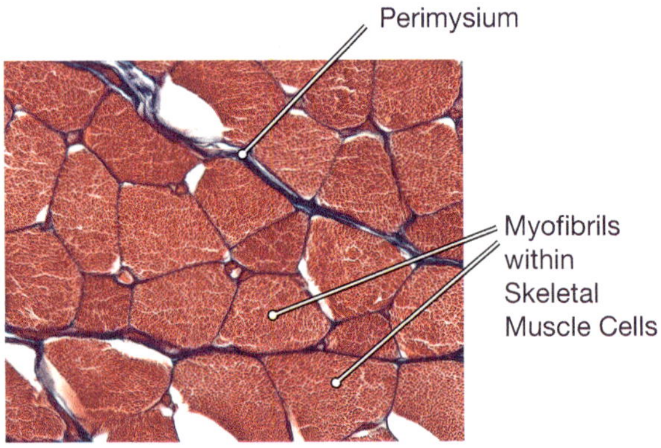

Figure 3.5: Skeletal muscle cells and myofibrils, cross section.

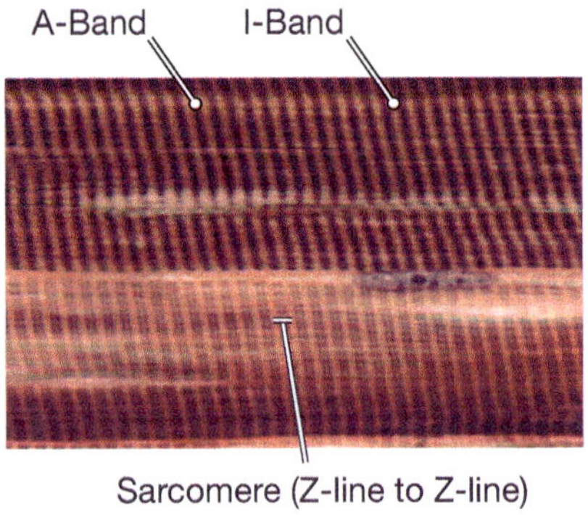

Figure 3.6: Skeletal muscle cells, longitudinal section.

Skeletal Muscle Cell

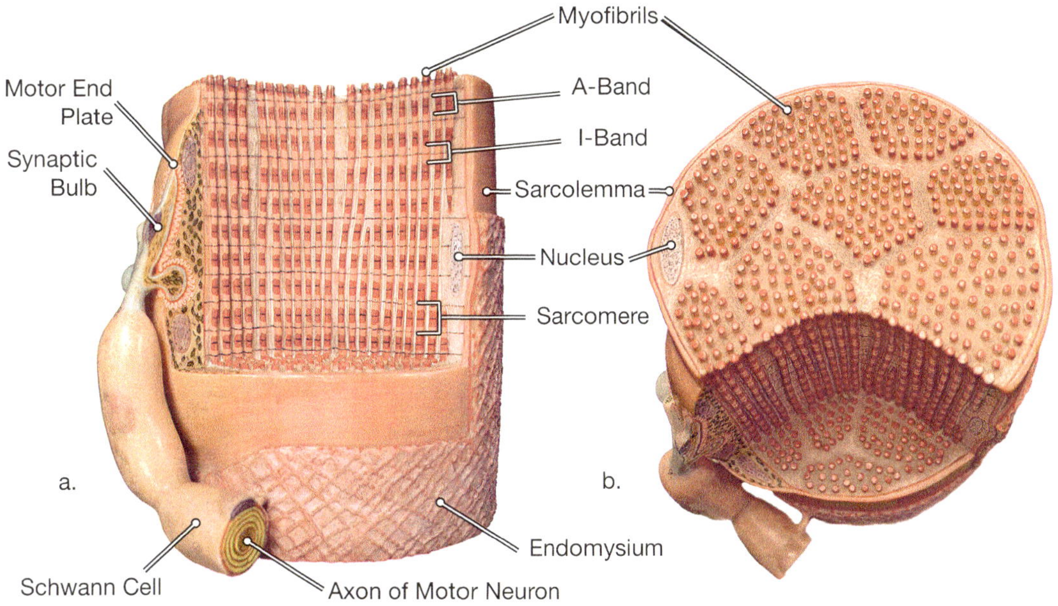

Figure 3.7: Skeletal muscle cells a. Longitudinal section. b. Cross section.

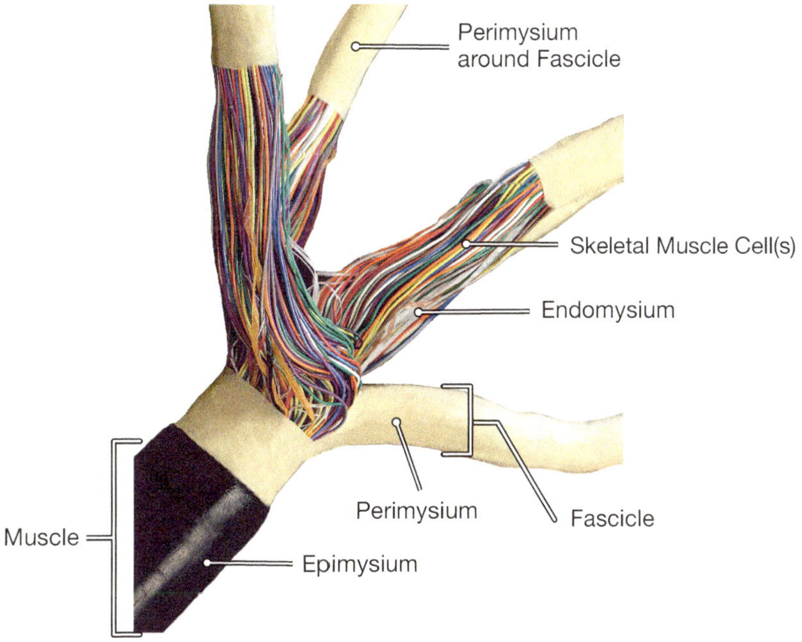

Figure 3.8: Telephone cable as skeletal muscle.

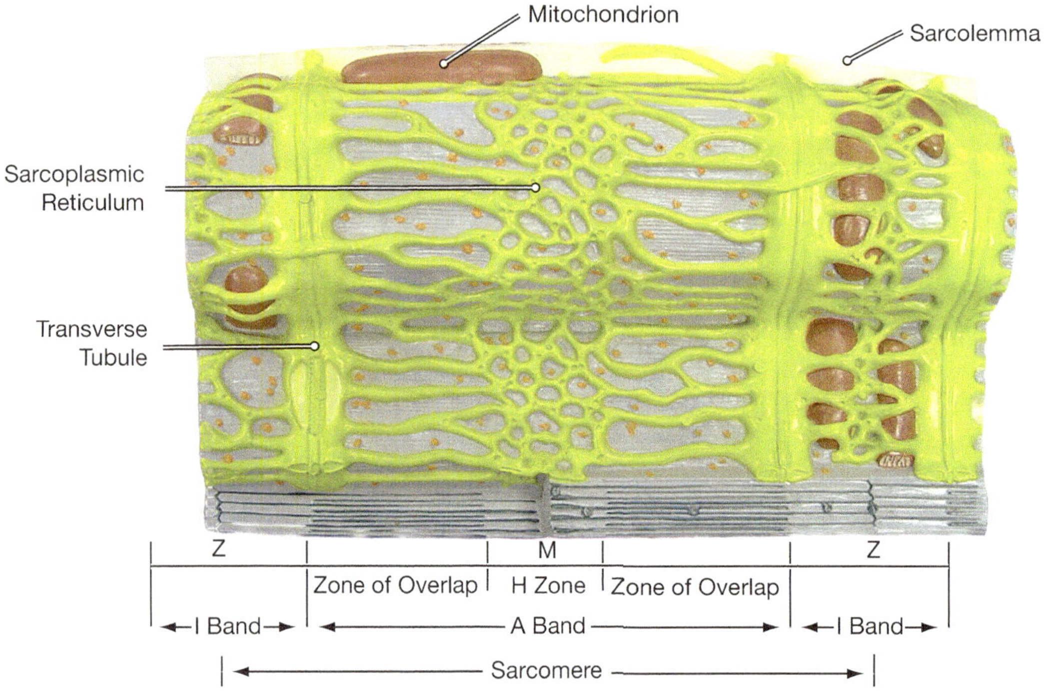

Figure 3.9: Sarcomere and sarcoplasmic reticulum.
© David G. Ward.

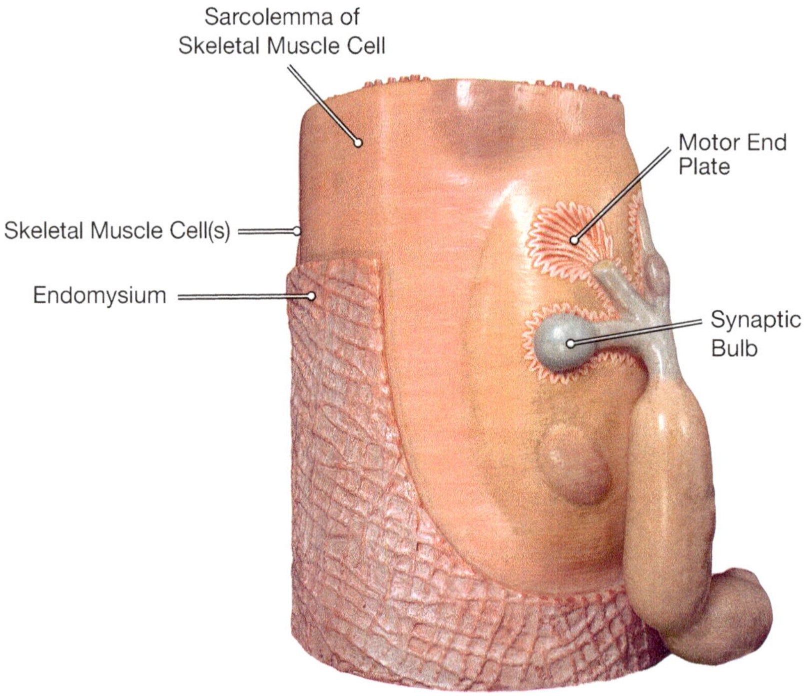

Figure 3.10: Skeletal muscle cell.
© David G. Ward.

Superficial Skeletal Muscles

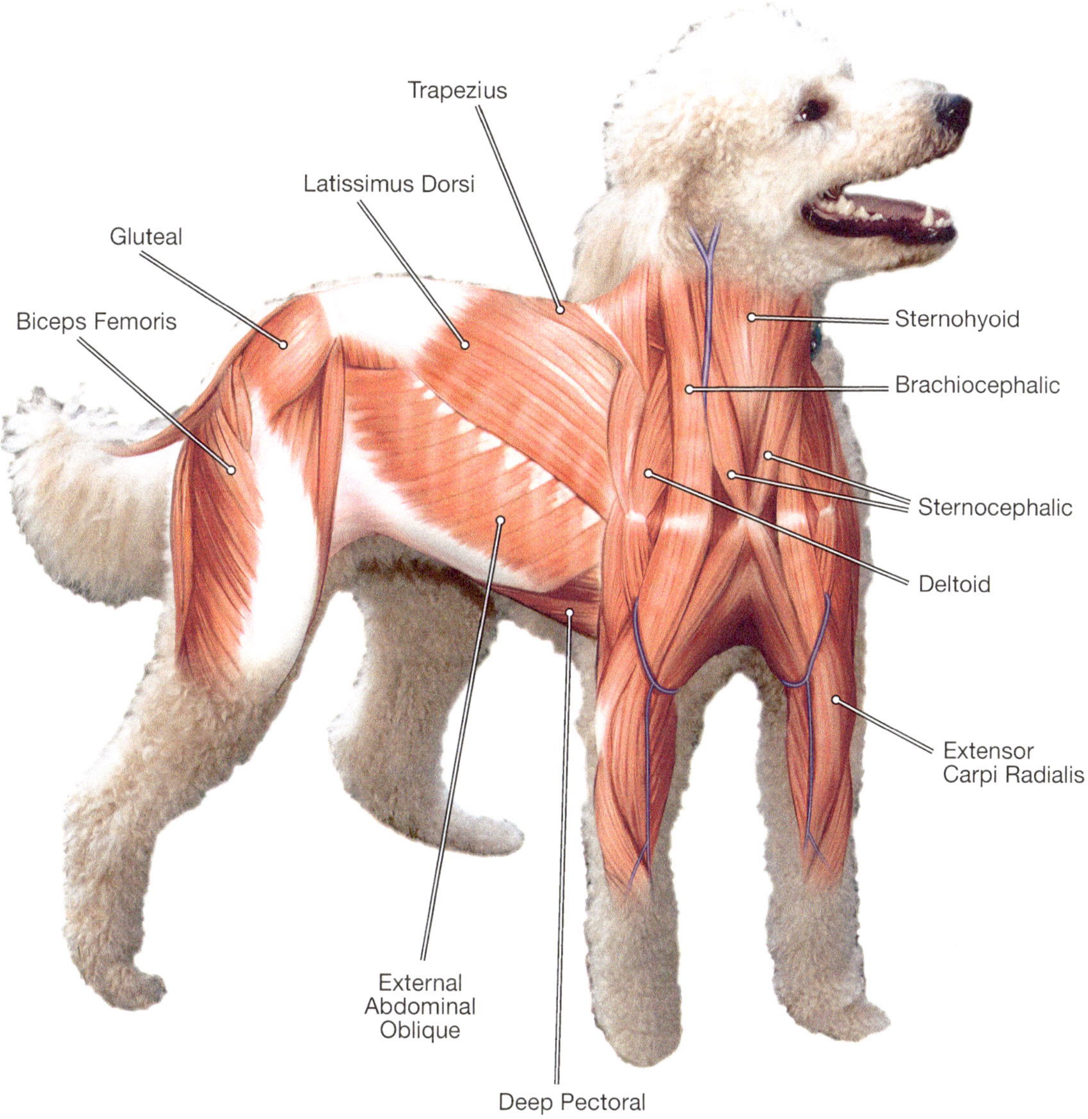

Figure 3.11: Skeletal muscles.
© NCSU.

Deep Skeletal Muscles

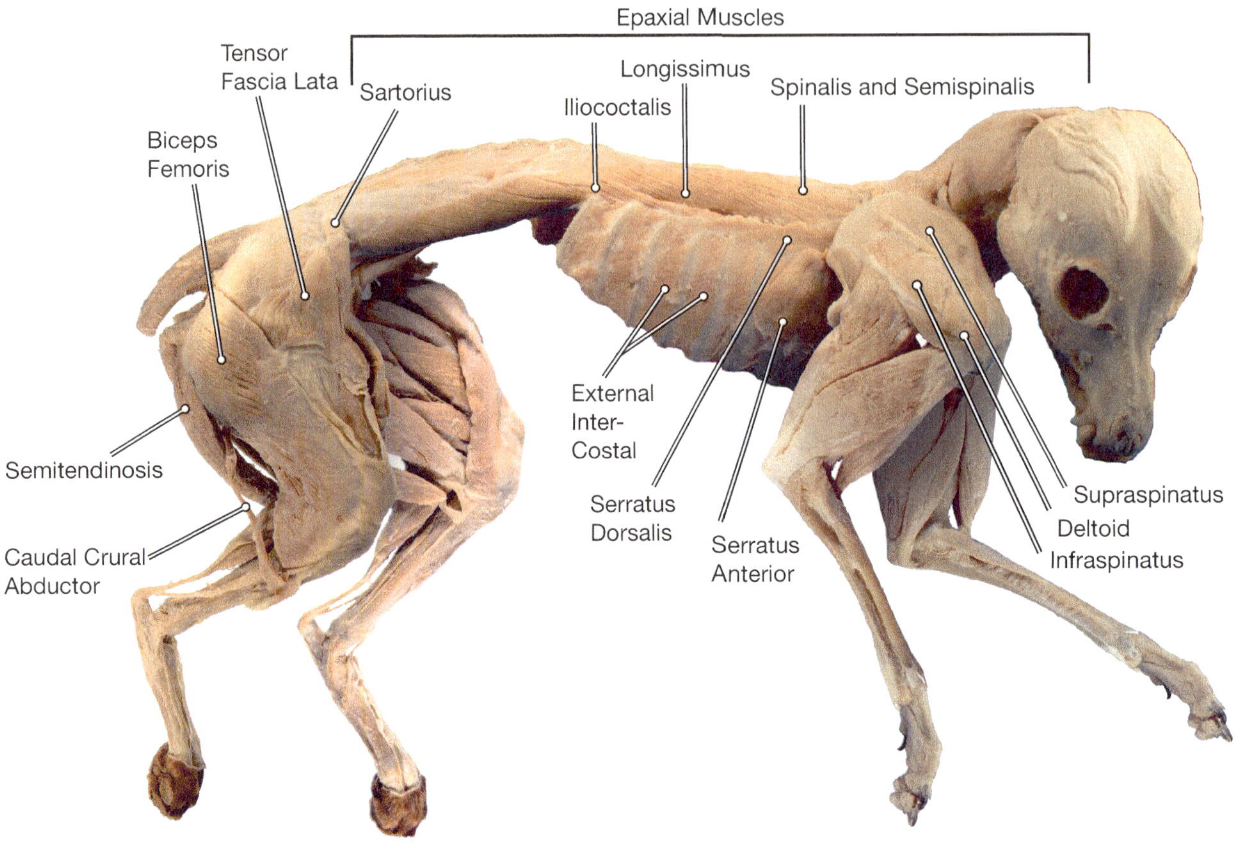

Figure 3.12: Muscles of the trunk.
© NCSU.

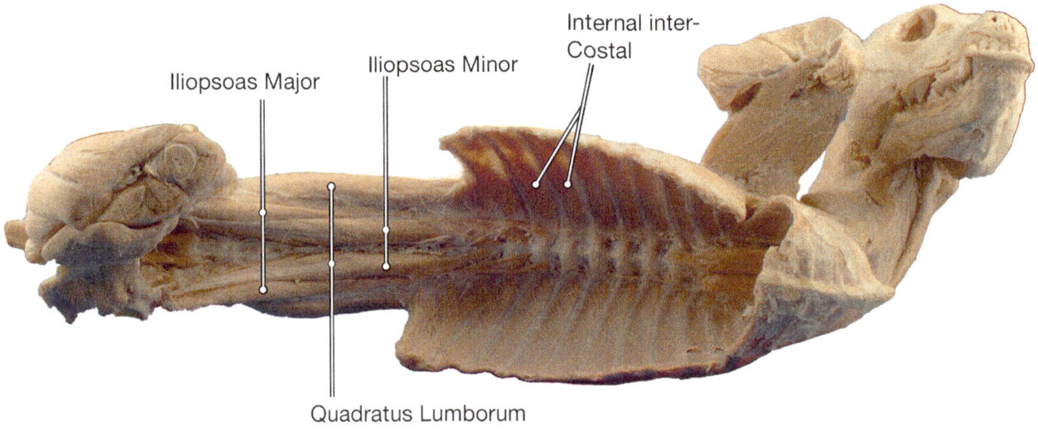

Figure 3.13: Interior view of the trunk muscles.
© NCSU.

Muscles: Thoracic Limb

Figure 3.14: Thoracic limb highlighted, lateral view.
© NCSU.

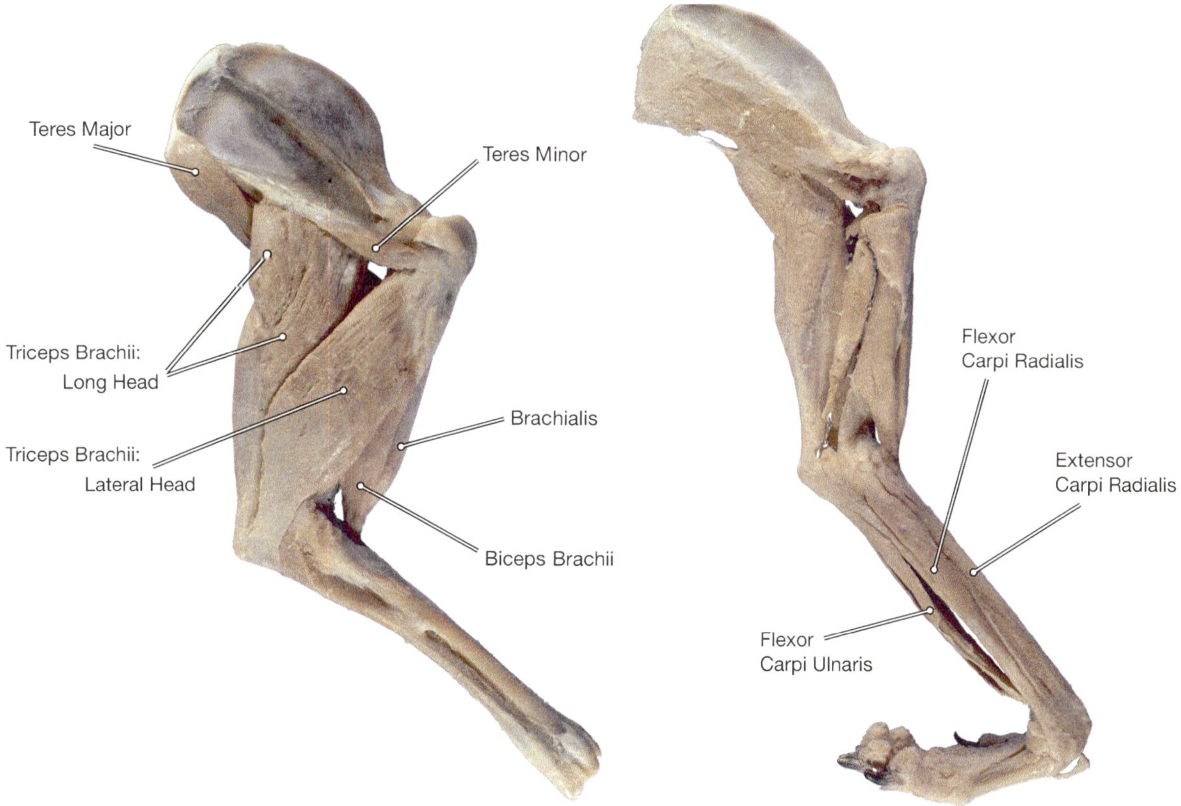

Figure 3.15: Thoracic limb, lateral views.
© NCSU.

Figure 3.16: Thoracic limb highlighted, medial view.

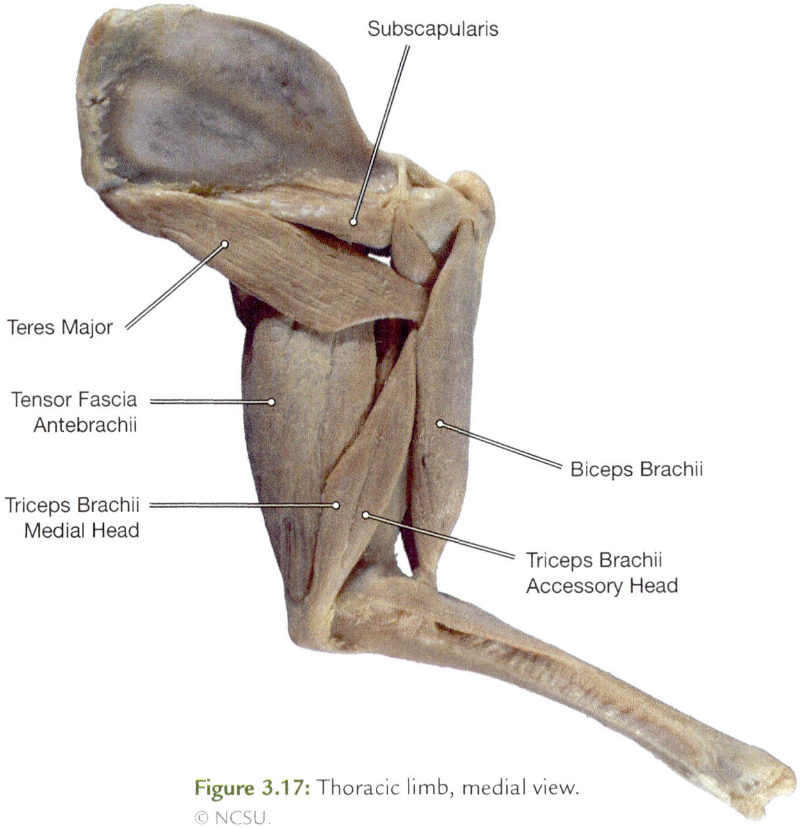

Figure 3.17: Thoracic limb, medial view.

Muscles: Pelvic Limb

Figure 3.18: Pelvic limb highlighted, medial view.

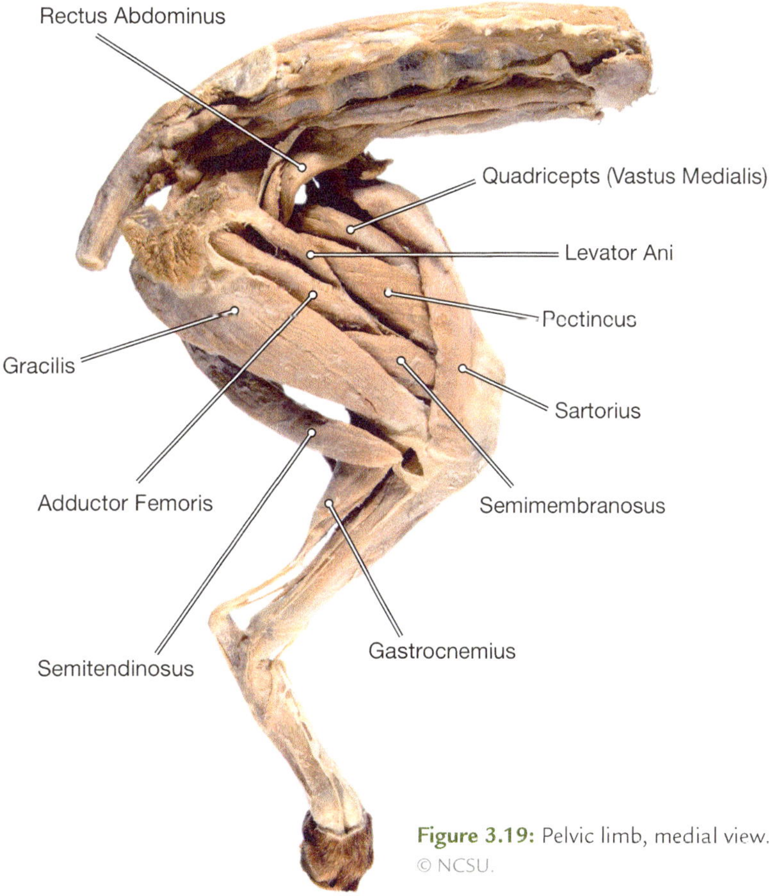

Figure 3.19: Pelvic limb, medial view.

Figure 3.20: Pelvic limb highlighted, lateral view.
© NCSU.

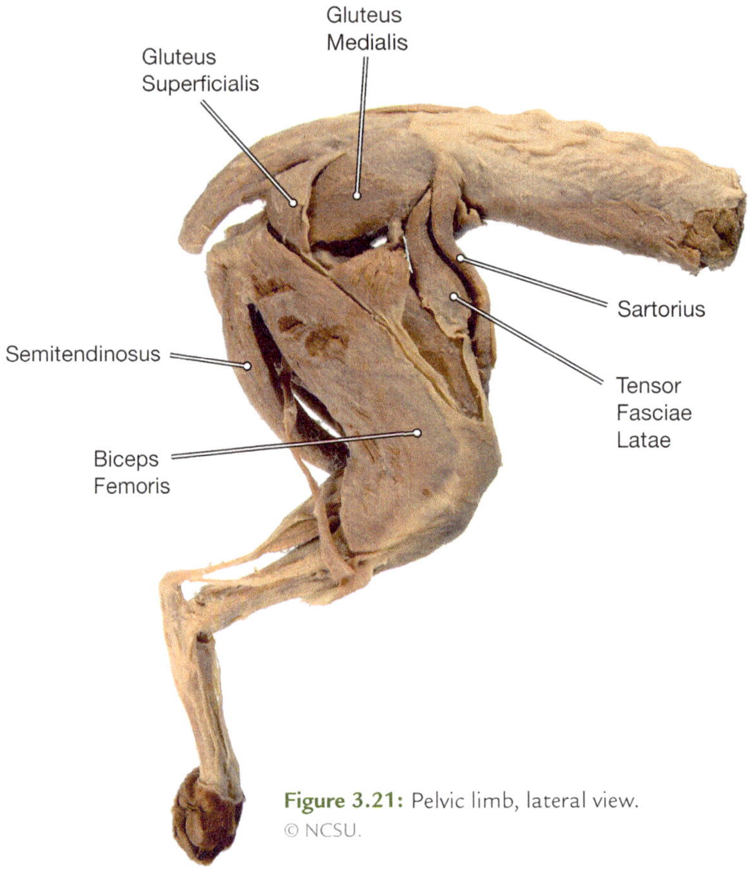

Figure 3.21: Pelvic limb, lateral view.
© NCSU.

Facial Muscles

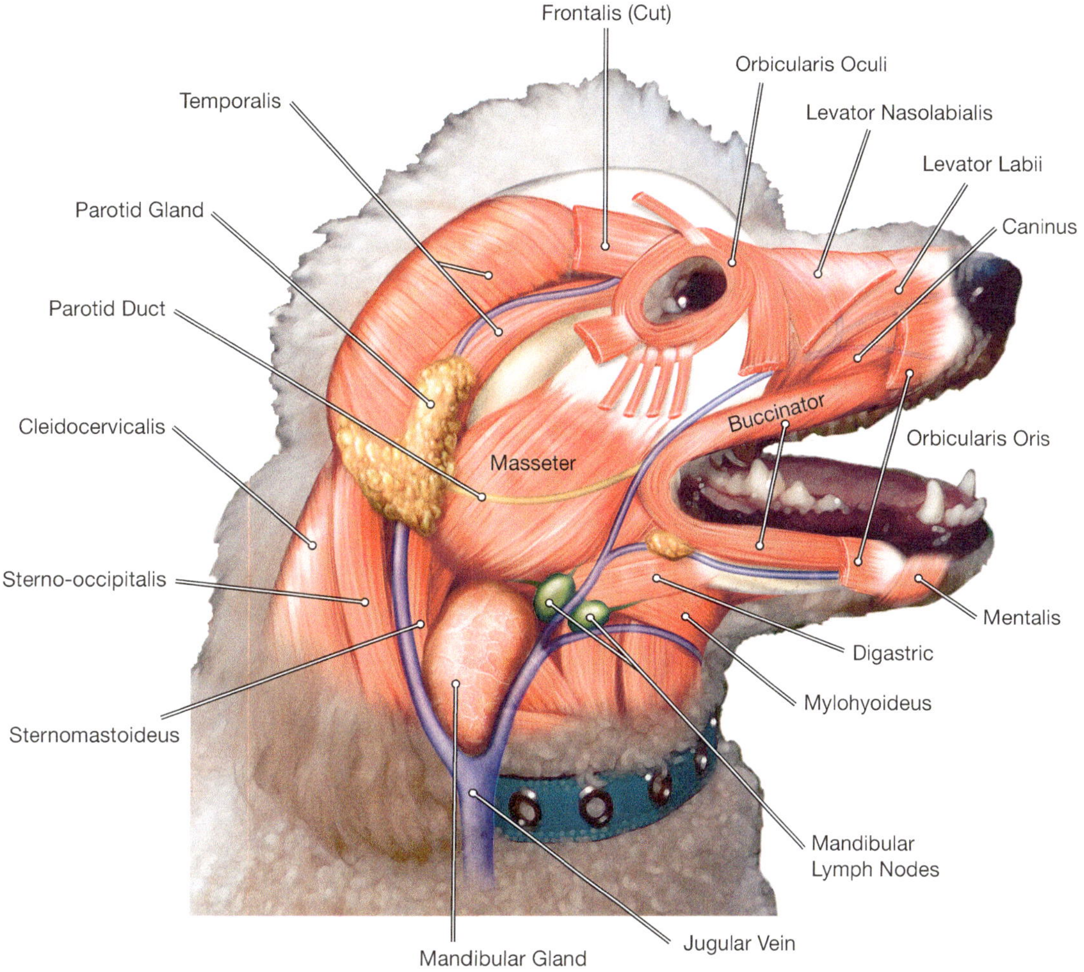

Figure 3.22: Facial muscles and related structures.
© NCSU.

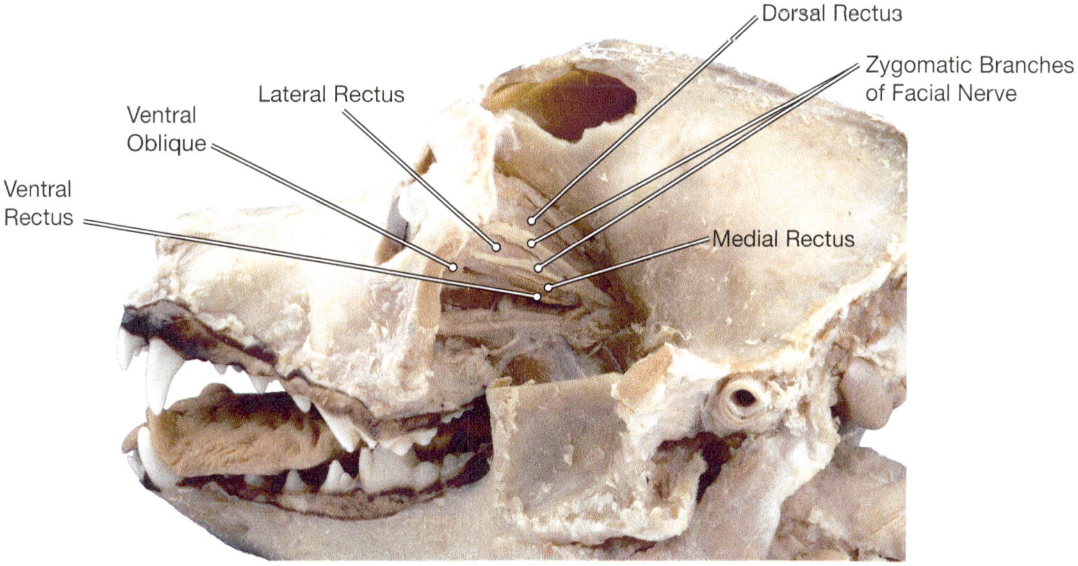

Figure 3.23: Facial muscles.
© NCSU.

Review: Labeling Activity

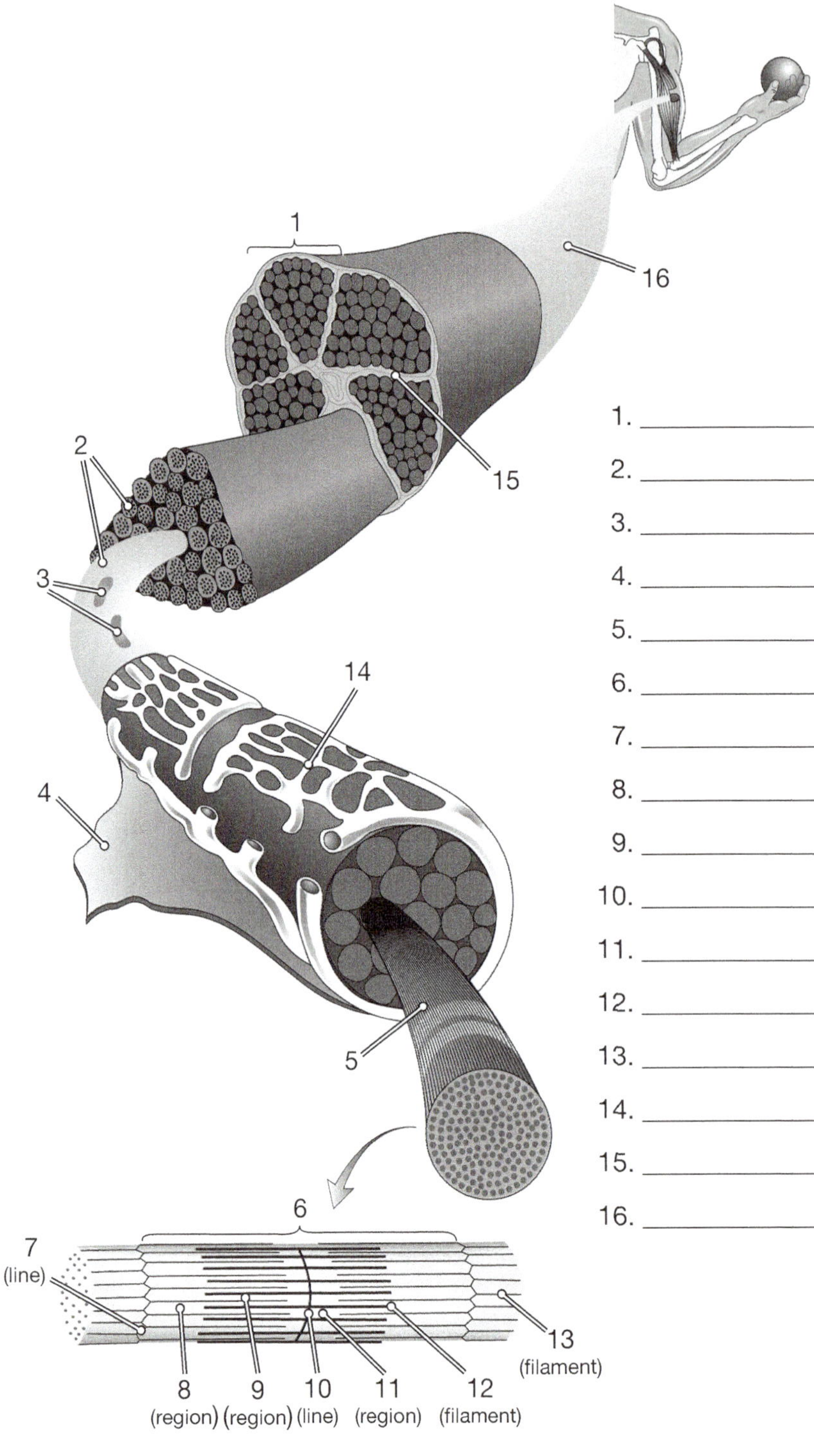

1. ______________________
2. ______________________
3. ______________________
4. ______________________
5. ______________________
6. ______________________
7. ______________________
8. ______________________
9. ______________________
10. ______________________
11. ______________________
12. ______________________
13. ______________________
14. ______________________
15. ______________________
16. ______________________

Figure 3.24: Muscle labeling activity.

Review: Labeling Activity

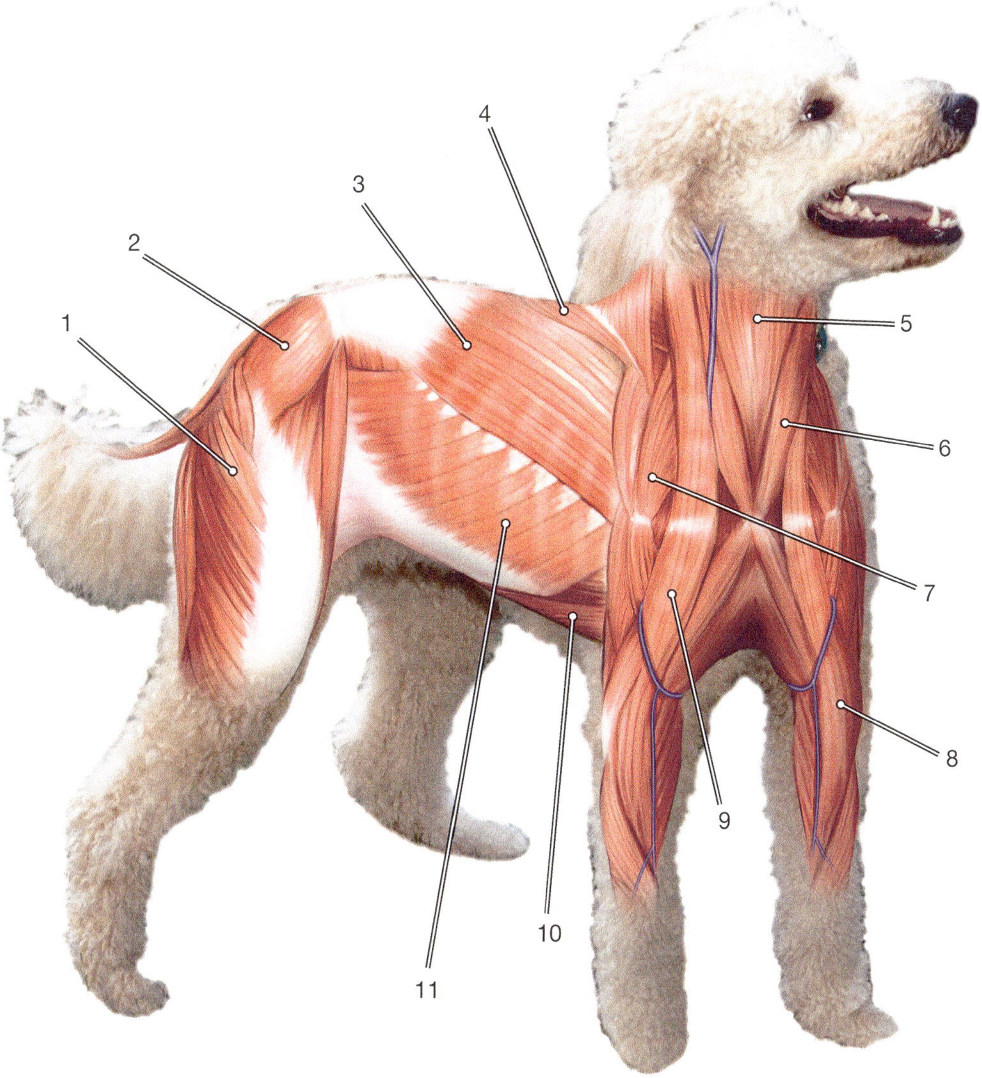

Figure 3.25: Skeletal muscles labeling activity.
© NCSU.

1. ______________________
2. ______________________
3. ______________________
4. ______________________
5. ______________________
6. ______________________
7. ______________________
8. ______________________
9. ______________________
10. ______________________
11. ______________________

Review: Labeling Activity

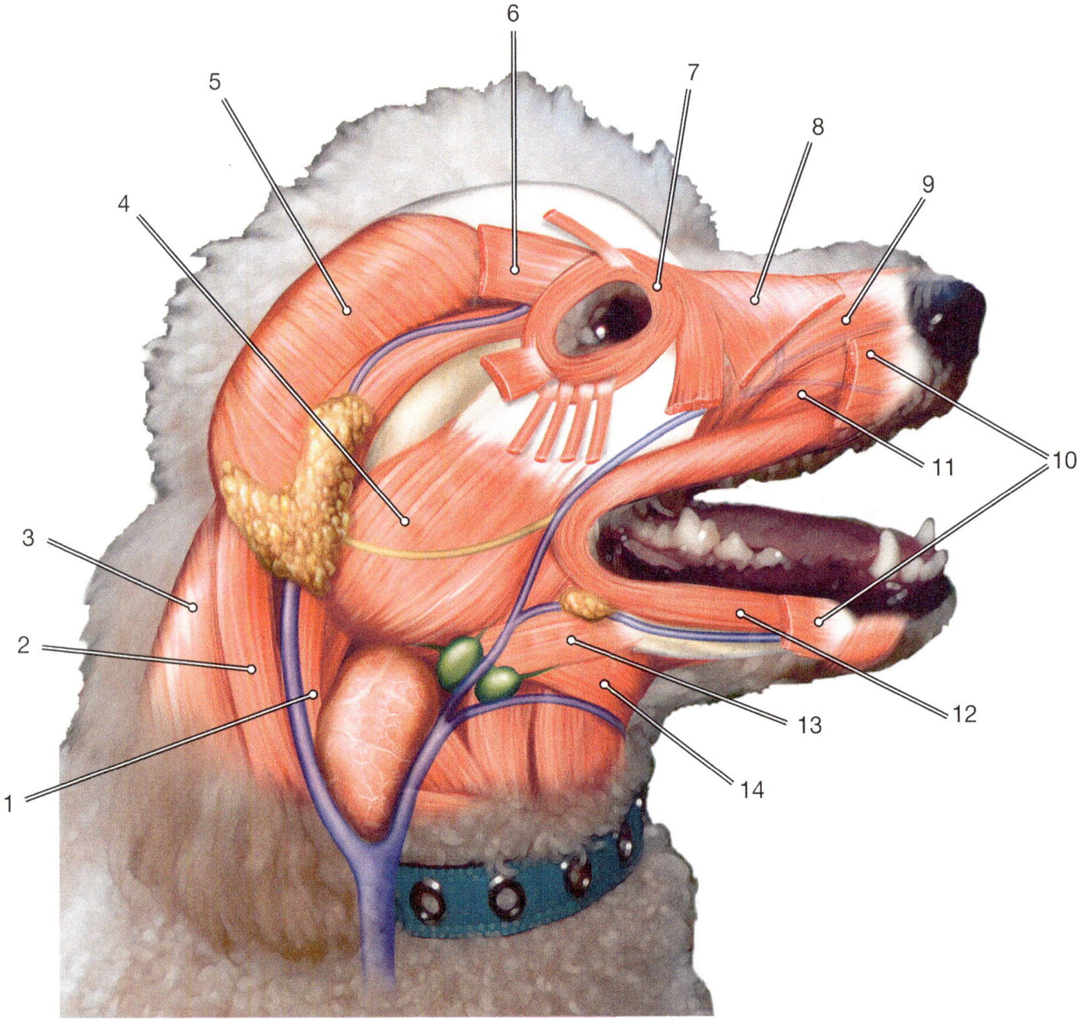

Figure 3.26: Facial muscles labeling activity.
© NCSU.

1. ____________________
2. ____________________
3. ____________________
4. ____________________
5. ____________________
6. ____________________
7. ____________________
8. ____________________
9. ____________________
10. ____________________
11. ____________________
12. ____________________
13. ____________________
14. ____________________

Review: Crossword

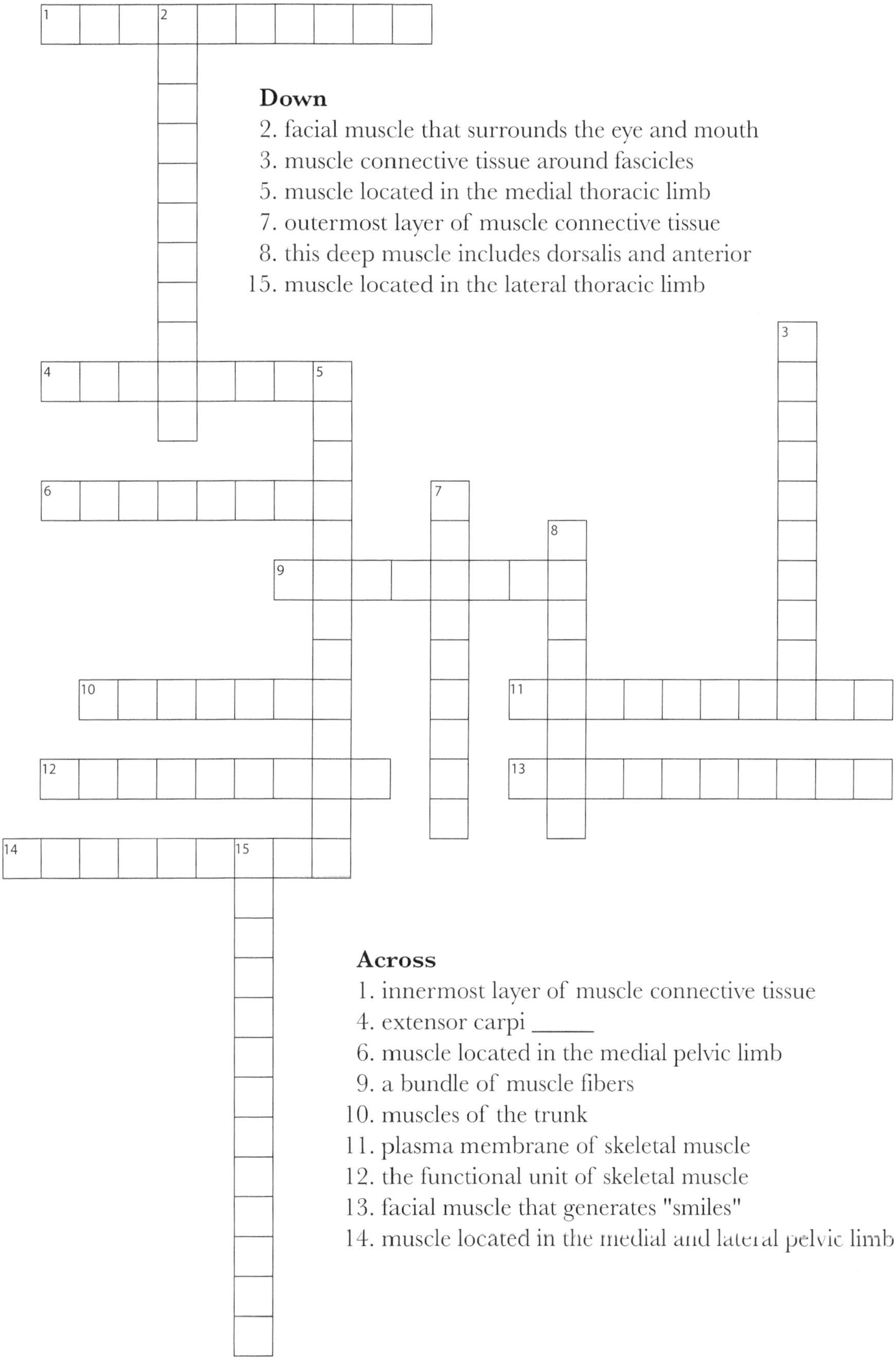

Down

2. facial muscle that surrounds the eye and mouth
3. muscle connective tissue around fascicles
5. muscle located in the medial thoracic limb
7. outermost layer of muscle connective tissue
8. this deep muscle includes dorsalis and anterior
15. muscle located in the lateral thoracic limb

Across

1. innermost layer of muscle connective tissue
4. extensor carpi _____
6. muscle located in the medial pelvic limb
9. a bundle of muscle fibers
10. muscles of the trunk
11. plasma membrane of skeletal muscle
12. the functional unit of skeletal muscle
13. facial muscle that generates "smiles"
14. muscle located in the medial and lateral pelvic limb

Chapter 4

Central Nervous System and Sensory Organs

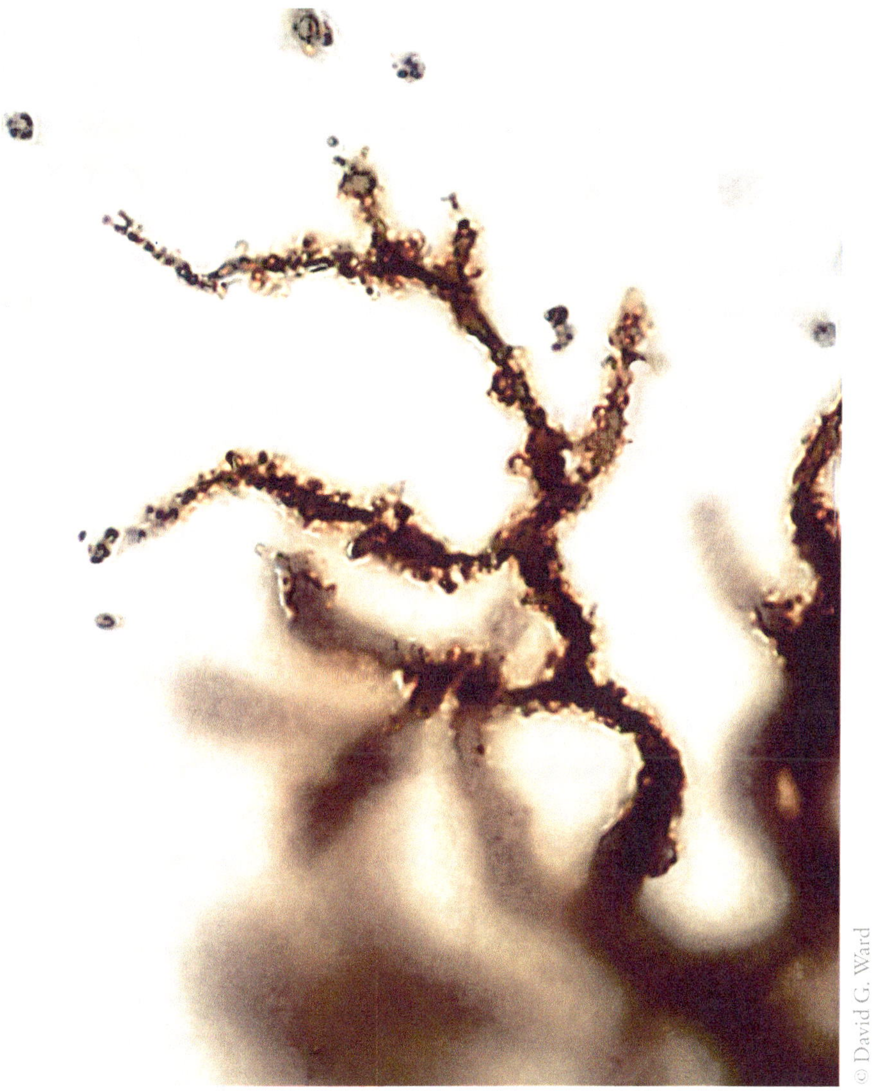

Dendrites of Neuron Showing Dendritic Spines (Brightfield, Silver, x1880)

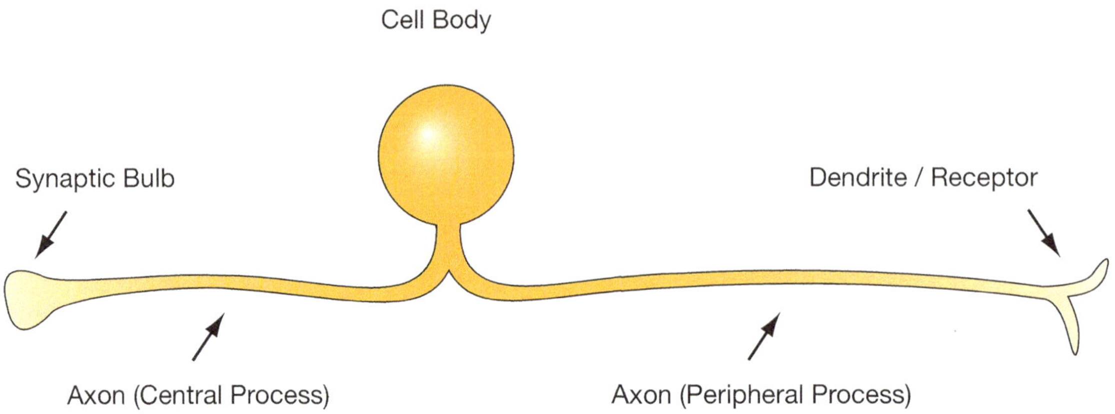

Figure 4.1: Unipolar neuron (sensory).

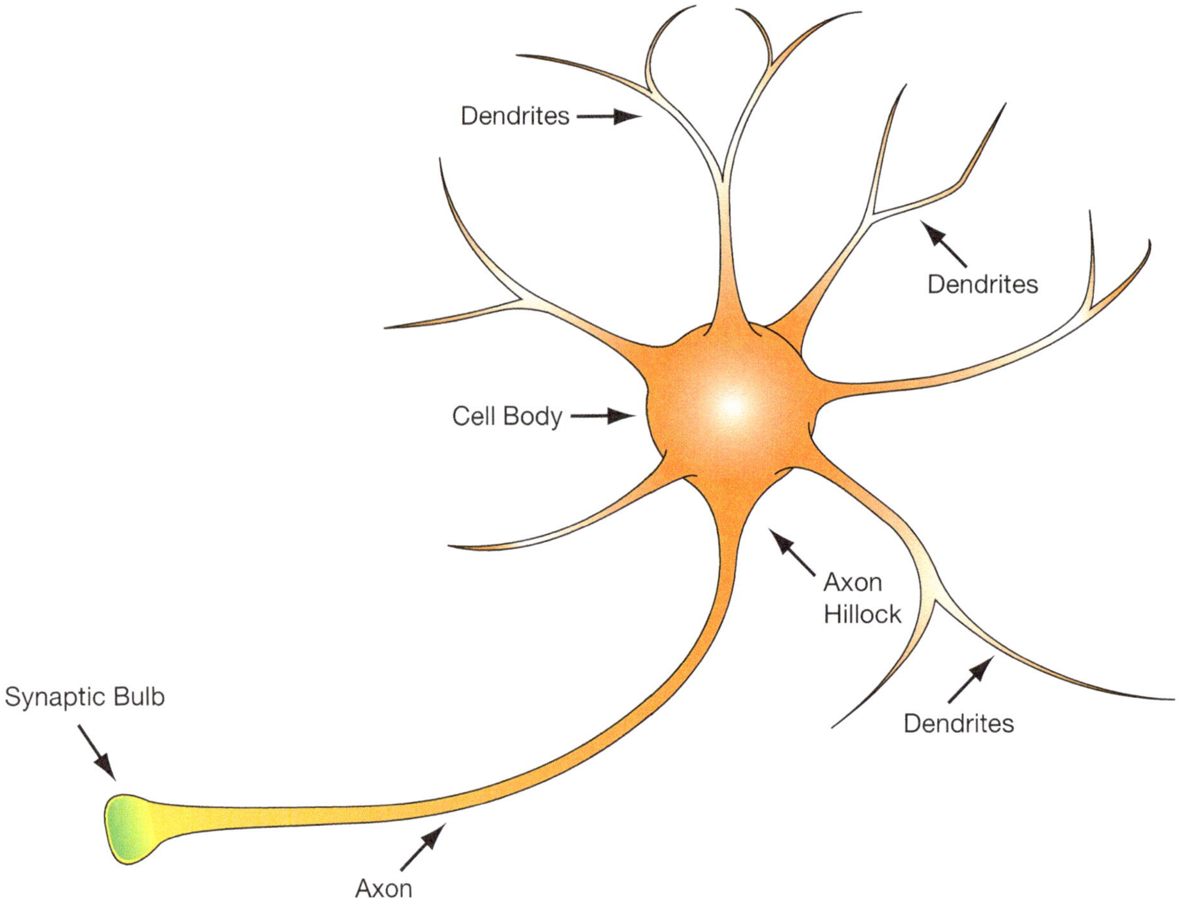

Figure 4.2: Multipolar neuron (motor).

Communication Between Neurons

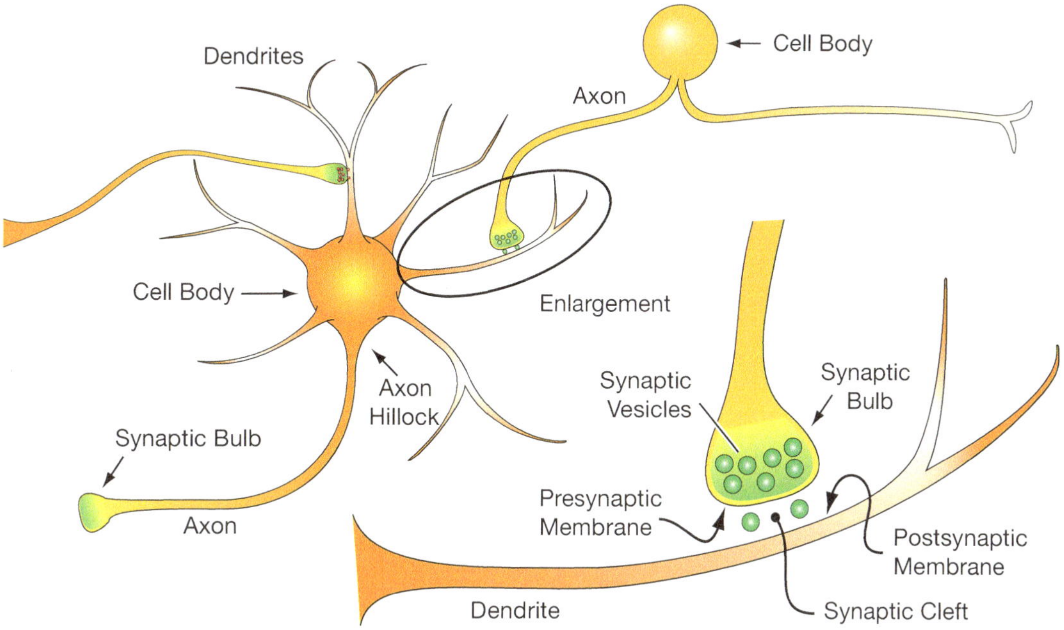

Figure 4.3: Synaptic communication.
© David G. Ward.

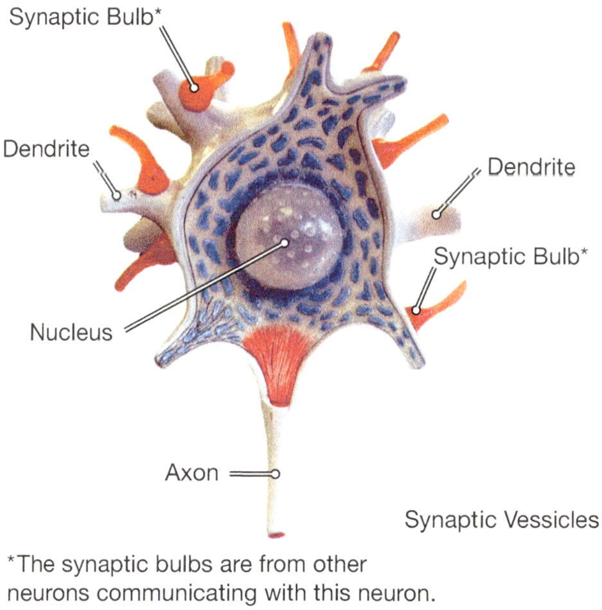

Figure 4.4: Mutlipolar neuron.
© David G. Ward.

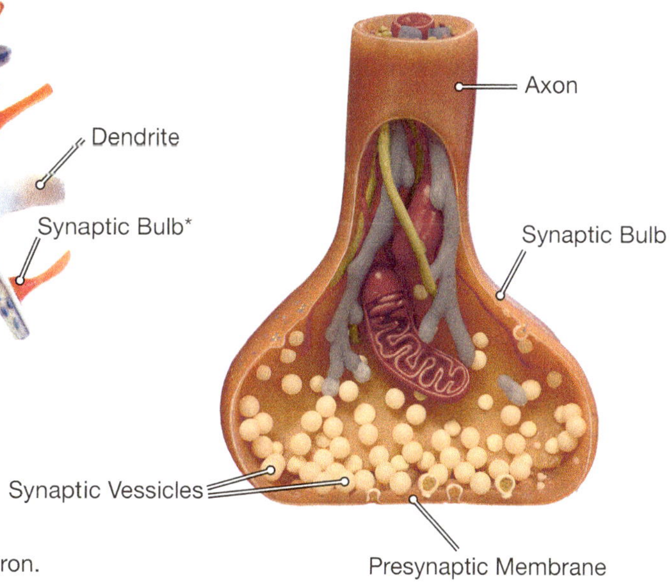

Figure 4.5: Synaptic bulb.
© David G. Ward.

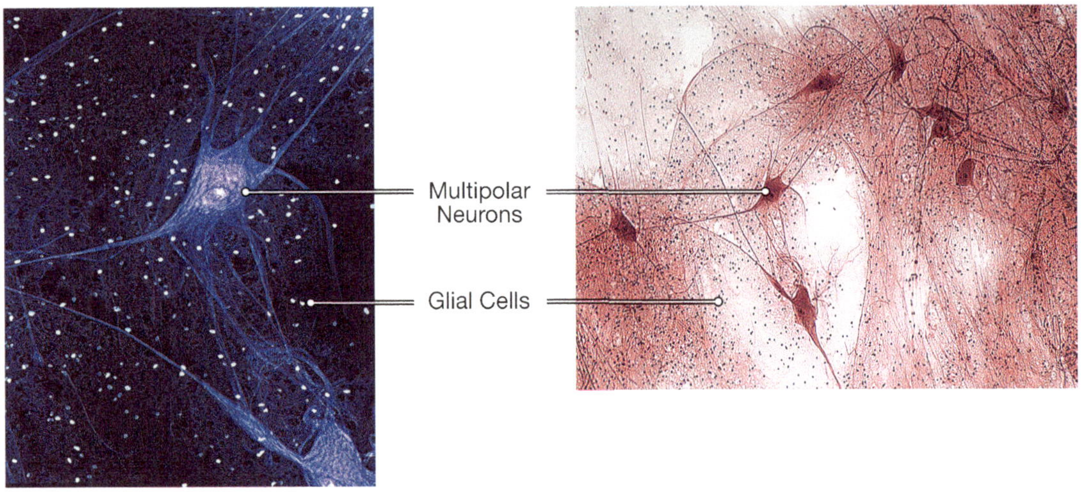

Figure 4.6: Spinal multipolar neurons.
© David G. Ward.

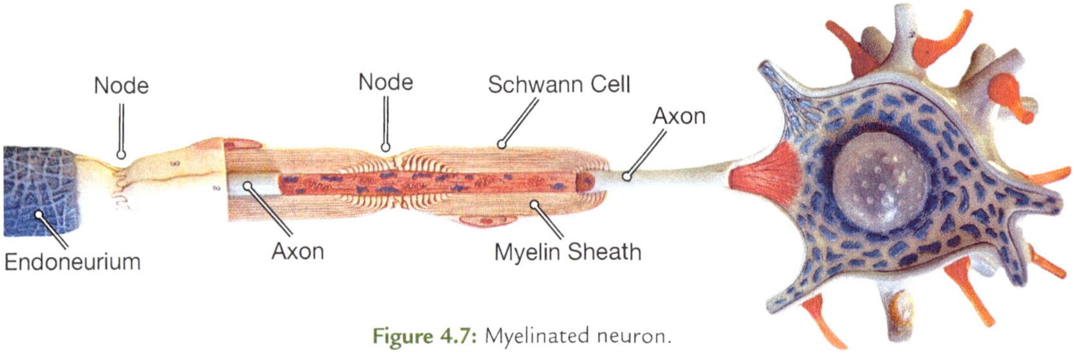

Figure 4.7: Myelinated neuron.
© David G. Ward.

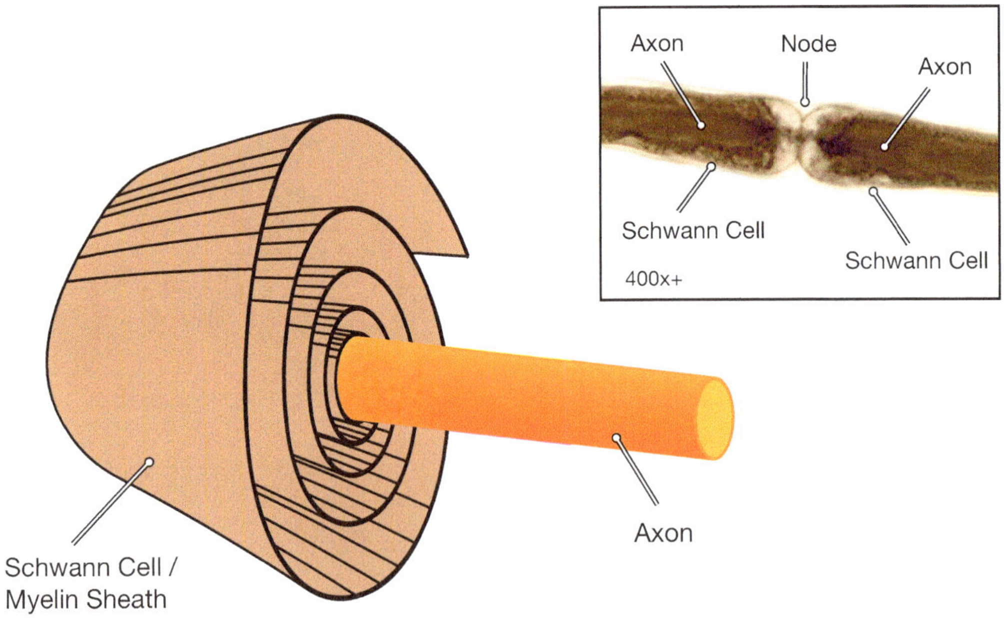

Figure 4.8: Schwann cell.
© David G. Ward.

Nerve and Schwann Cells

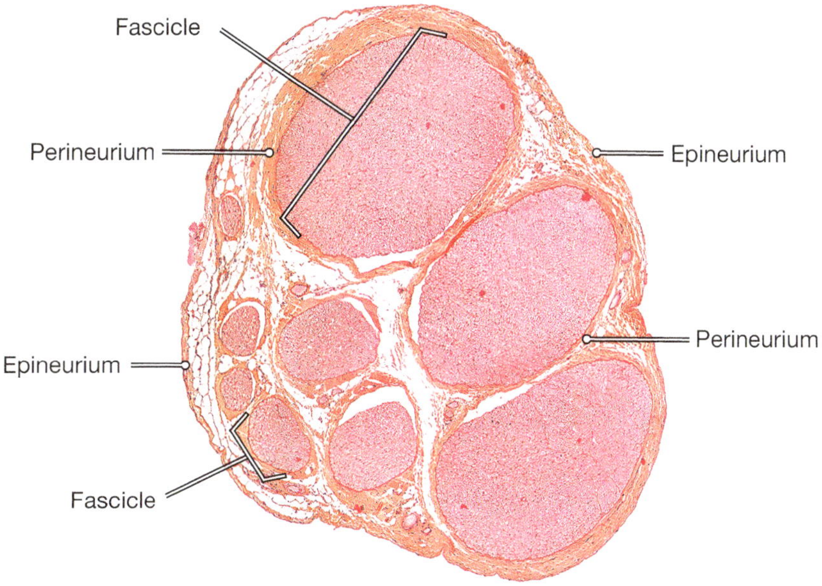

Figure 4.9: Nerve cell, histology.
© David G. Ward.

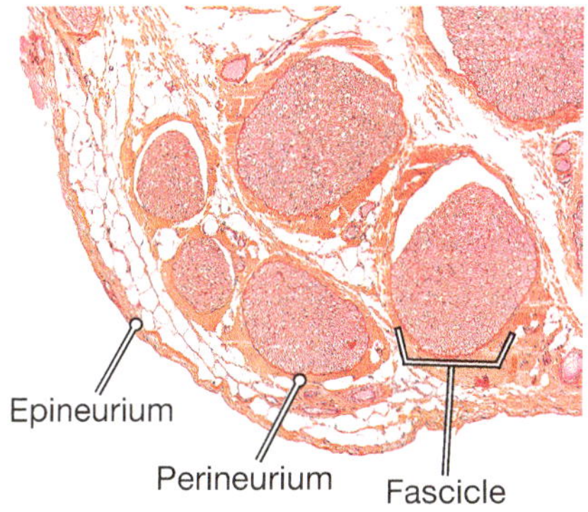

Figure 4.10: Nerve.
© David G. Ward.

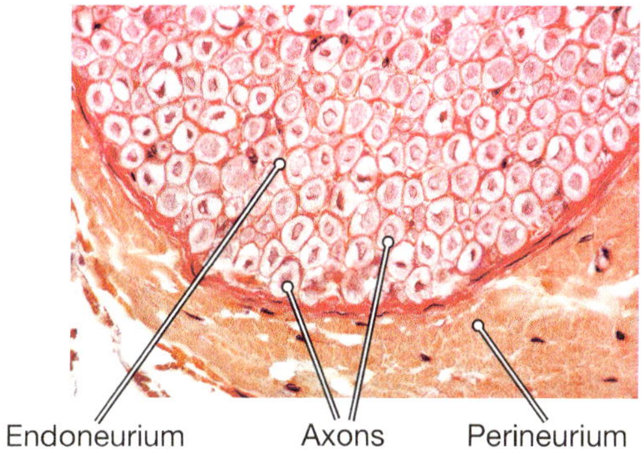

Figure 4.11: Nerve fascicle.
© David G. Ward.

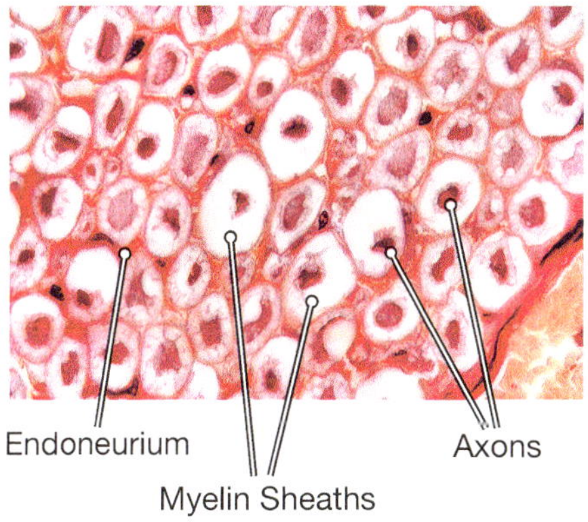

Figure 4.12: Schwann cells and axons.
© David G. Ward.

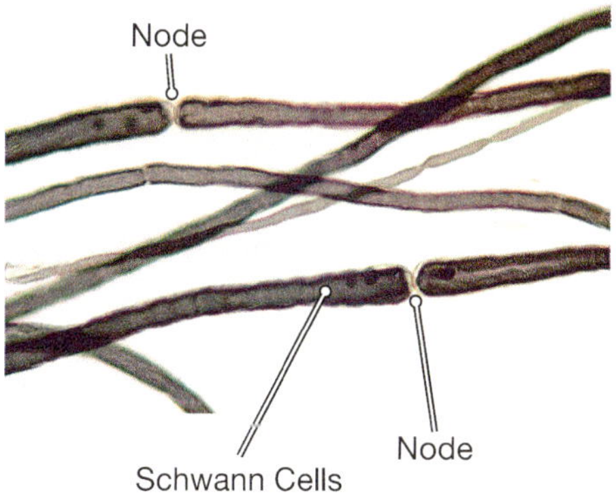

Figure 4.13: Schwann cells.
© David G. Ward.

Neuromuscular Junctions and Motor Units

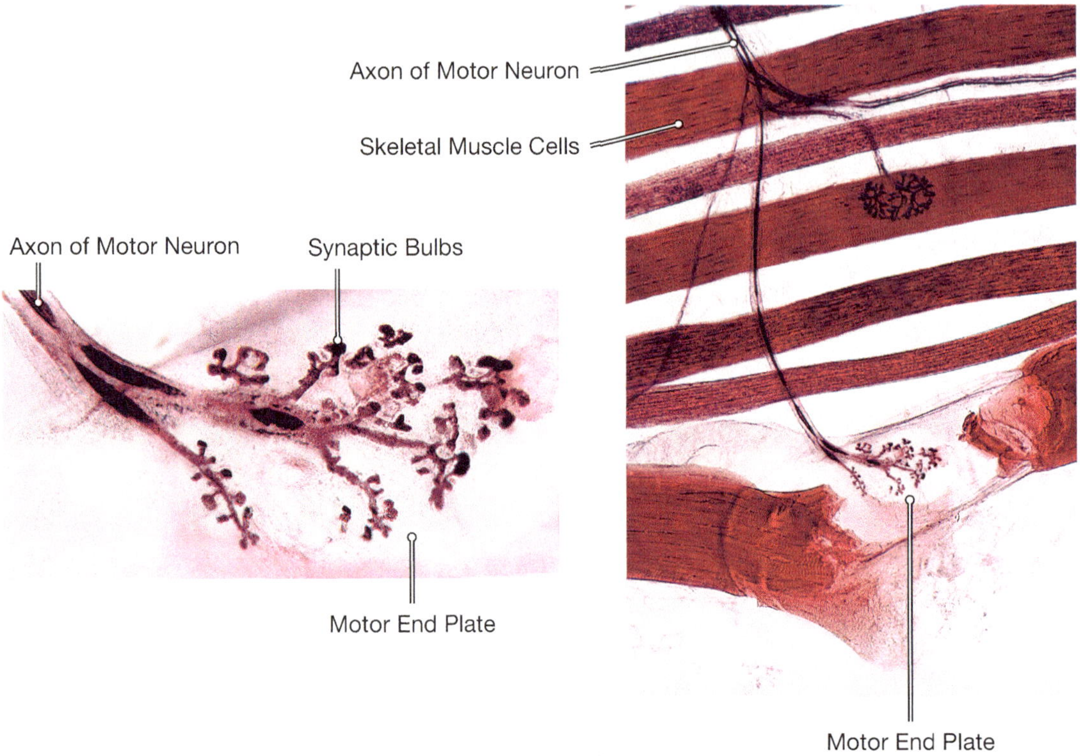

Figure 4.14: Neuromuscular junctions.
© David G. Ward.

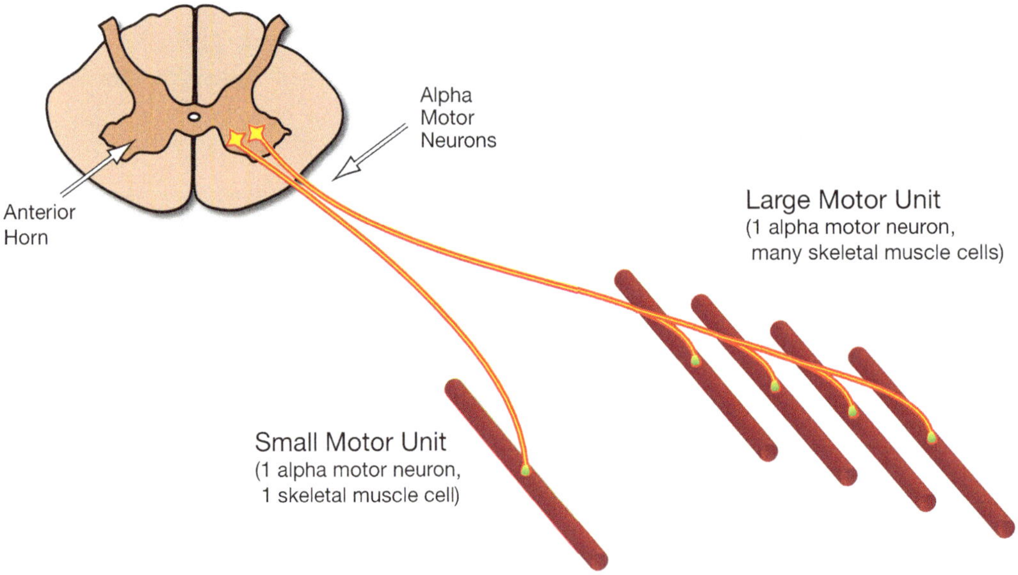

Figure 4.15: Motor units.
© David G. Ward.

Spinal Cord

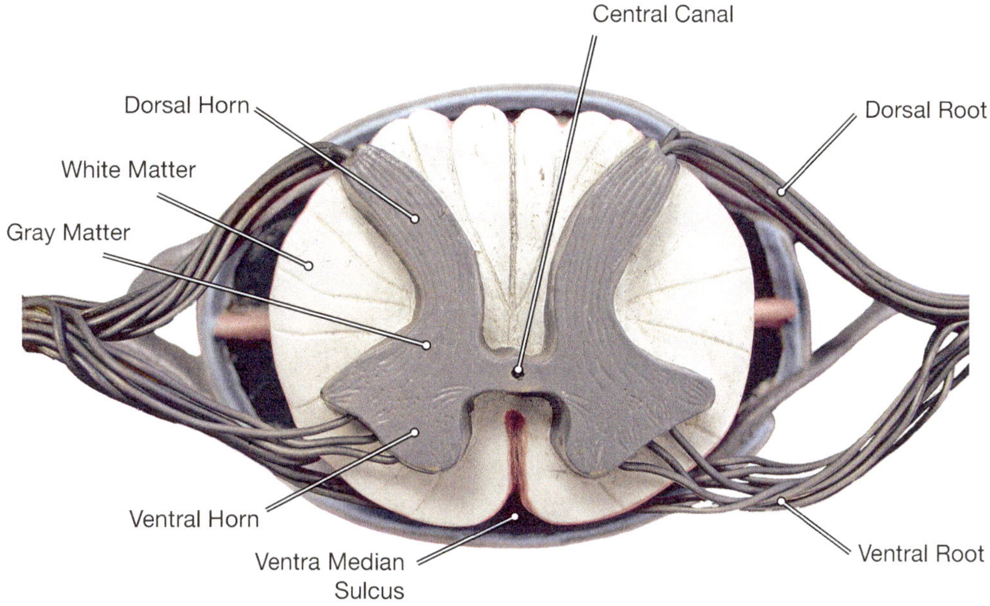

Figure 4.16: Cross section with meninges.
© David G. Ward.

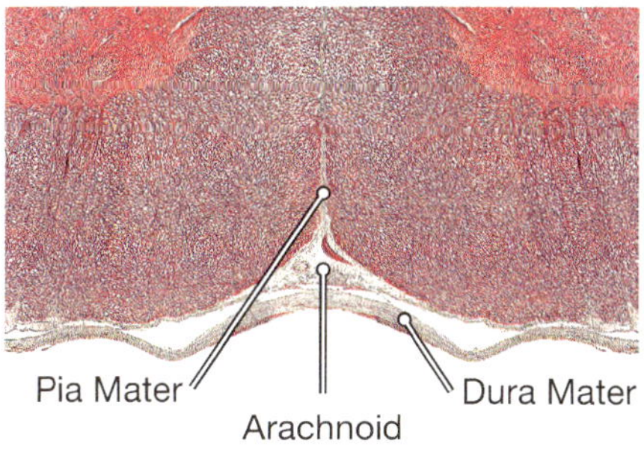

Figure 4.17: Meninges.
© David G. Ward.

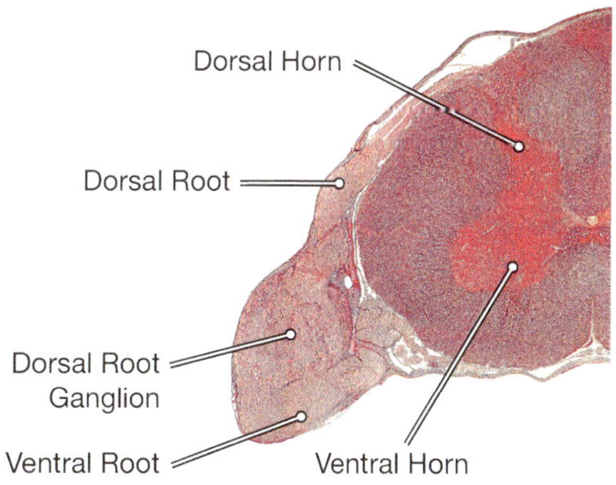

Figure 4.18: Horns and roots.
© David G. Ward.

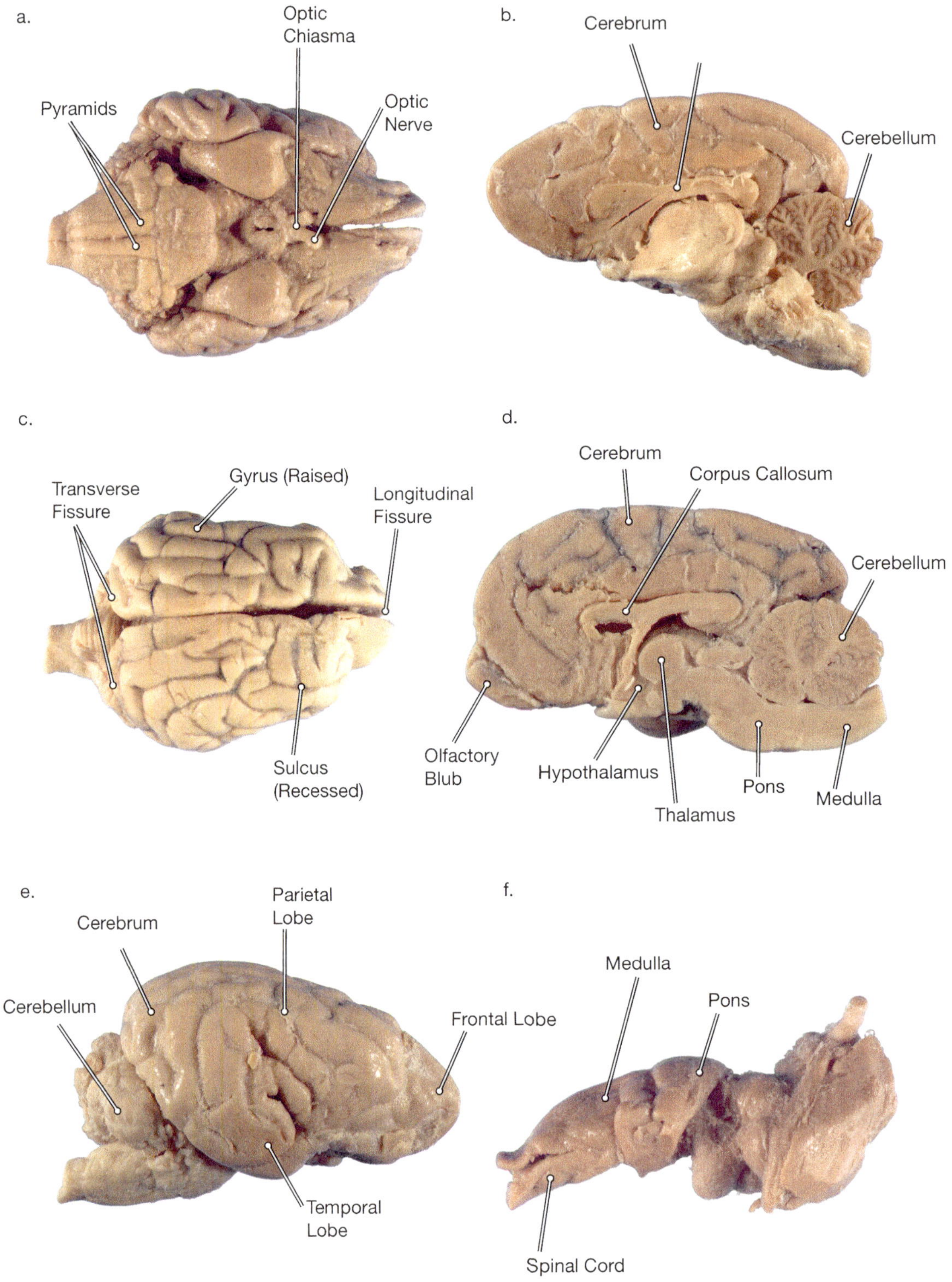

Figure 4.19: Brain.

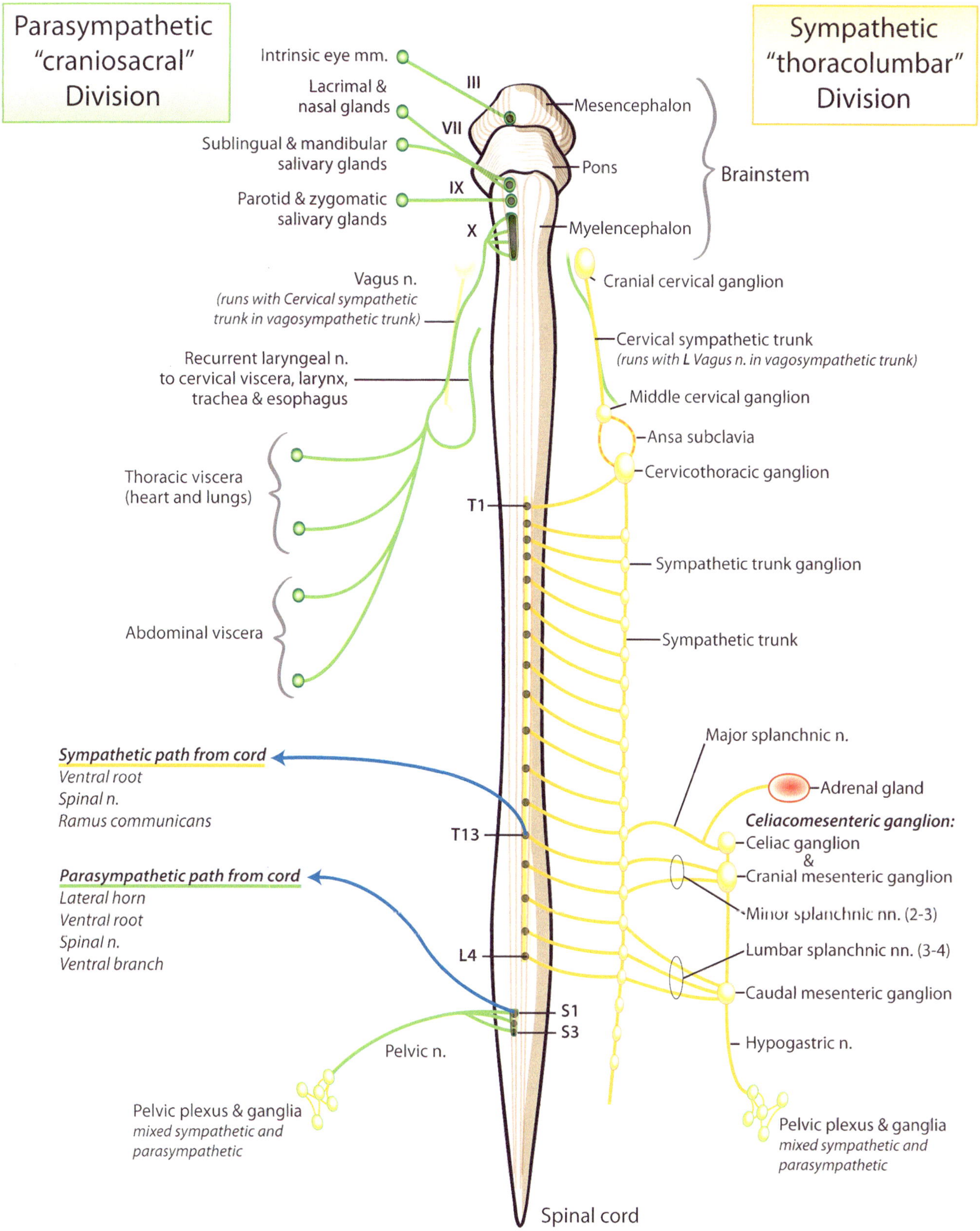

Figure 4.20: Spinal cord divisions.

Parasympathetic Ganglia and Neurotransmitters

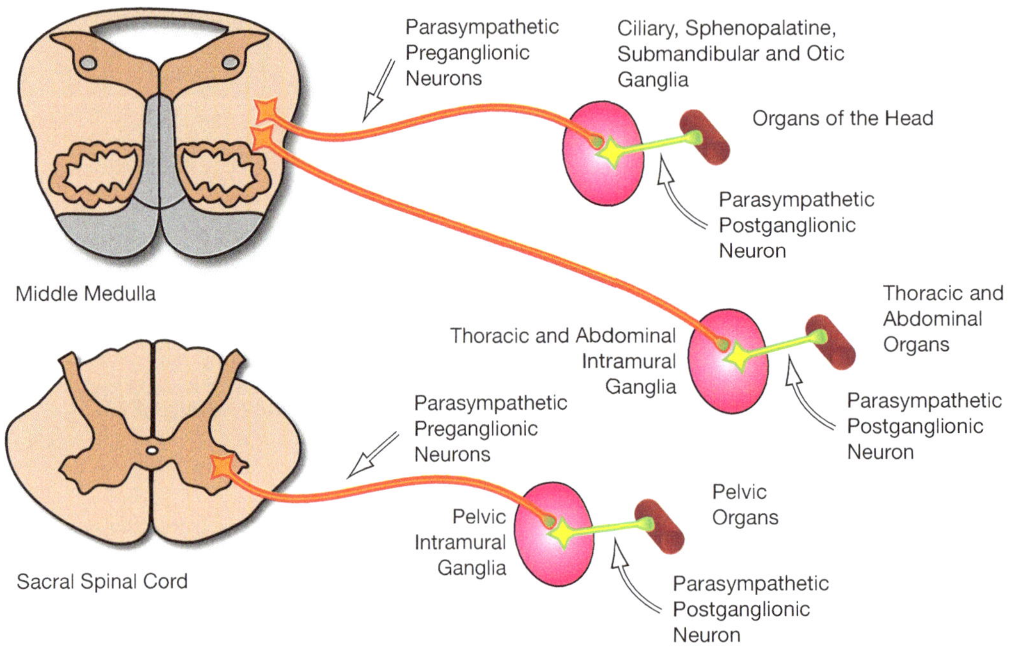

Figure 4.21: Parasympathetic ganglia.

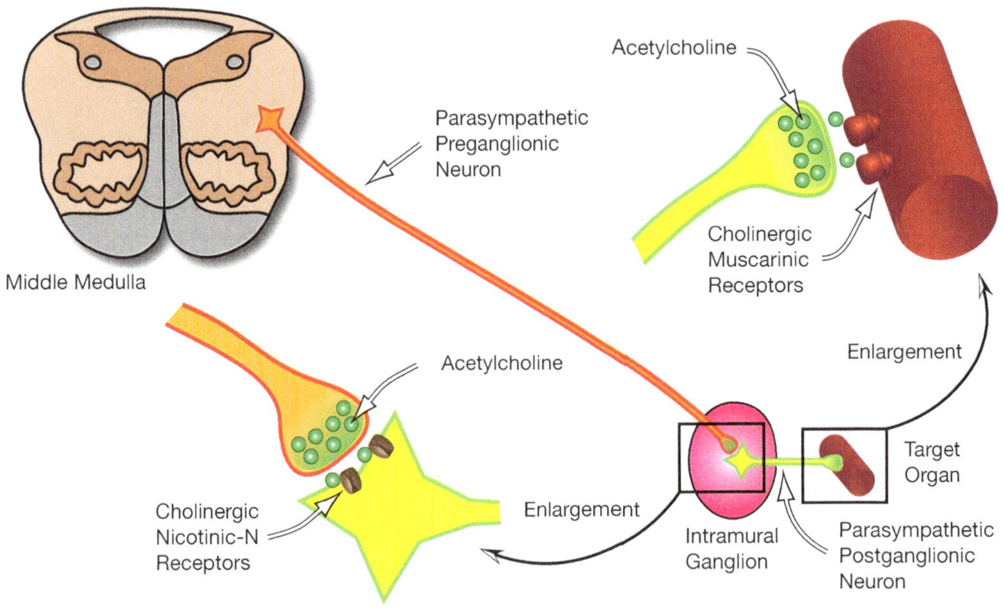

Figure 4.22: Parasympathetic neurotransmitters.

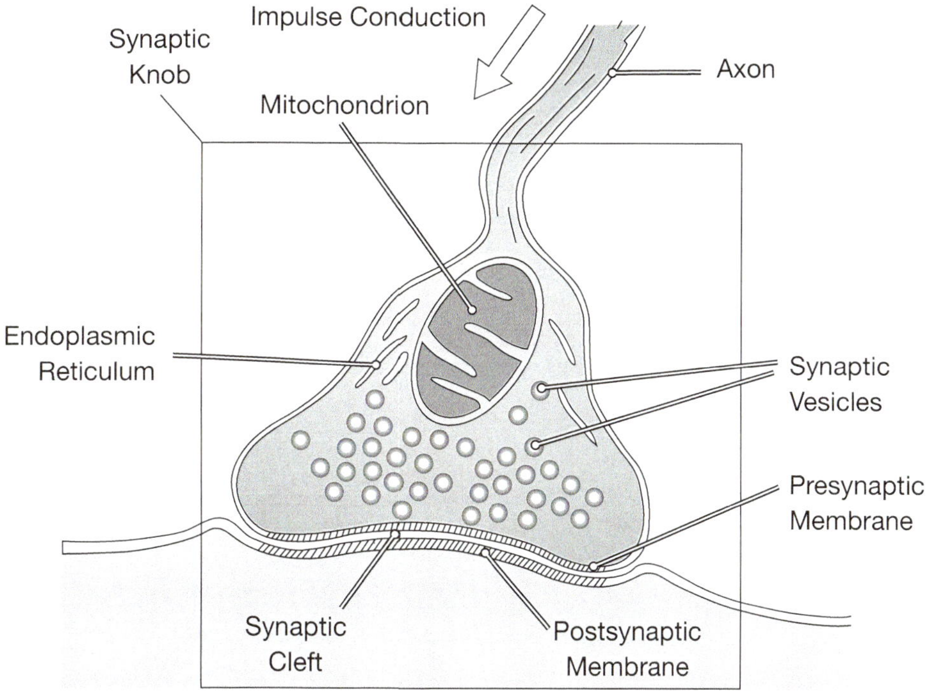

Figure 4.23: Synapse.

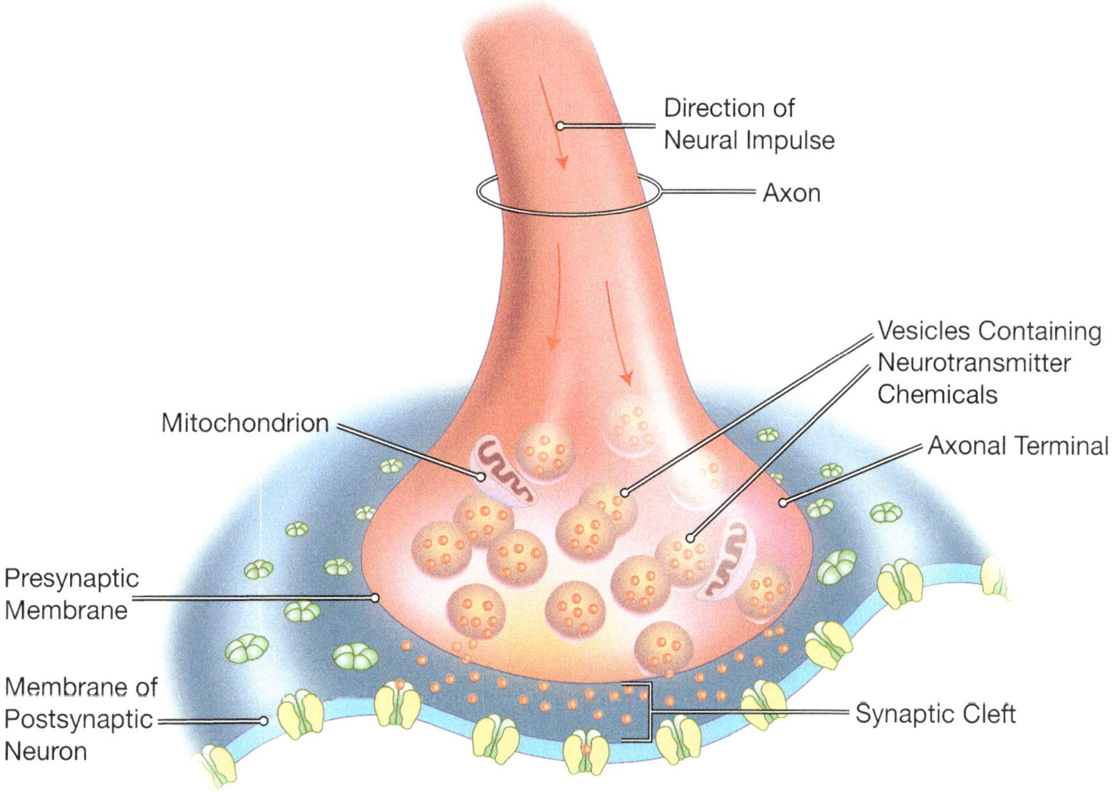

Figure 4.24: Synapse structure.

Sympathetic Ganglia and Neurotransmitters

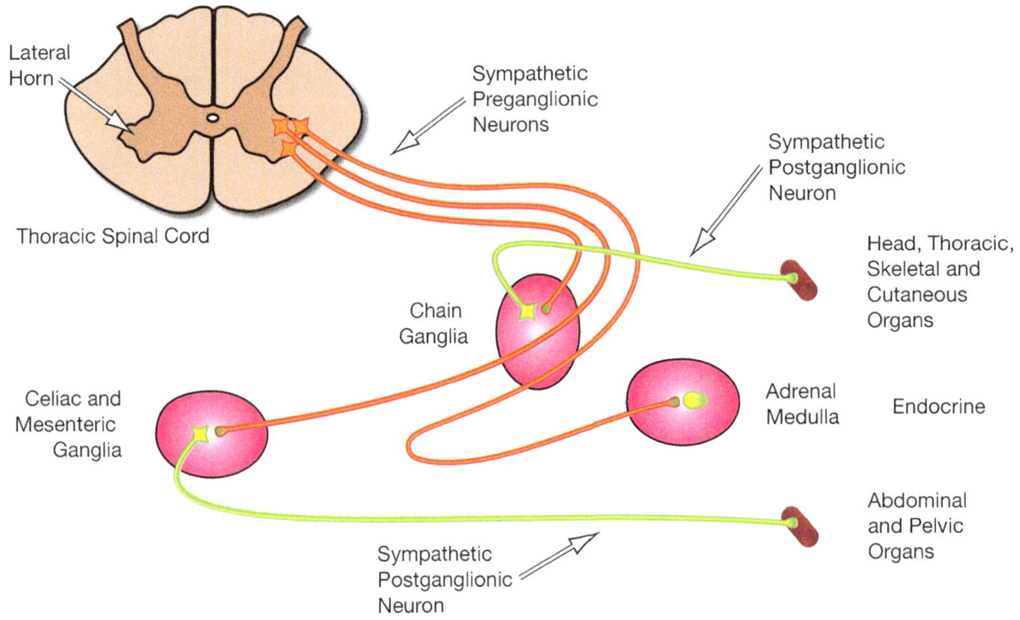

Figure 4.25: Sympathetic ganglia.
© David G. Ward.

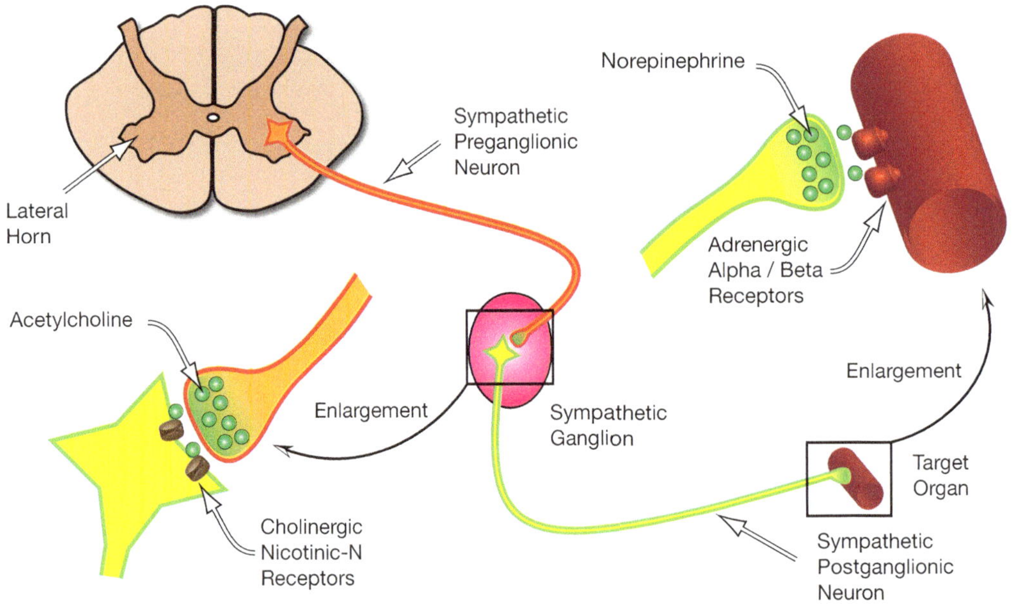

Figure 4.26: Sympathetic neurotransmitters.
© David G. Ward.

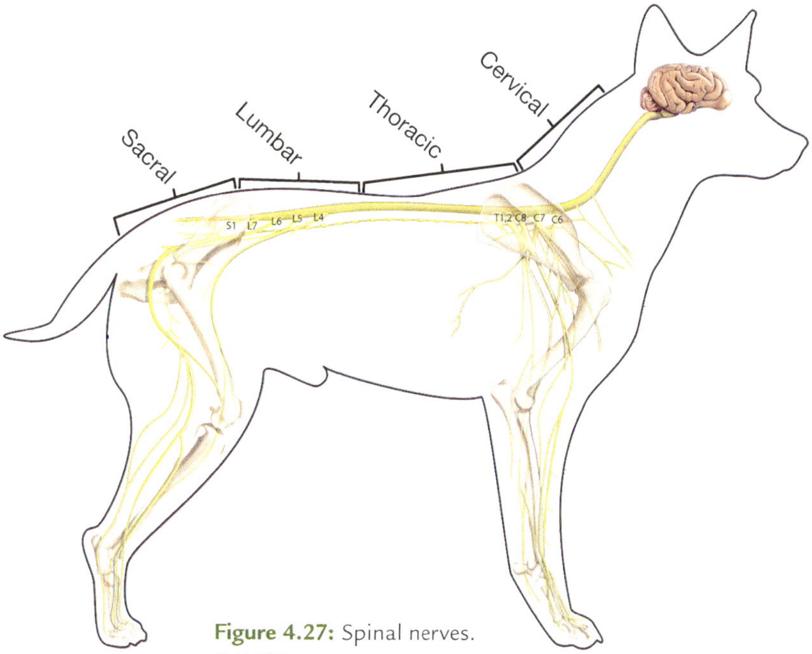

Figure 4.27: Spinal nerves.

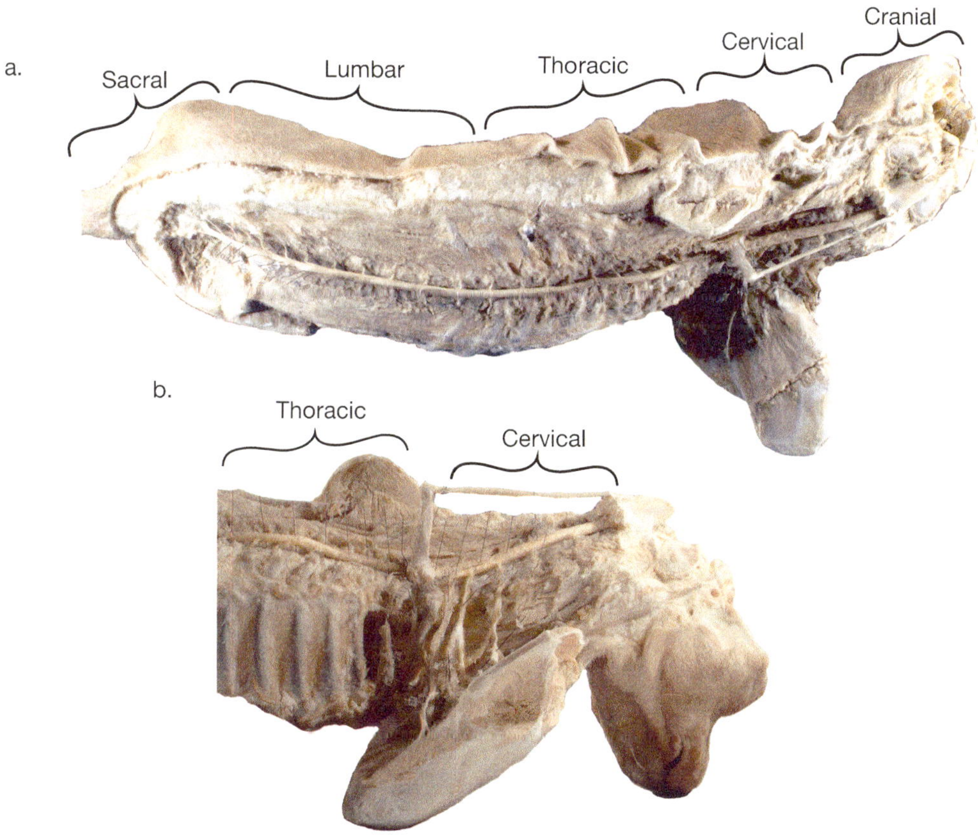

Figure 4.28: Spinal nerves.

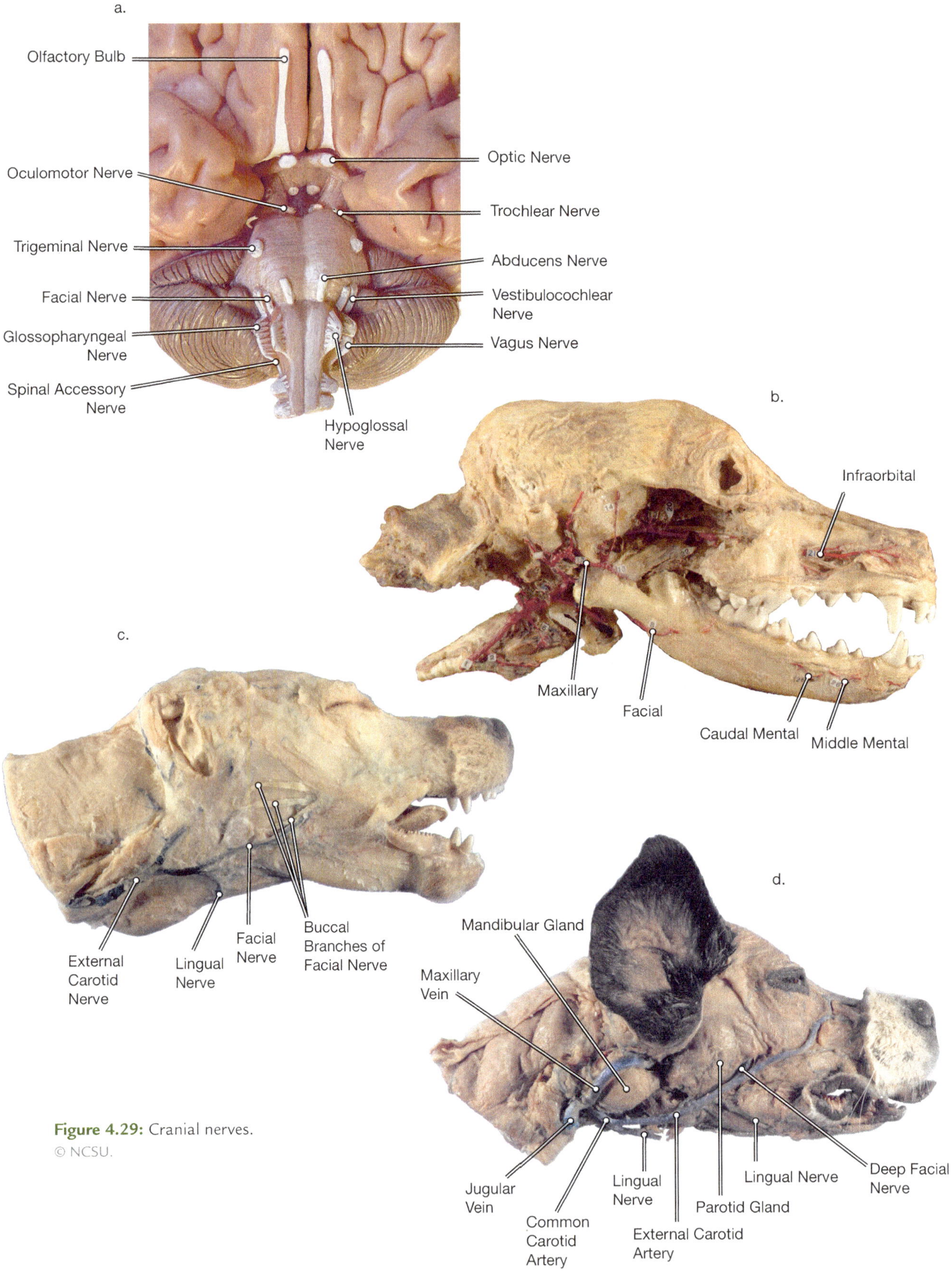

Figure 4.29: Cranial nerves.

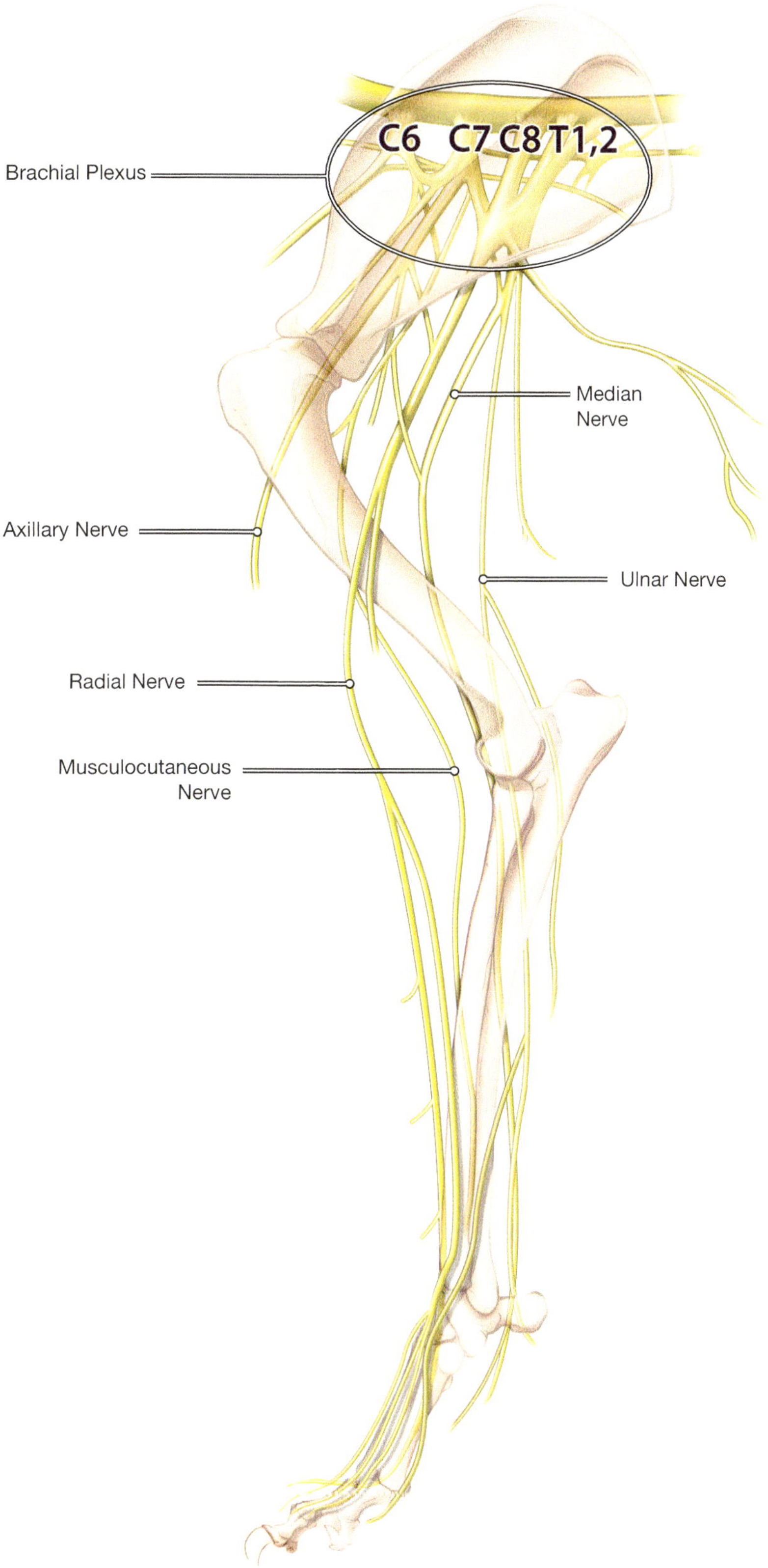

Figure 4.30: Nerves of the thoracic limb, lateral view.

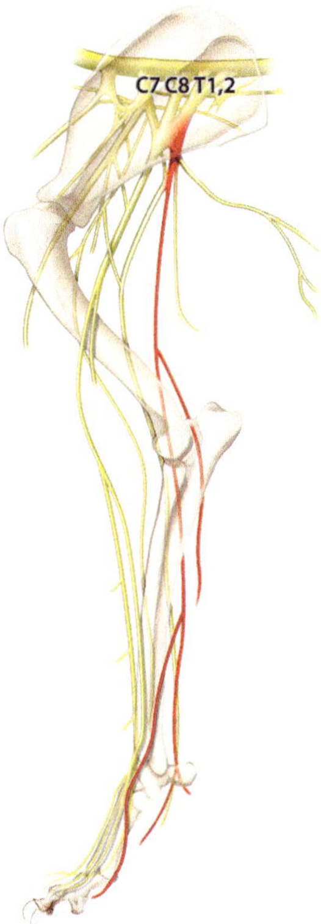

Figure 4.31: Ulnar nerve.

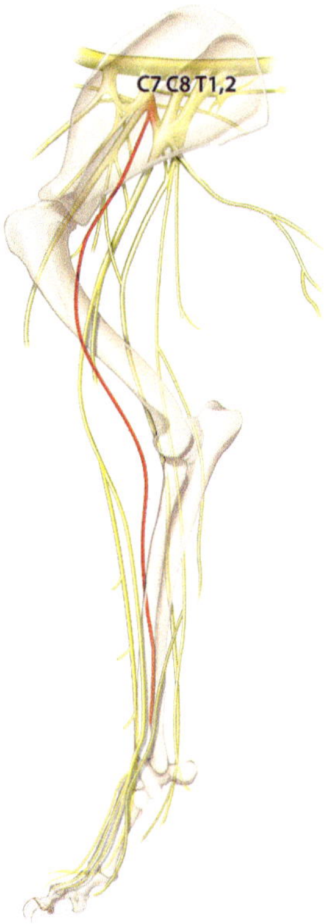

Figure 4.32: Musculocutaneous nerve.

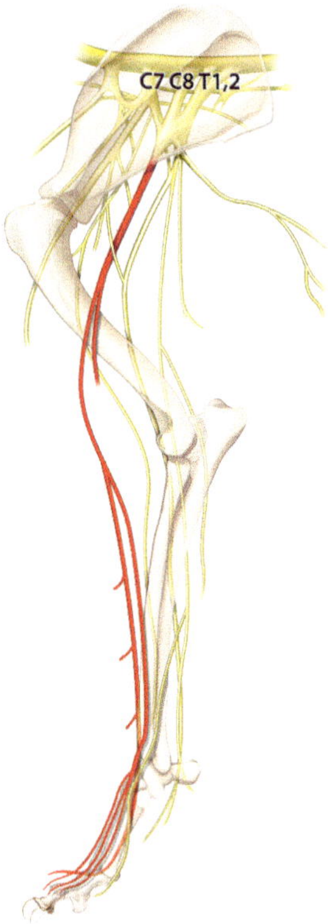

Figure 4.33: Radial nerve.

Figure 4.34: Axillary nerve.

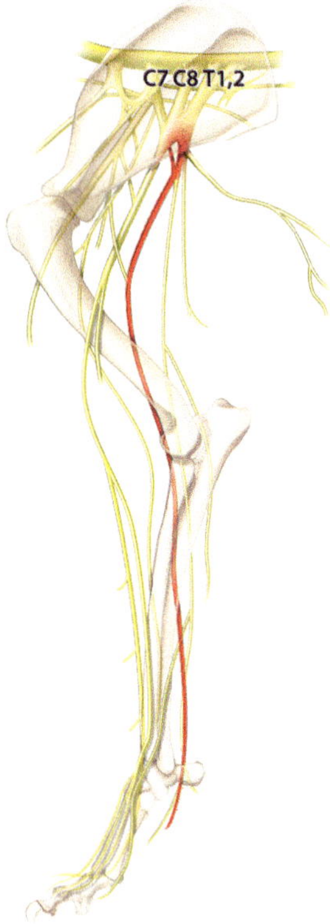

Figure 4.35: Median nerve.

Nerves of the Lumbosacral Plexus

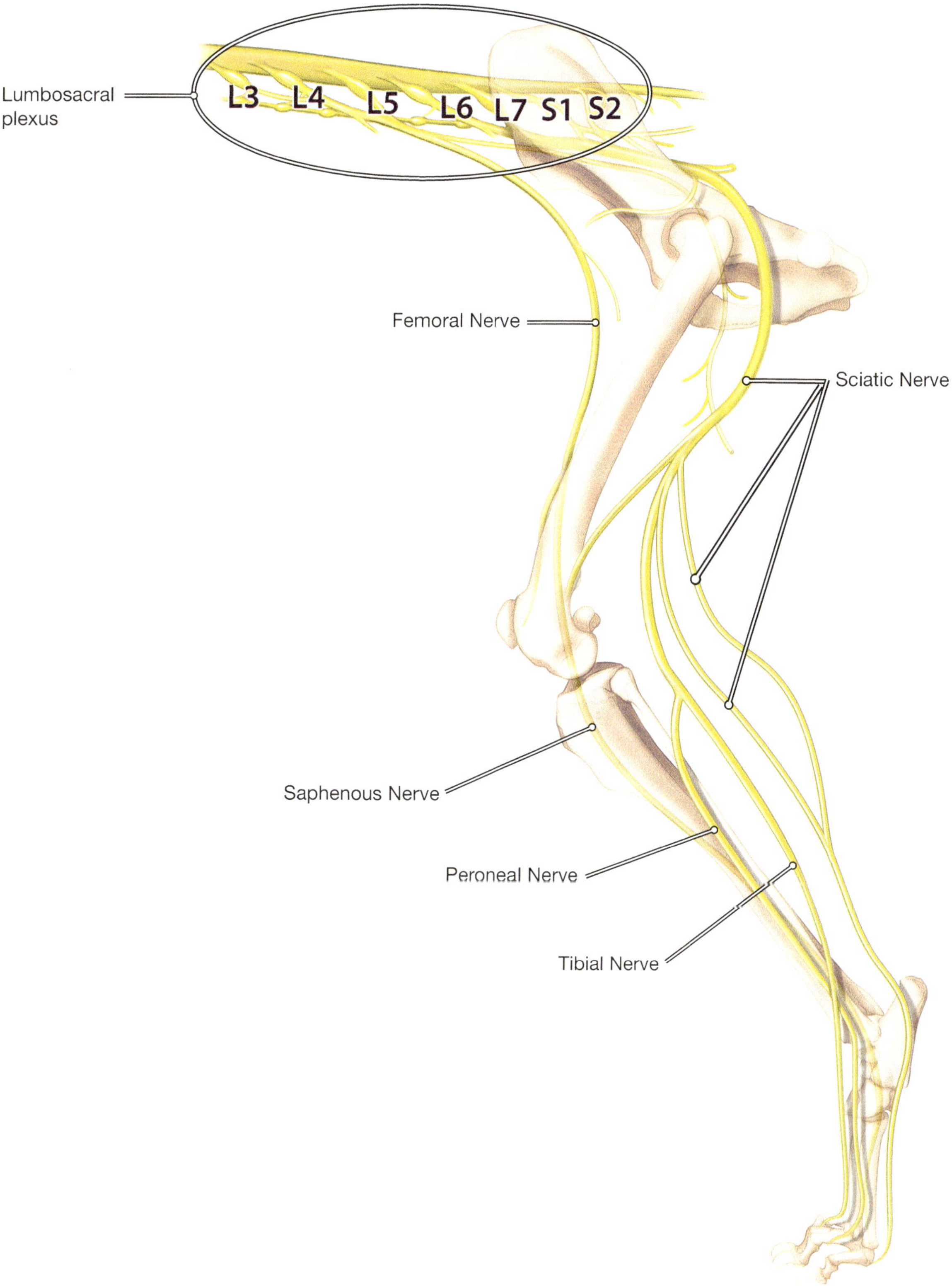

Figure 4.36: Nerves of lumbosacral plexes, lateral view.

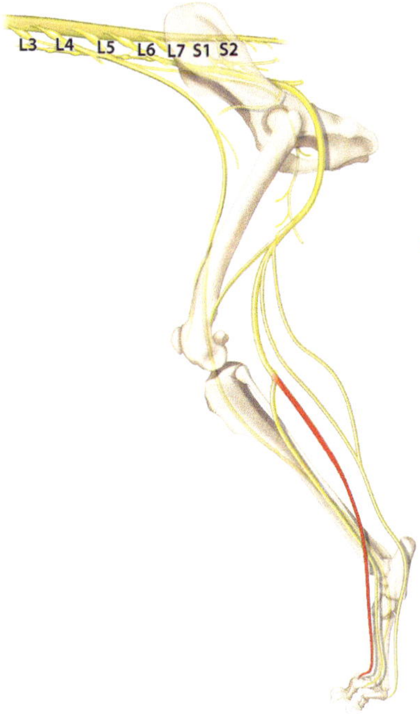

Figure 4.37: Tibial nerve.
© NCSU.

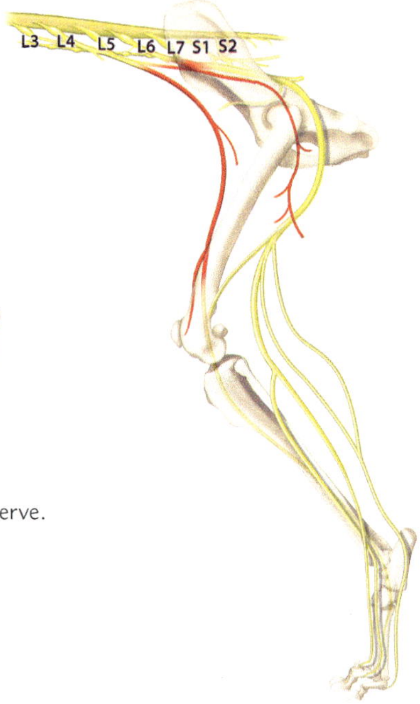

Figure 4.38: Femoral nerve.
© NCSU.

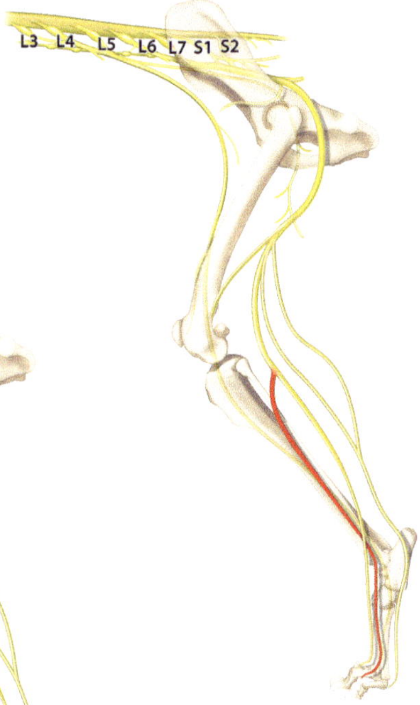

Figure 4.39: Peroneal (fibular) nerve.
© NCSU.

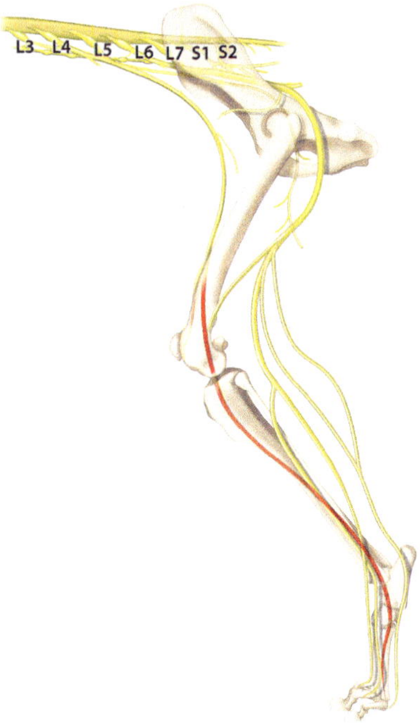

Figure 4.40: Saphenous nerve.
© NCSU.

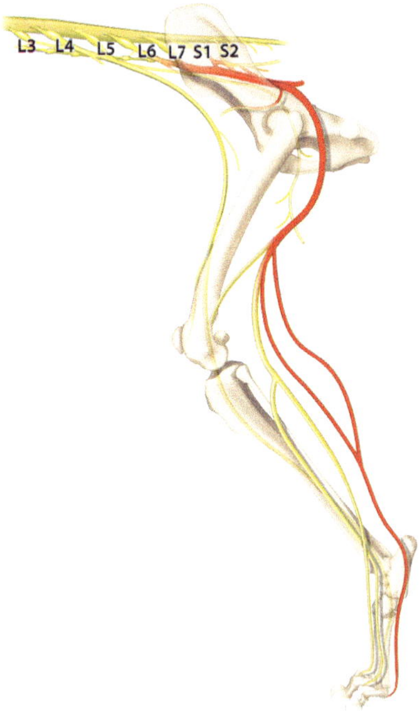

Figure 4.41: Sciatic nerve.
© NCSU.

Dermatomes and Stretch Reflex

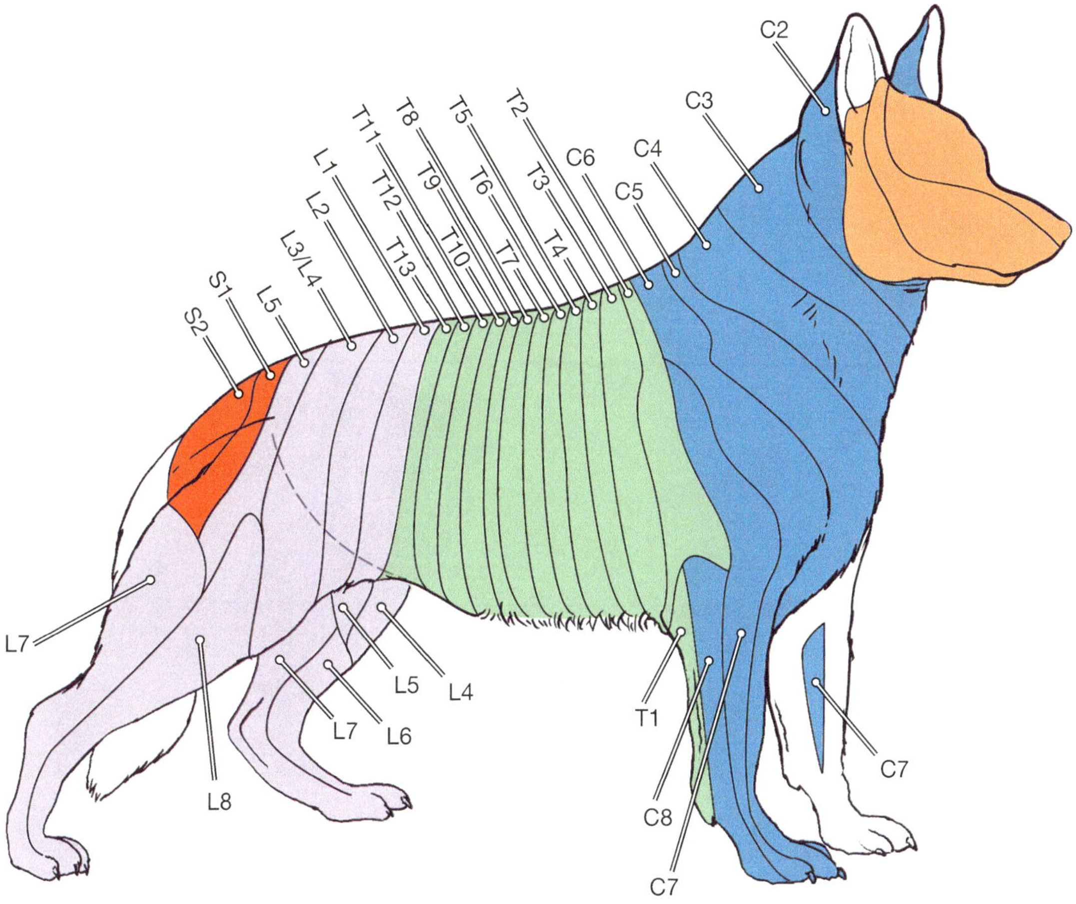

Figure 4.42: Dermatomes.

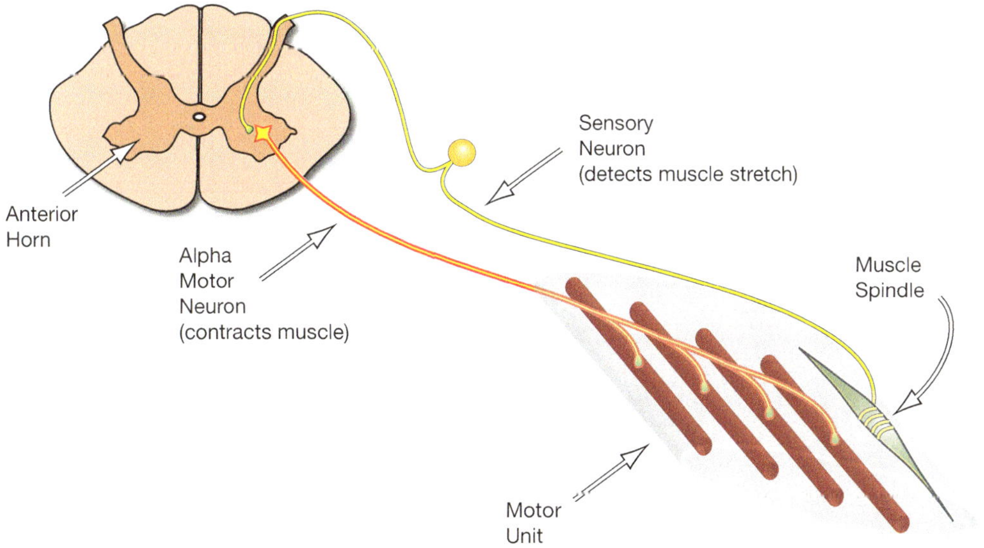

Figure 4.43: Spinal stretch reflex.

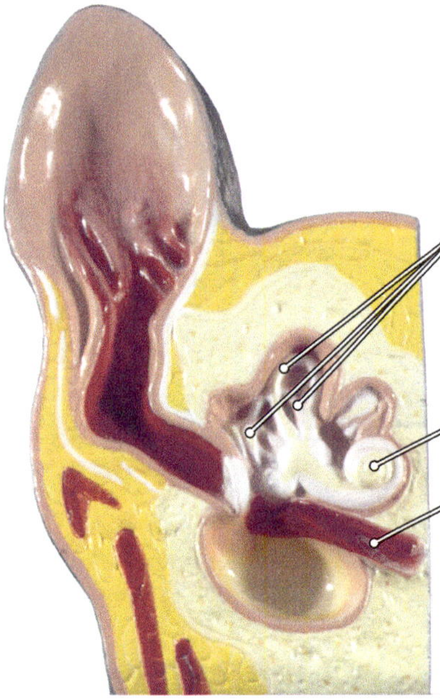

Figure 4.44: Inner ear model.
© NCSU.

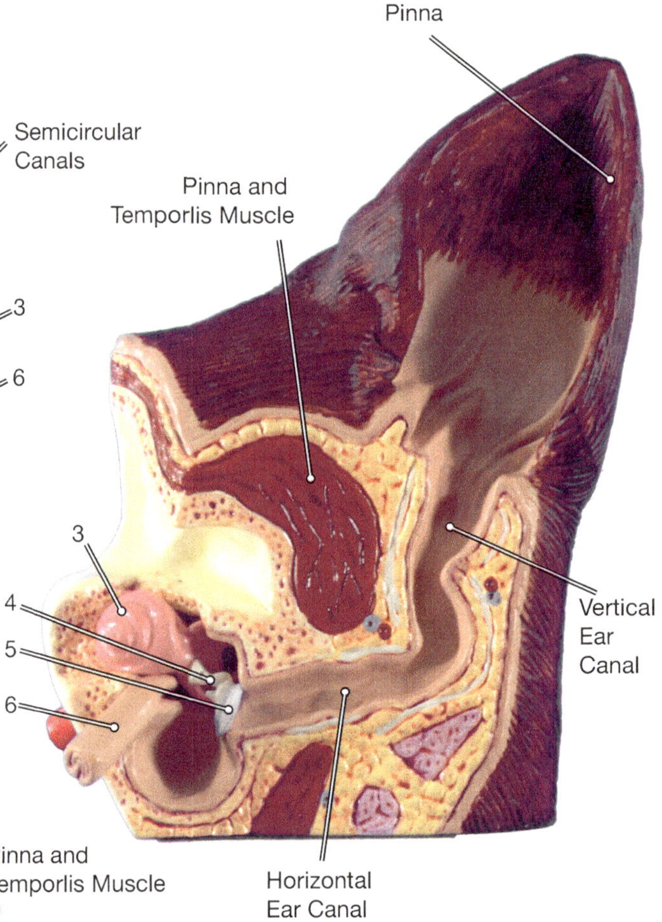

Figure 4.45: Heathly ear model.
© NCSU.

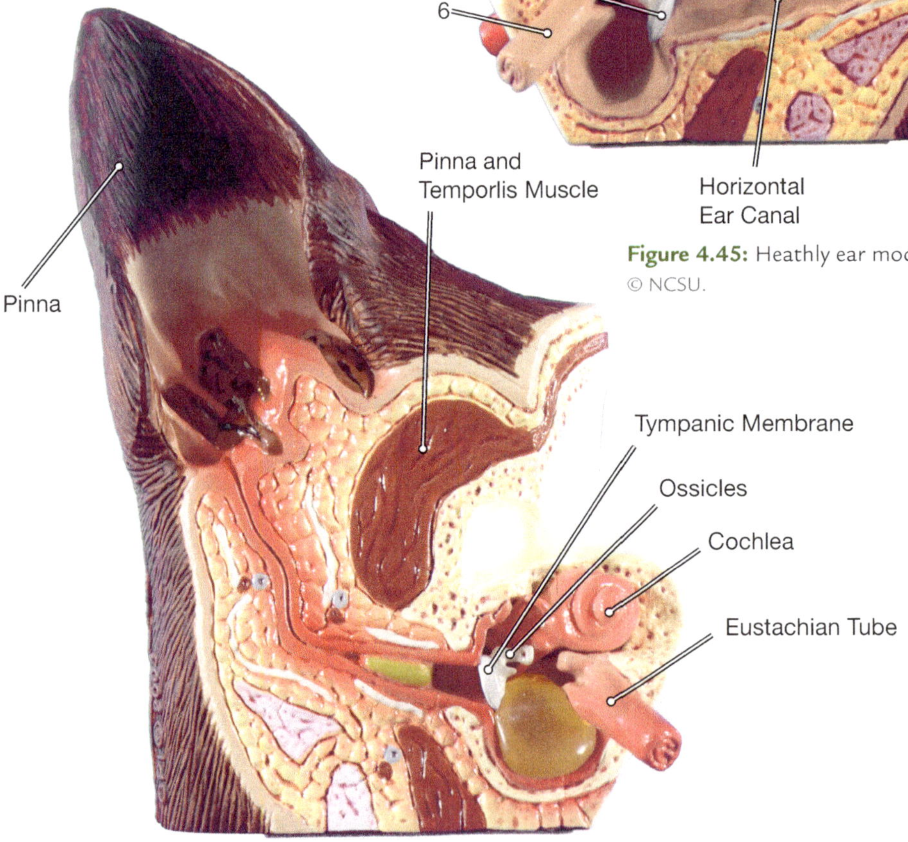

Figure 4.46: Diseased ear model.
© NCSU.

Cochlea

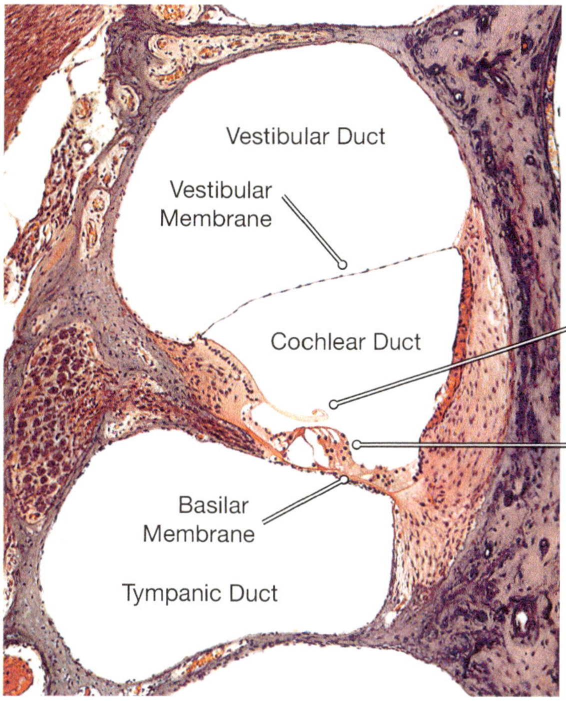

Figure 4.47: Cochlea.

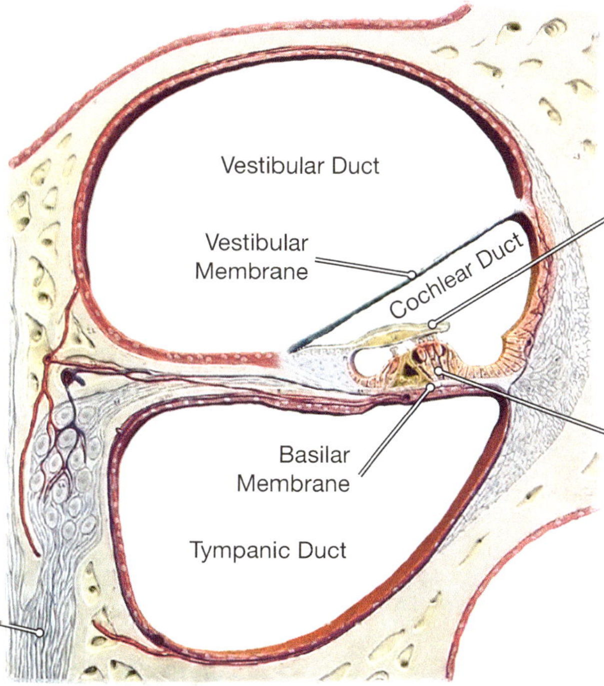

Figure 4.48: Cochlea.

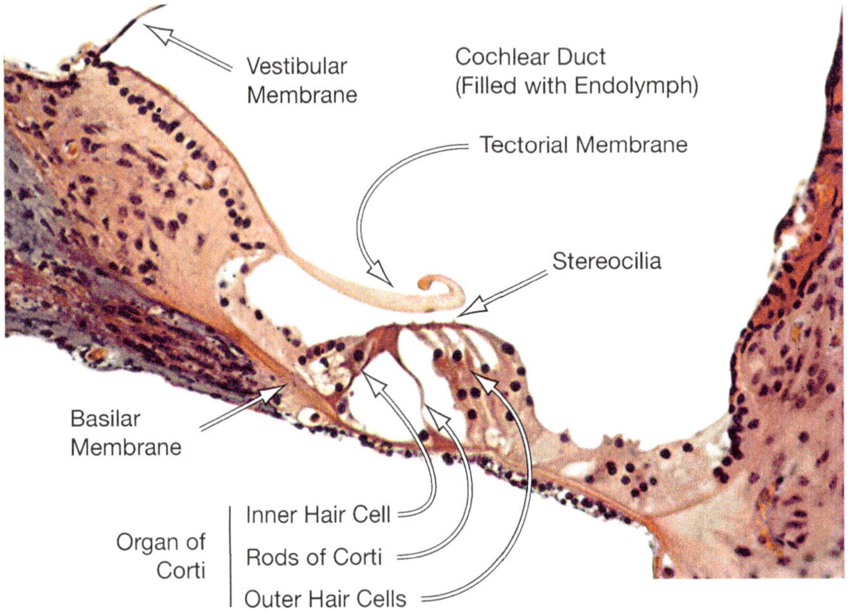

Figure 4.49: Organ of corti.
© David G. Ward.

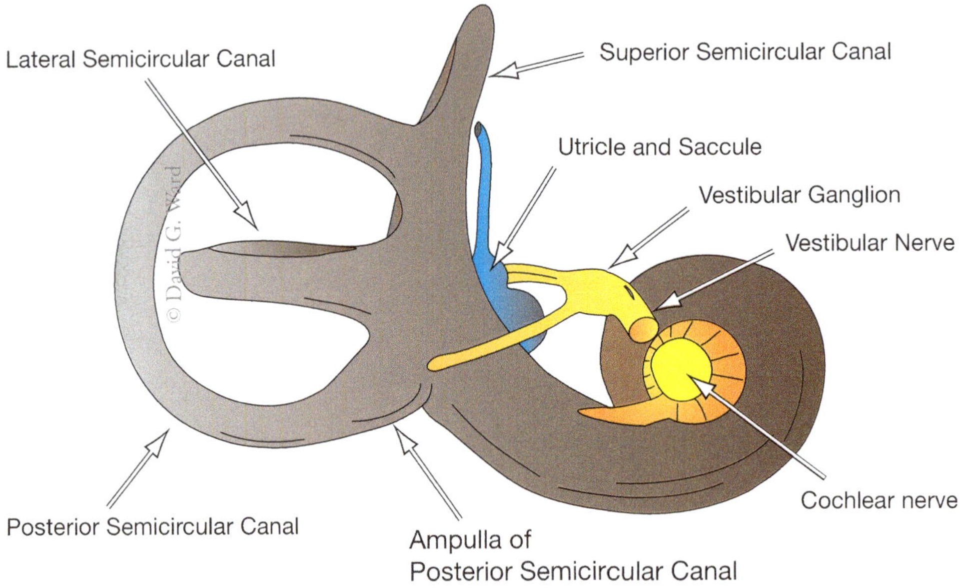

Figure 4.50: Vestibular apparatus and cochlea, posterior view.
© David G. Ward.

Sensory Organs: Eye

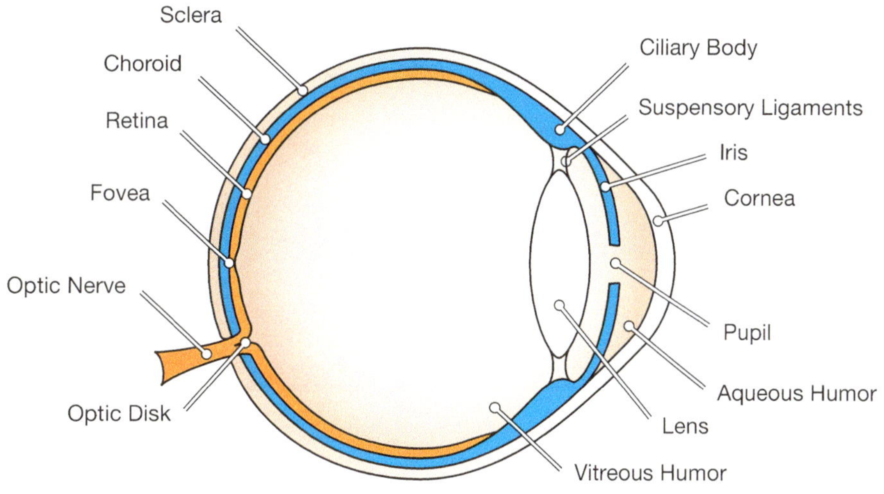

Figure 4.51: Eye, midsagittal.
© David G. Ward.

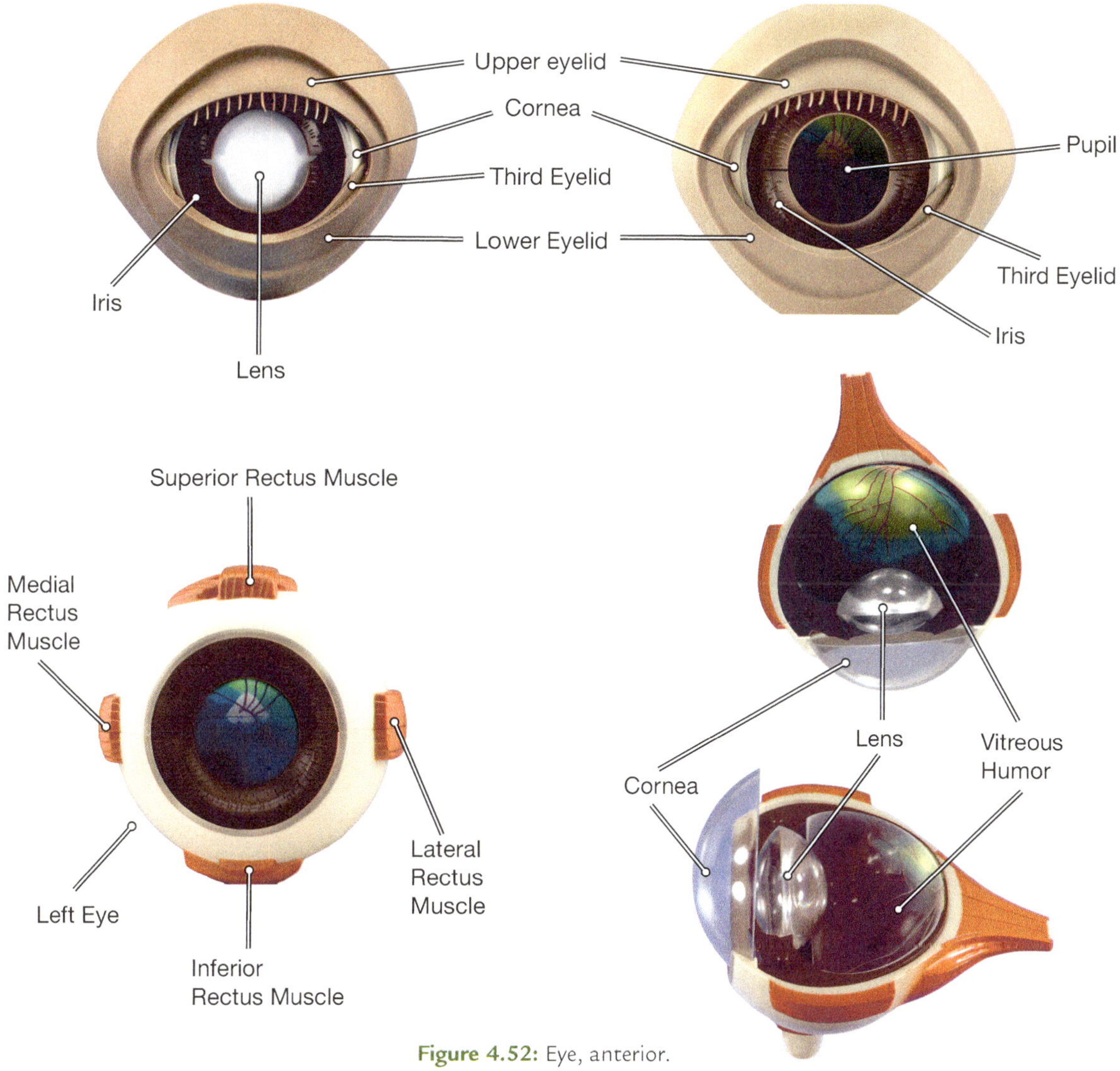

Figure 4.52: Eye, anterior.
© NCSU.

Eye and Retina

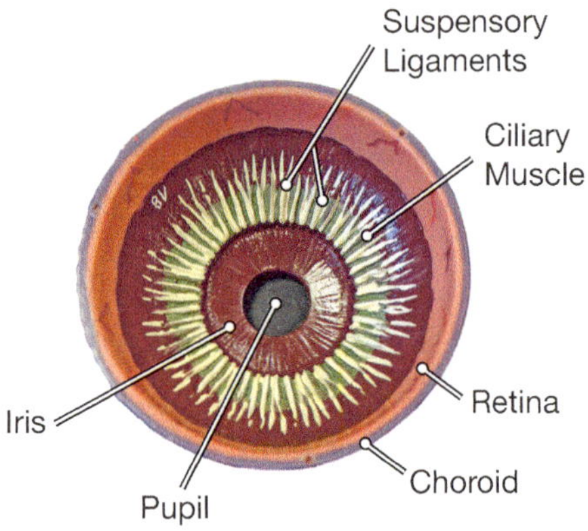

Figure 4.53: Ciliary body.

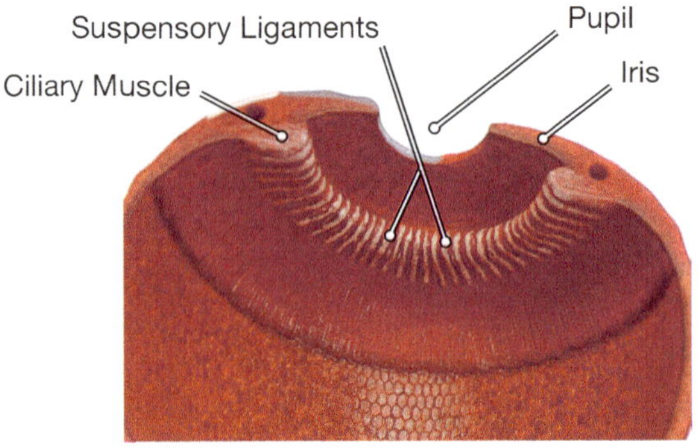

Figure 4.54: Ciliary body.

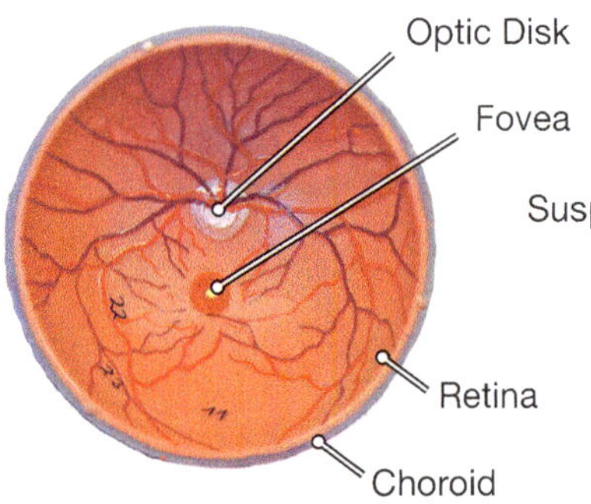

Figure 4.55: Retina.

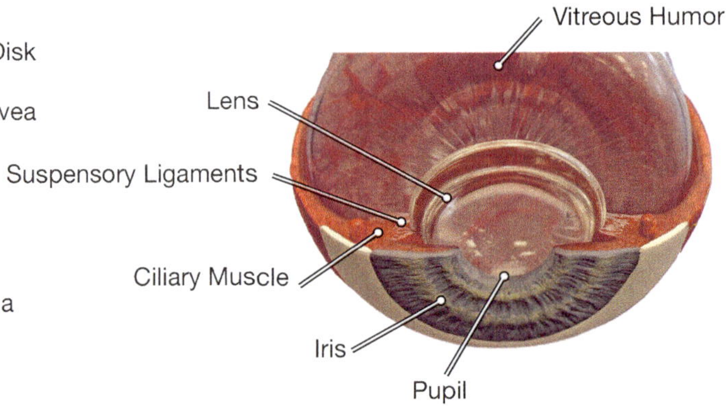

Figure 4.56: Lens.

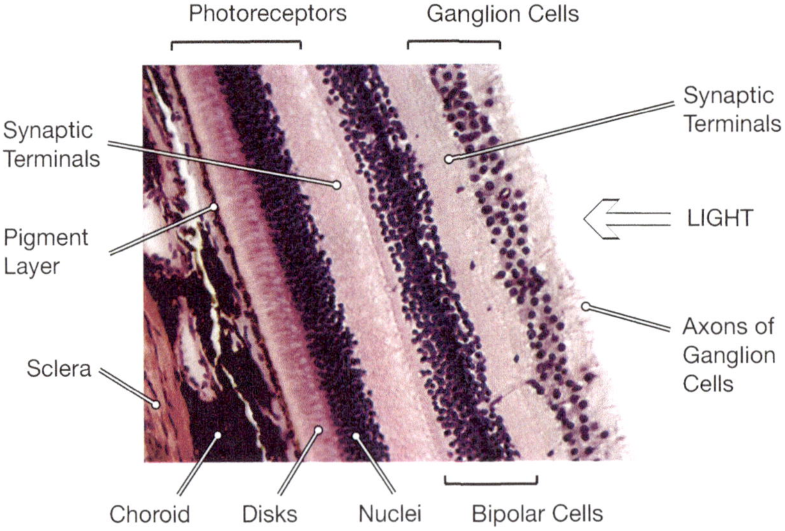

Figure 4.57: Retina.

Review: Coloring Activity

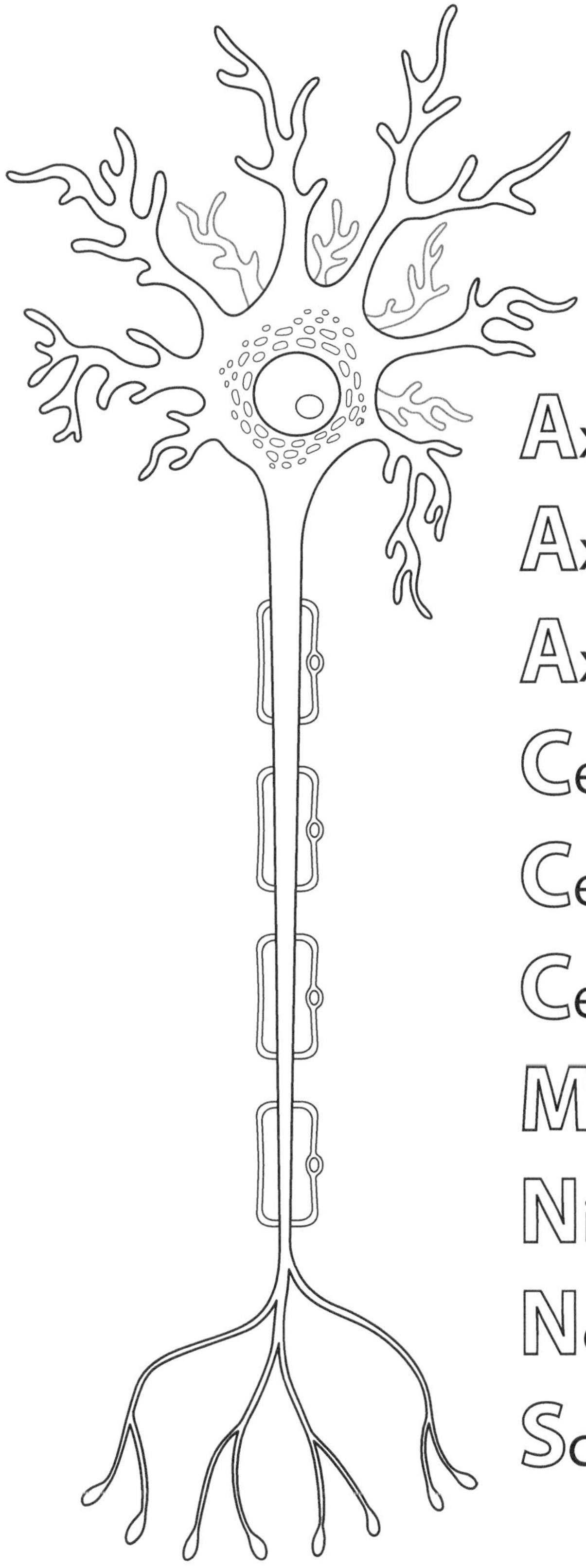

Axon

Axon hillock

Axon terminals

Cell body

Cell body nucleolus

Cell body nucleus

Myelin sheath

Nissl bodies

Node of Ranvier

Schwann cell nucleus

Figure 4.58: Nerve coloring activity.

Review: Word Search

N	I	V	P	L	W	C	D	V	R	X	O	C	M	H	F	E	U	B	T	D
V	O	Z	I	S	A	N	Q	E	E	P	U	U	B	L	J	N	E	M	Y	I
T	V	I	R	T	P	T	T	F	T	R	S	P	I	A	M	A	T	E	R	O
M	Y	H	T	Y	R	A	N	I	D	O	T	Q	L	Z	T	R	S	W	E	N
A	U	C	J	C	M	E	C	O	L	F	Y	I	O	I	H	B	F	R	L	H
B	V	J	Y	A	N	C	O	L	Z	Y	H	R	C	E	V	M	S	A	Y	C
N	W	A	R	O	H	U	A	U	X	I	E	A	B	A	F	E	Y	D	Q	A
C	Q	U	H	I	X	C	J	D	S	G	R	O	B	Z	L	M	N	R	X	R
J	D	Q	A	U	S	L	I	R	M	H	G	O	Z	A	Q	C	A	E	E	A
Q	G	S	S	U	X	K	B	H	A	Q	U	G	H	N	V	I	P	T	I	P
P	M	L	P	F	W	B	N	S	K	L	Y	M	W	A	P	N	T	I	B	V
A	S	R	I	V	V	W	O	K	I	W	U	H	O	A	N	A	I	N	O	O
J	O	X	L	L	E	C	N	N	A	W	H	C	S	R	D	P	C	A	A	U
C	P	T	L	T	Y	D	G	Q	L	P	C	Q	S	I	U	M	B	M	A	V
R	X	P	H	N	I	W	I	S	U	S	L	W	K	U	Y	Y	U	S	M	J
P	G	R	I	K	B	L	N	W	V	W	H	Y	Y	V	M	T	L	R	E	H
A	E	T	G	H	I	V	B	A	C	C	V	E	J	K	V	O	B	S	P	K
E	X	M	P	V	R	E	O	B	U	N	O	B	I	V	C	F	R	R	O	B
S	Y	N	A	P	T	I	C	C	L	E	F	T	Z	K	J	X	G	U	T	F
V	E	Z	Y	T	I	C	Y	W	A	C	Z	A	W	X	Q	Z	P	F	E	N
D	C	B	C	I	X	W	R	C	M	R	M	X	L	W	G	Z	G	P	W	N

1. terminal end of a neuron
2. space of communication between two neurons
3. responsible for creating myelin sheath
4. nervous and skeletal muscle communication
5. innermost meninge layer
6. center meninge layer
7. outermost meninge layer
8. where the optic nerve enters the brain and crosses over
9. structure that joins the left and right hemisphere of the cerebrum
10. the eardrum
11. where bipolar cells are located
12. the number of eyelids in a dog eye
13. the gelatinous substance posterior to the lens
14. the ear canal most proximal to the pinna
15. the ear canal most proximal to the tympanic membrane

Review: Crossword Puzzle

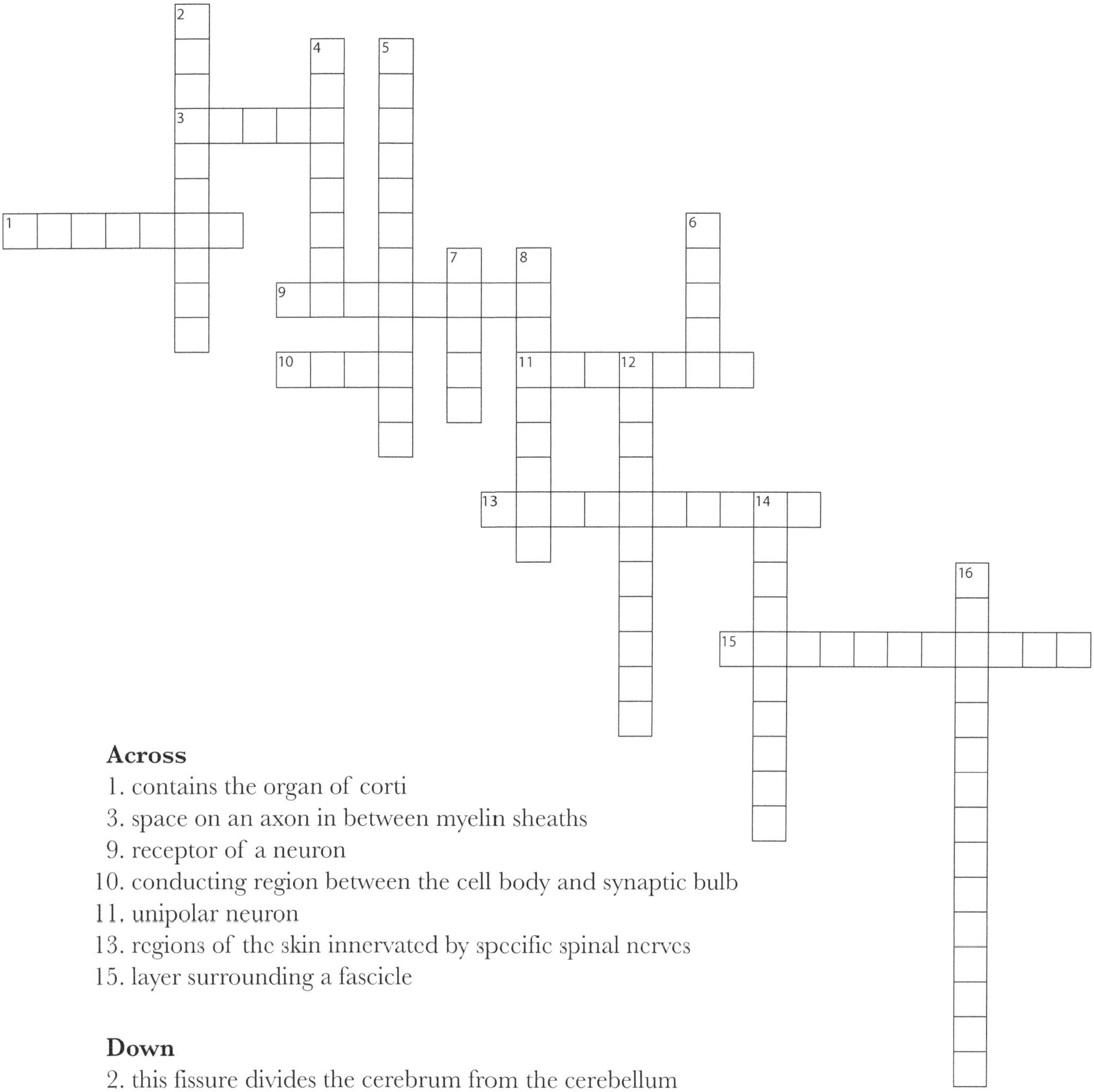

Across

1. contains the organ of corti
3. space on an axon in between myelin sheaths
9. receptor of a neuron
10. conducting region between the cell body and synaptic bulb
11. unipolar neuron
13. regions of the skin innervated by specific spinal nerves
15. layer surrounding a fascicle

Down

2. this fissure divides the cerebrum from the cerebellum
4. bundle of nerve fibers
5. this fissure divides the left from the right hemisphere of the cerebellum
6. multipolar neuron
7. dog's ear
8. where neurotransmitters are stored
12. thoracolumbar division
14. outermost layer of a nerve
16. craniosacral division

Chapter 5

Blood and the Cardiovascular System

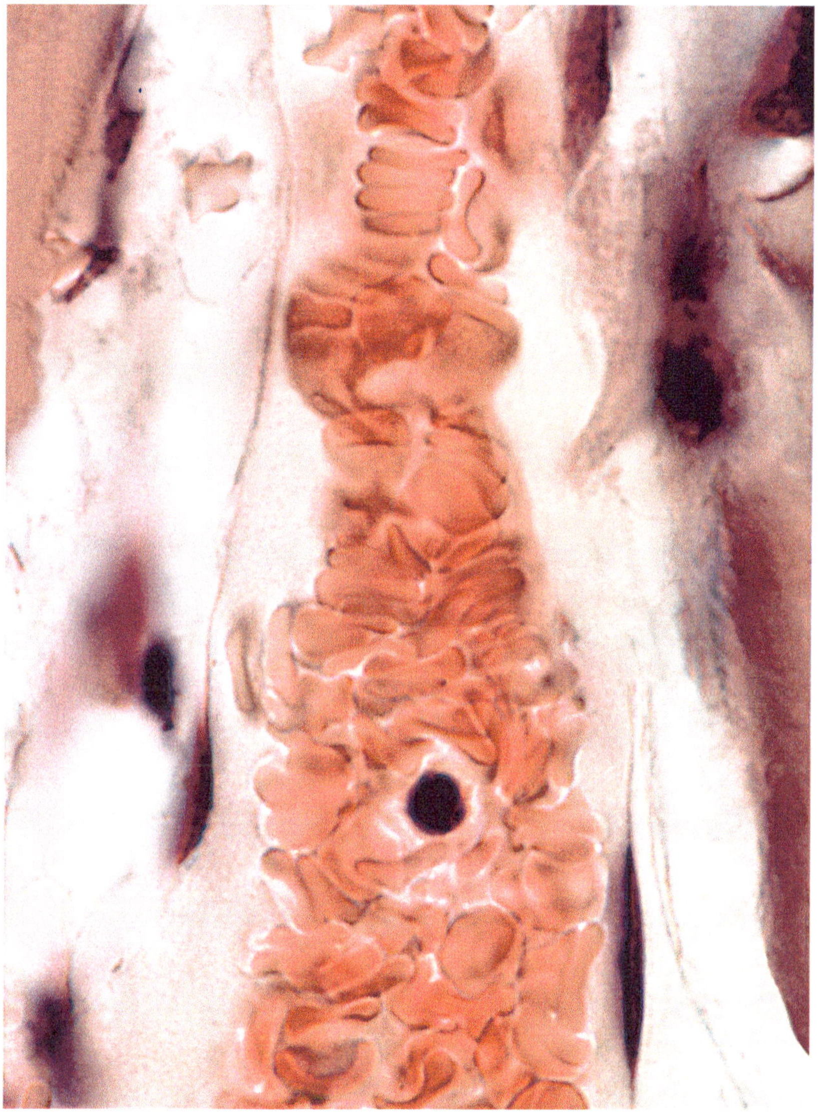

Erythrocytes in Blood Vessel in Spongy Bone (Brightfield, H&E; x1880)

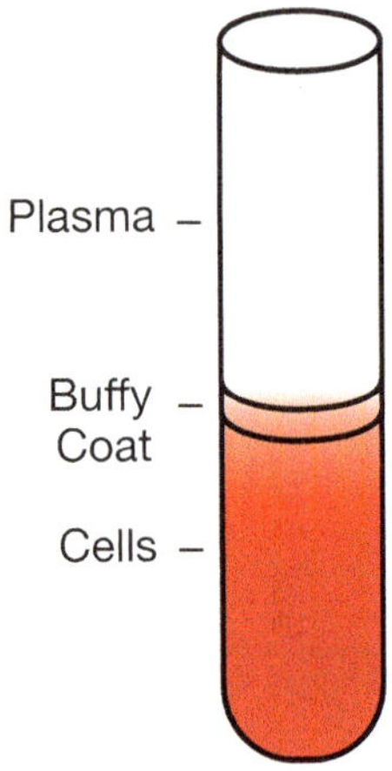

Platelets	150,000-400,000/mm^3	Hematocrit	35-55%
Leukocytes	4.0-12.0 x 10^3/mm^3	Hemoglobin	14-20 g/dL
Neutrophils	51-72%		
Lymphocytes	8-35%		
Monocytes	1-9%		
Eosinophils	0-9%		
Basophils	0-2%		
Erythrocytes	5.7-10.5 x 10^6/mm^3		

Figure 5.1: Blood composition.

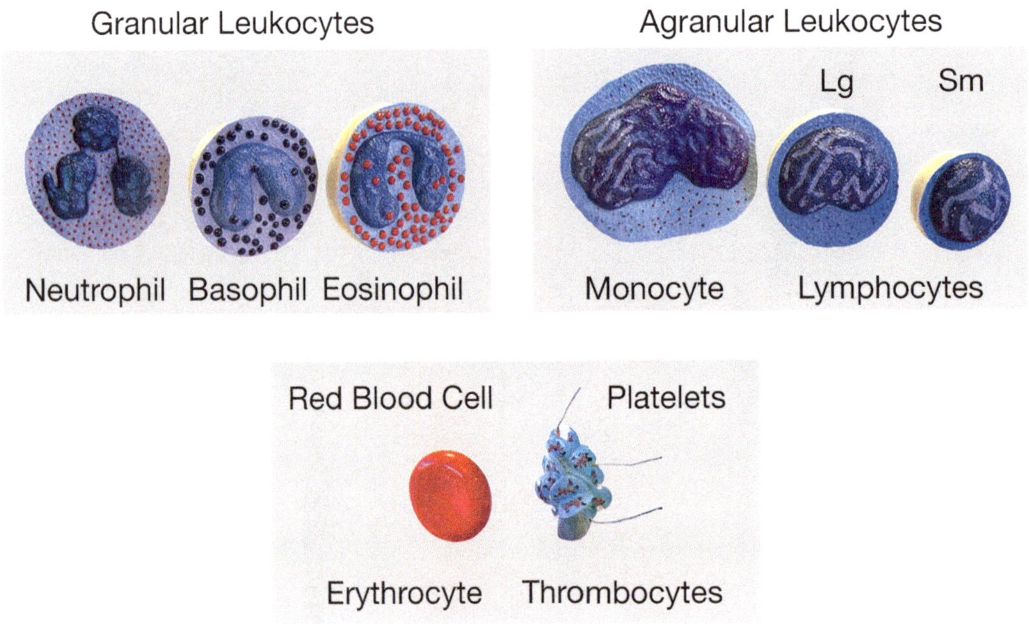

Figure 5.2: White blood cells.

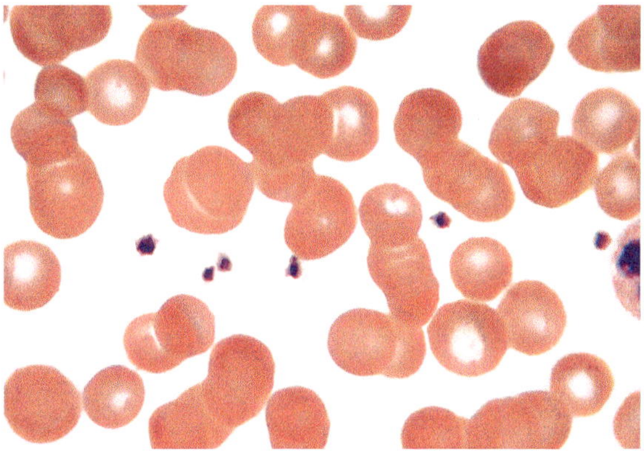

Figure 5.3: Erythrocytes and platelets.

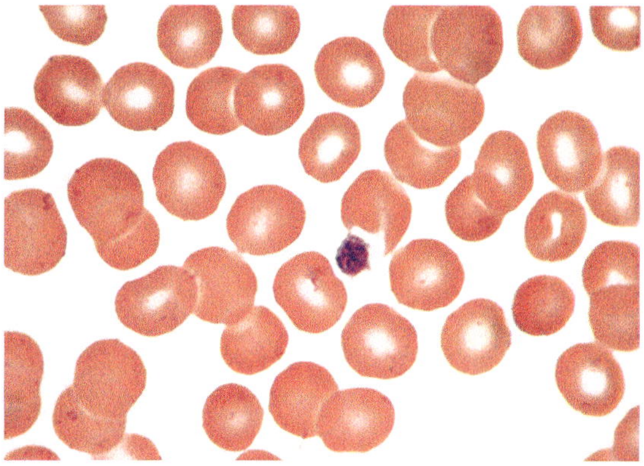

Figure 5.4: Erythrocytes and platelets.

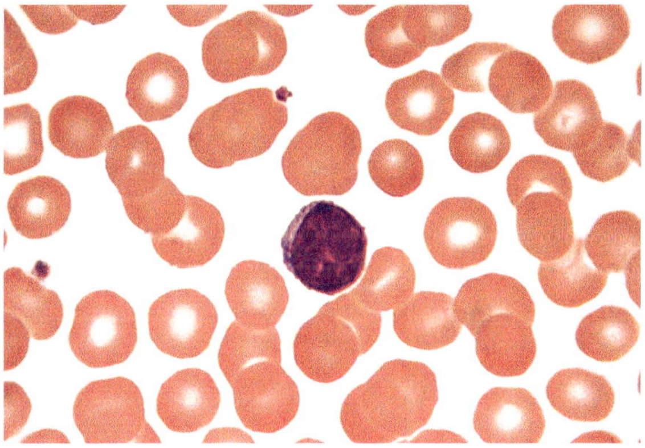

Figure 5.5: Small lymphocyte.

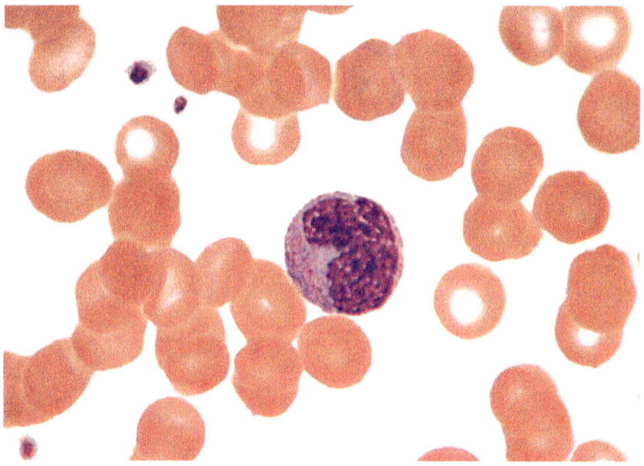

Figure 5.6: Large lymphocyte.

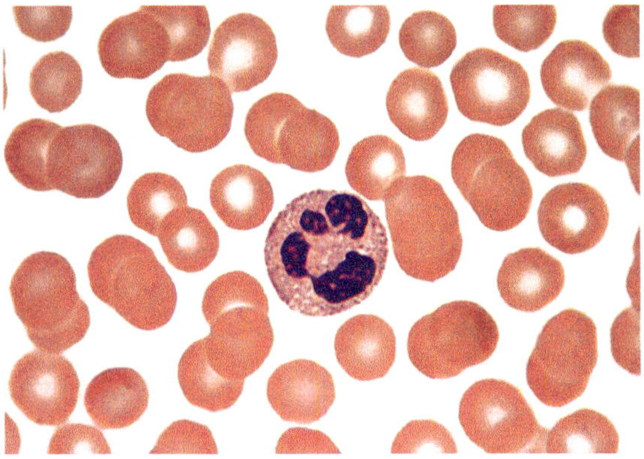

Figure 5.7: Neutrophil.

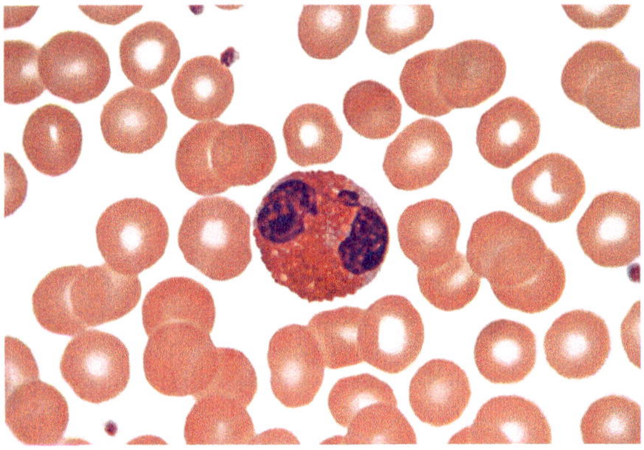

Figure 5.8: Eosinophil.

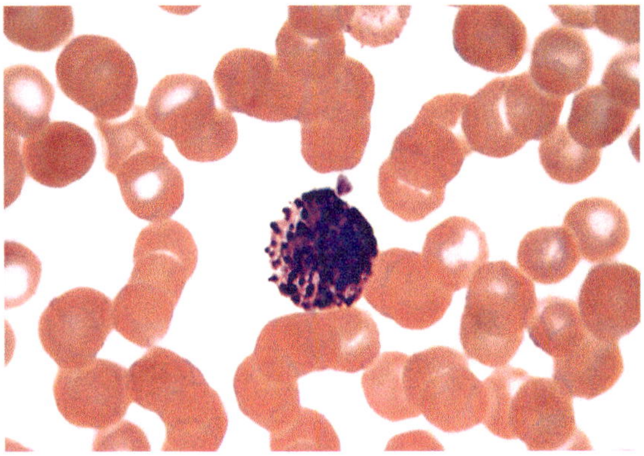

Figure 5.9: Basophil.

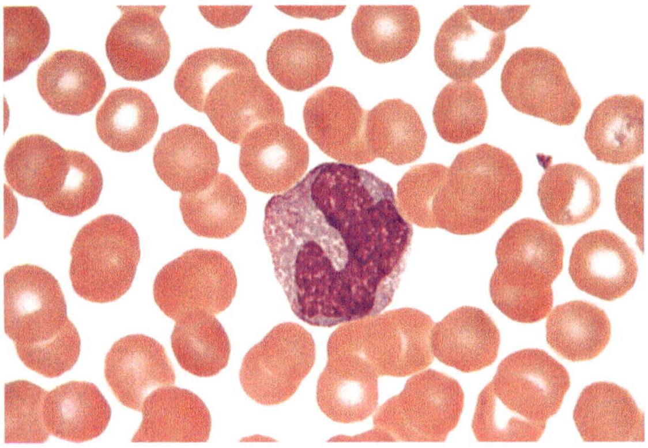

Figure 5.10: Monocyte.

Heart Wall, Valves and Cardiac Muscle

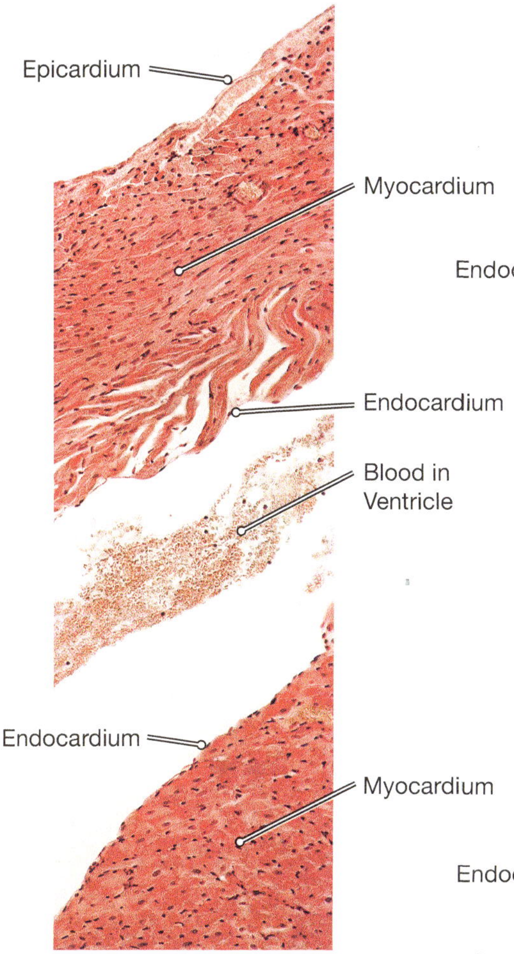

Figure 5.11: Heart wall.

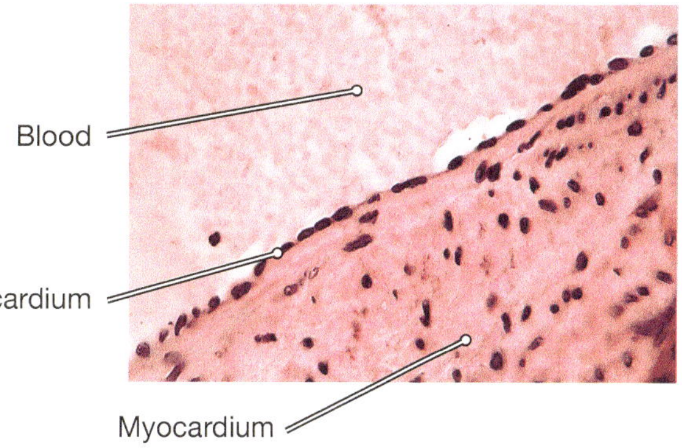

Figure 5.12: Endocardium.

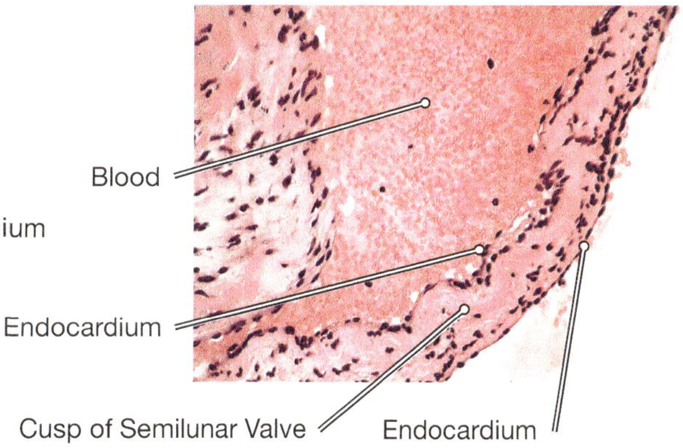

Figure 5.13: Cusp of semilunar valve.

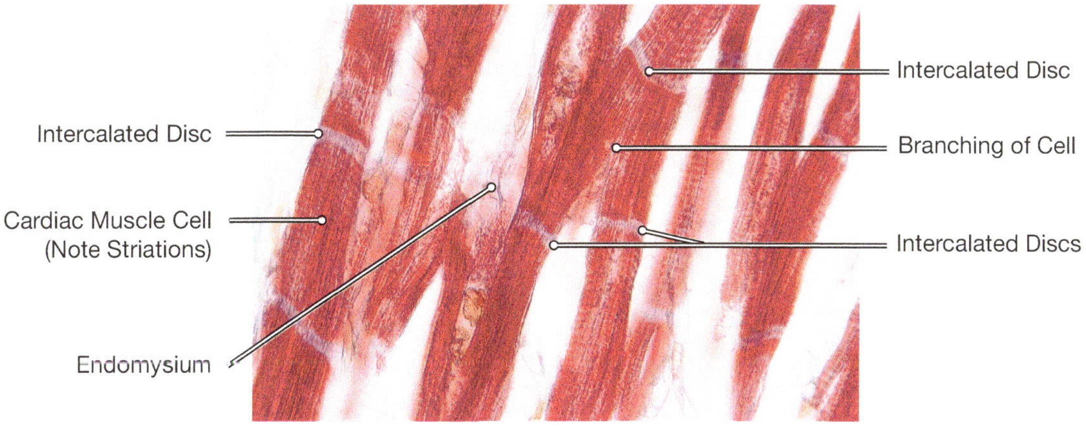

Figure 5.14: Cardiac muscle.

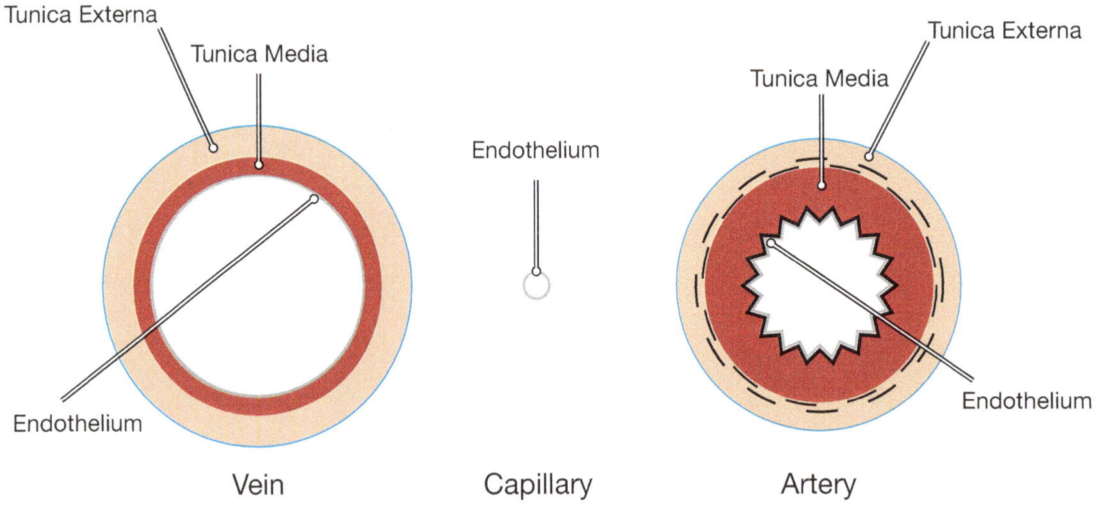

Figure 5.15: Arteries, capillaries and veins.
© David G. Ward.

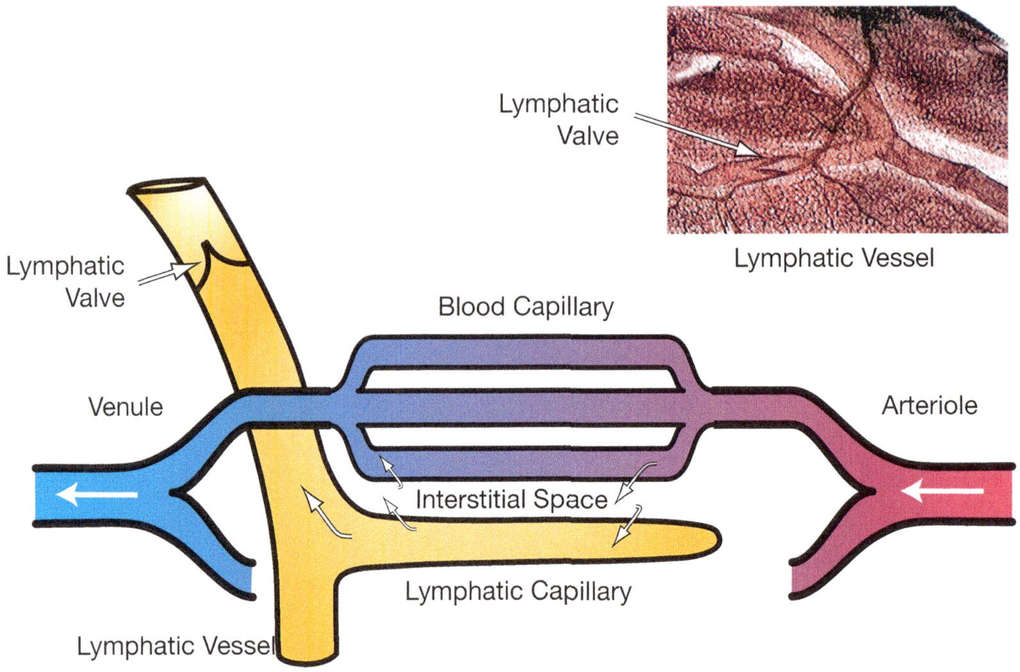

Figure 5.16: Microcirculation.
© David G. Ward.

Arteries, Veins and Smooth Muscle

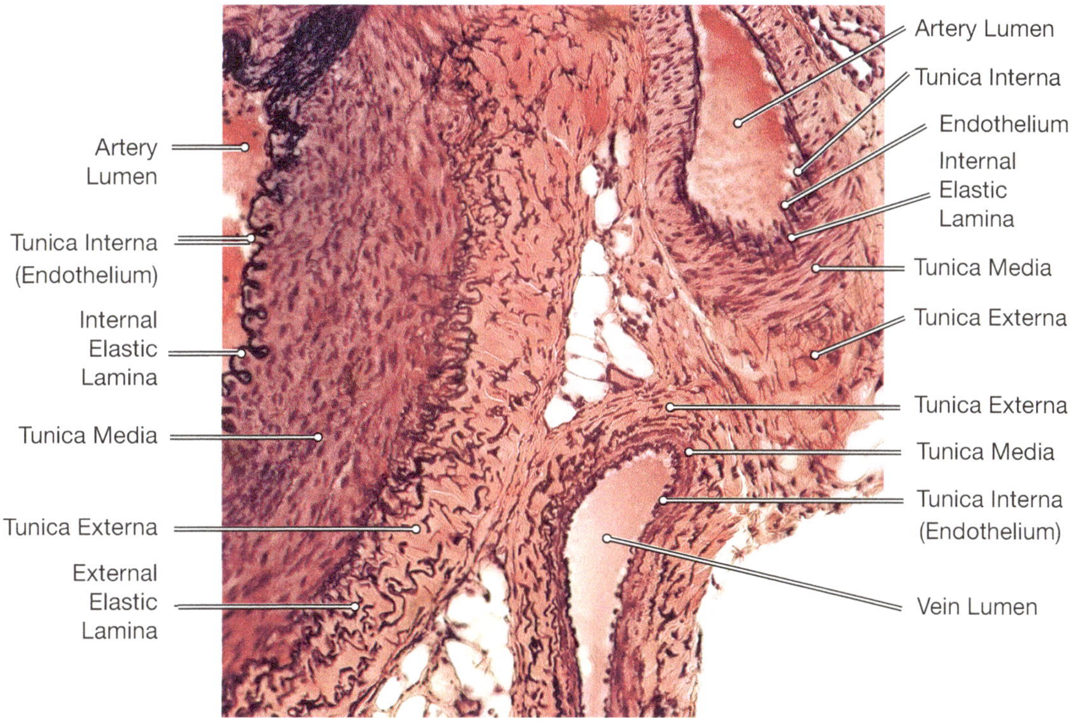

Figure 5.17: Artery and vein.
© David G. Ward.

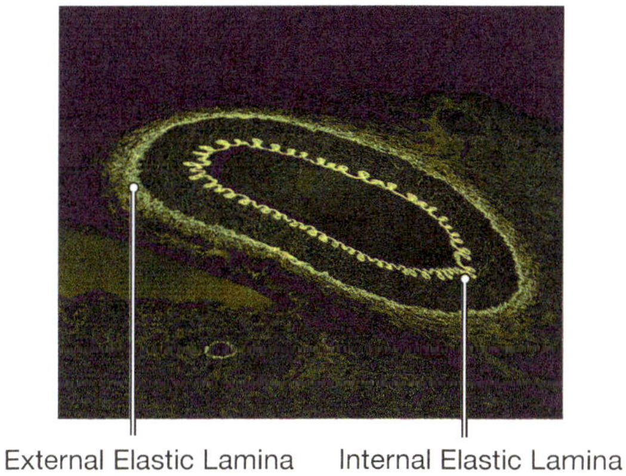

Figure 5.18: Artery, elastic lamina (fluorescence).
© David G. Ward.

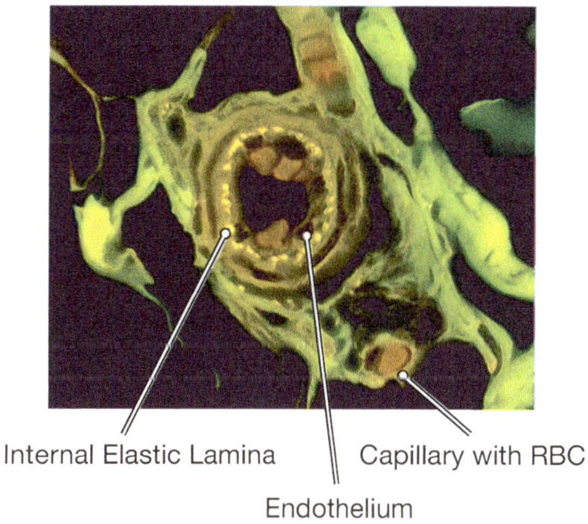

Figure 5.19: Arteriole (autofluorescence).
© David G. Ward.

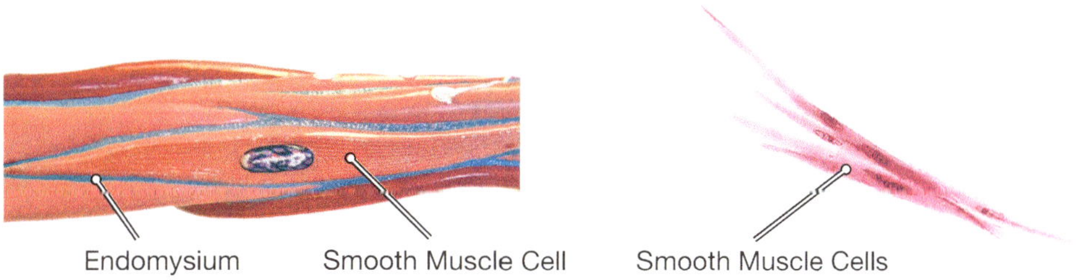

Figure 5.20: Smooth muscle.
© David G. Ward.

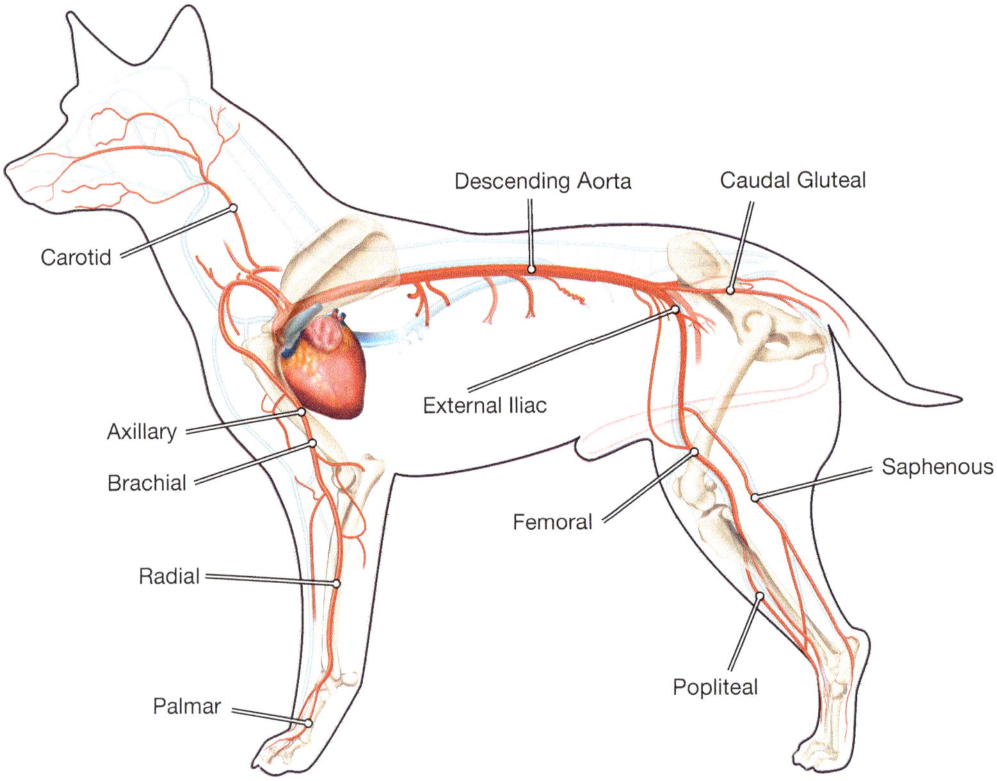

Figure 5.21: Arteries.
© NCSU.

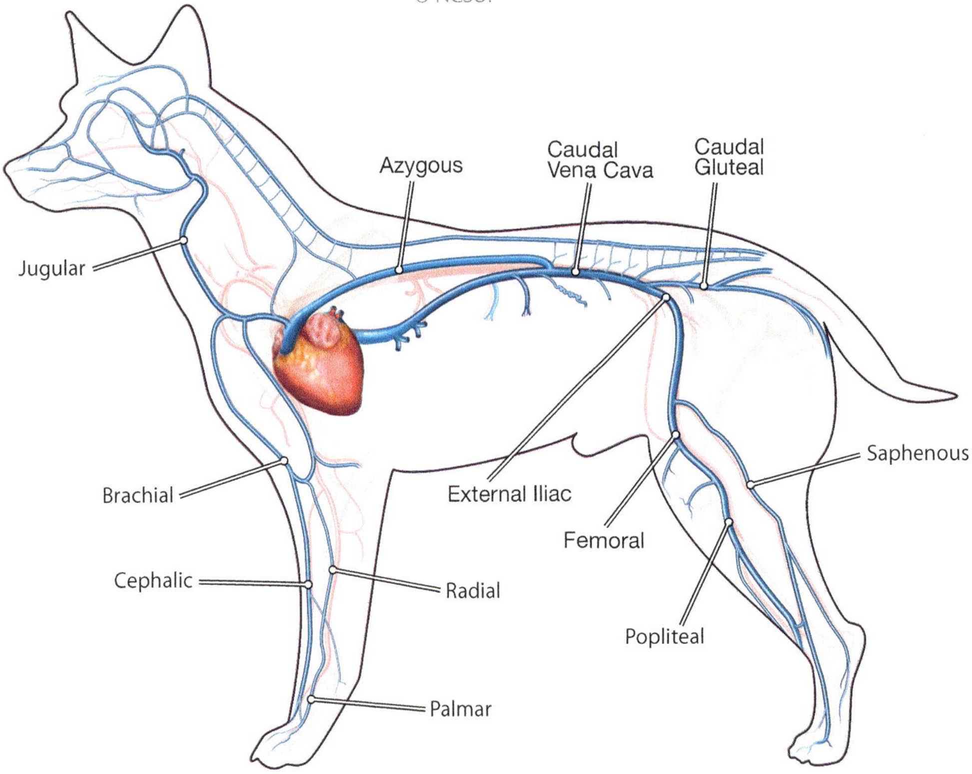

Figure 5.22: Veins.
© NCSU.

Venipuncture: Overview of Commonly-Used Veins

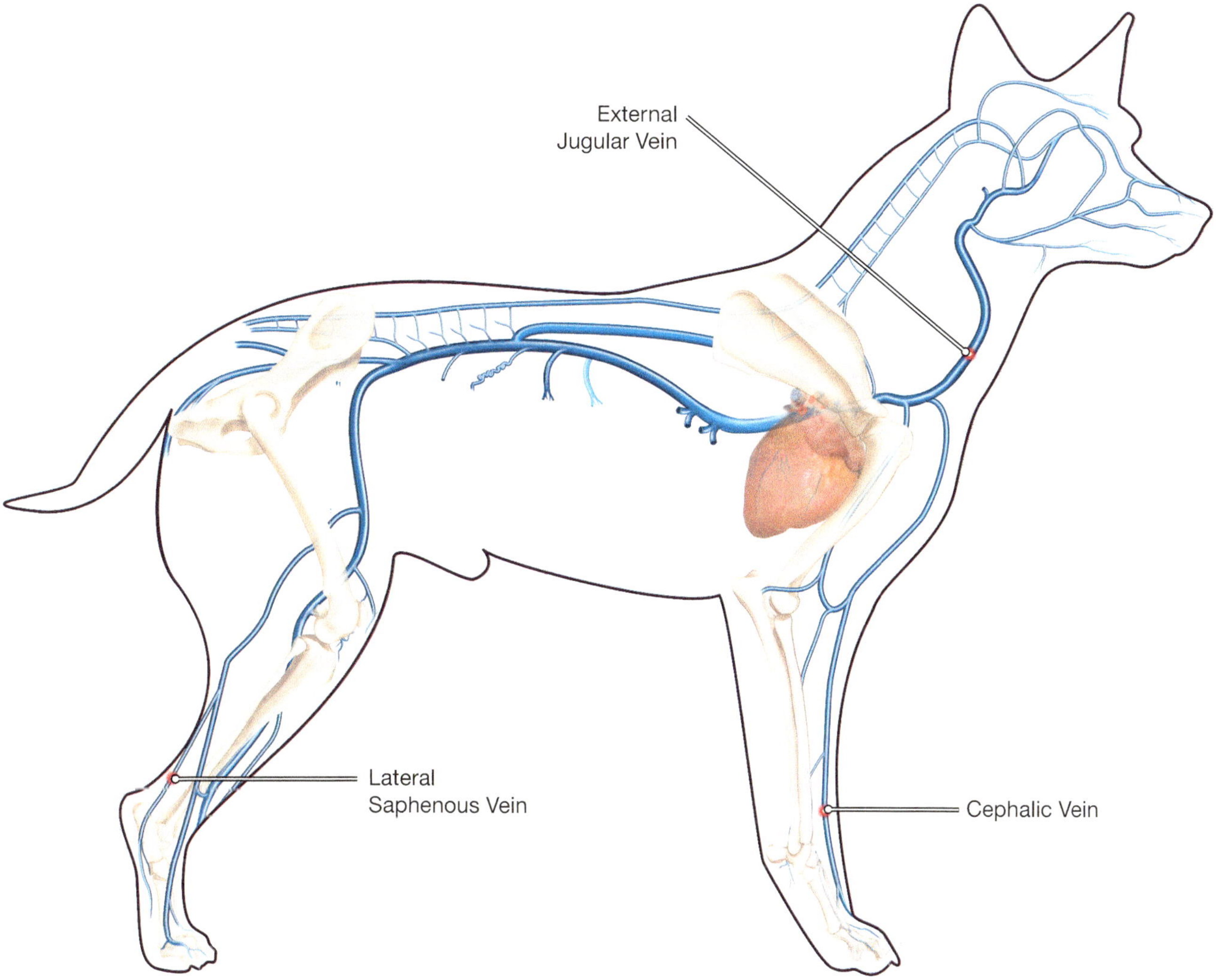

Figure 5.23: Commonly-used veins.
© NCSU.

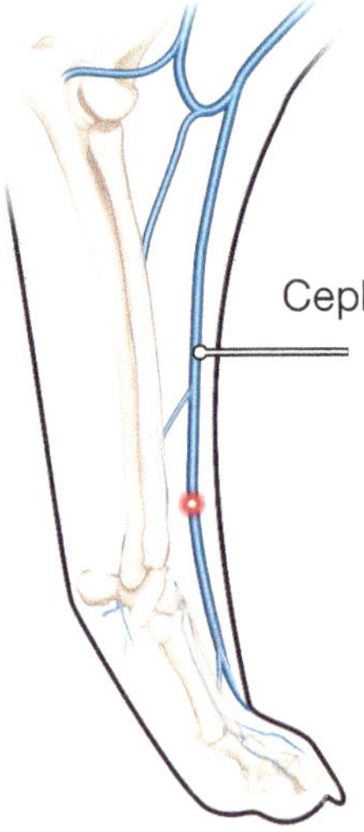

Figure 5.24: Cephalic vien.
© NCSU.

Figure 5.25: Cephalic vein accessed on the cranial aspect of the thoracic limb.
© NCSU.

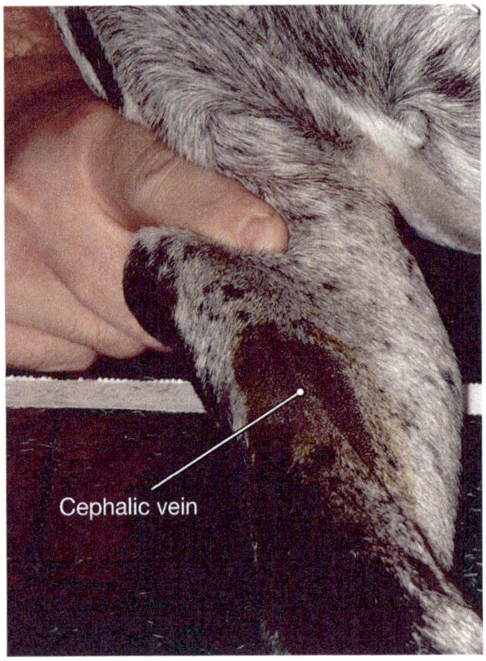

Figure 5.26: Vein is occluded distally with finger.
© NCSU.

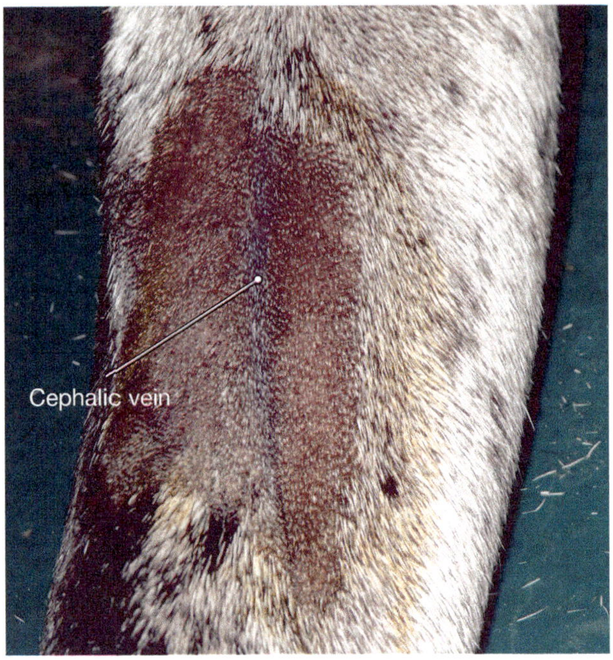

Figure 5.27: After occlusion, the target vessel is more prominent with accumulated blood.
© NCSU.

Venipuncture: External Jugular

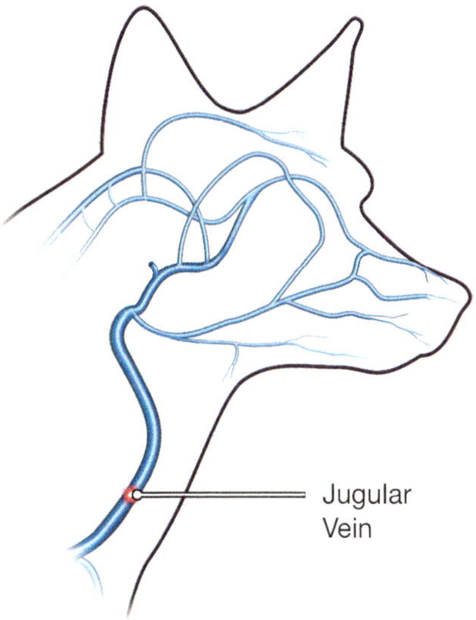

Figure 5.28: Jugular vein.
© NCSU.

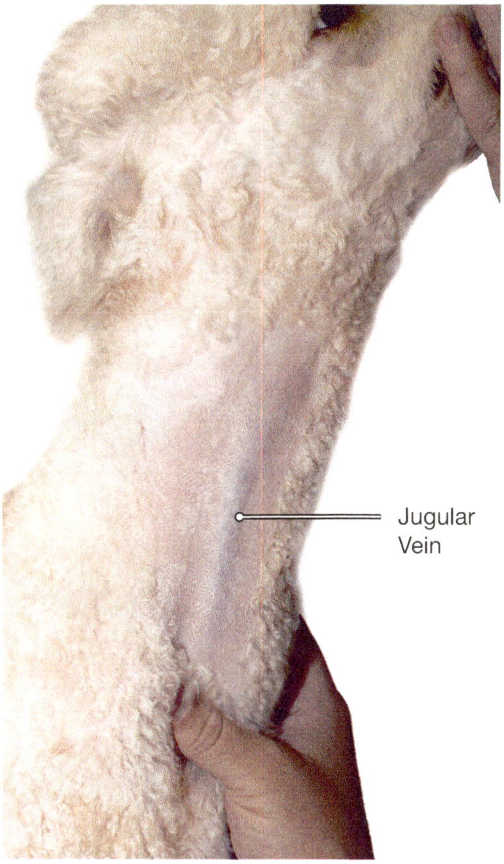

Figure 5.29: Jugular vein.
© NCSU.

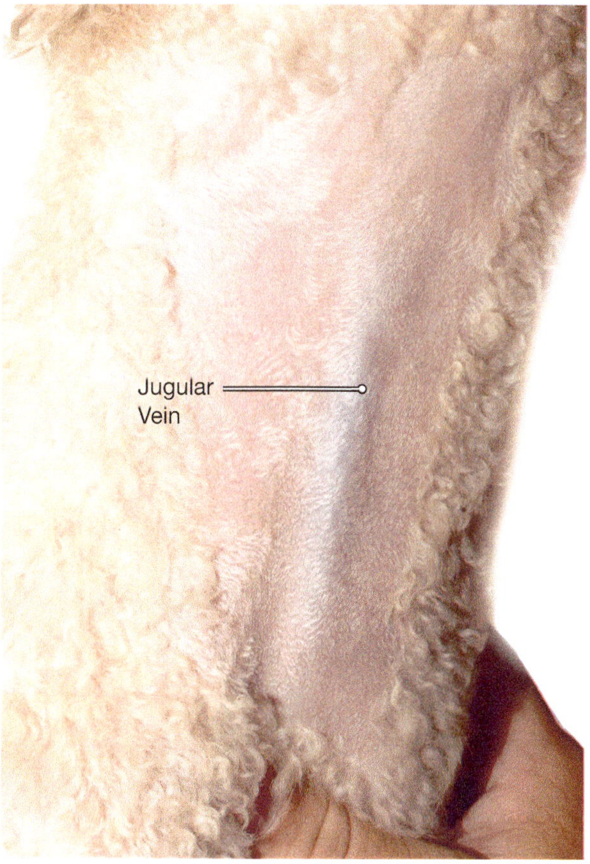

Figure 5.30: Jugular vein, occluded with pressure.
© NCSU.

Venipuncture: Saphenous

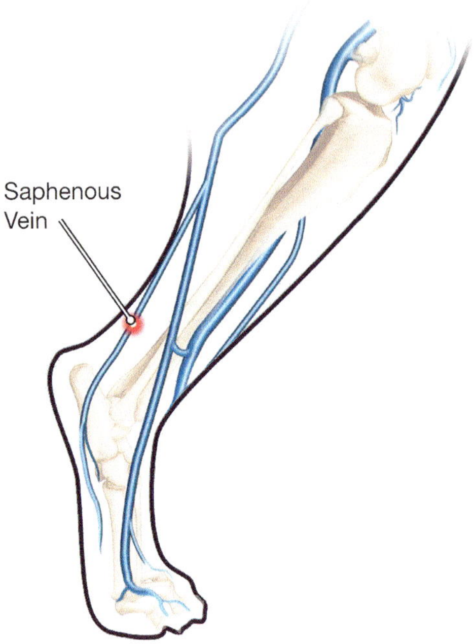

Figure 5.31: Saphenous vein accessed on the lateral aspect of the pelvic limb.
© NCSU.

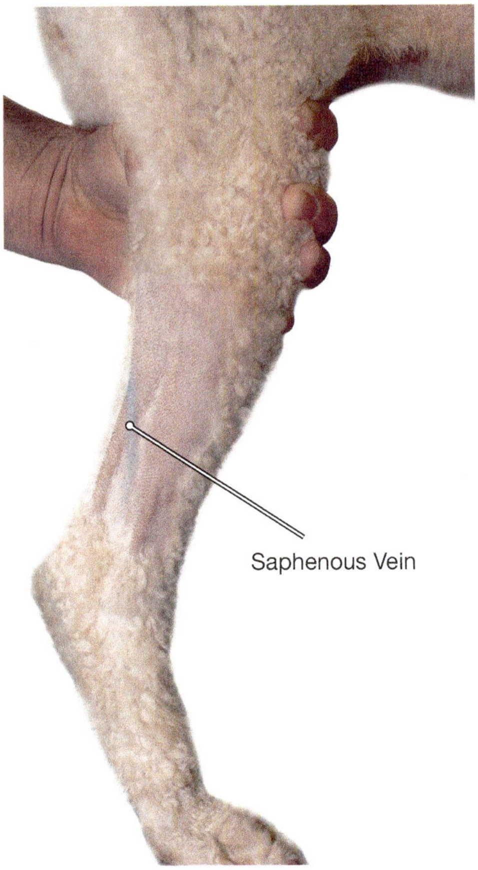

Figure 5.32: Saphenous vein, occluded with pressure. © NCSU.

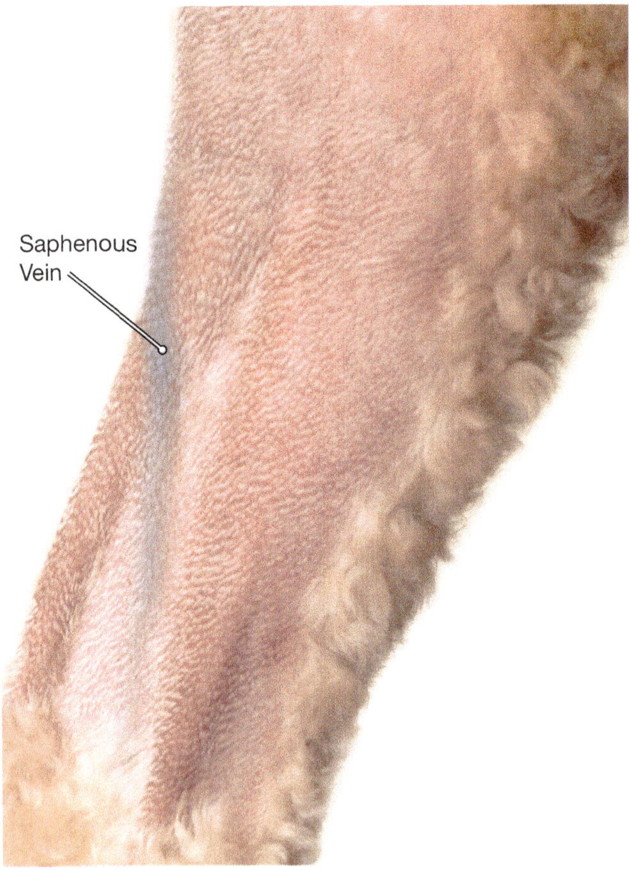

Figure 5.33: Saphenous vein.
© NCSU.

Lymph Nodes

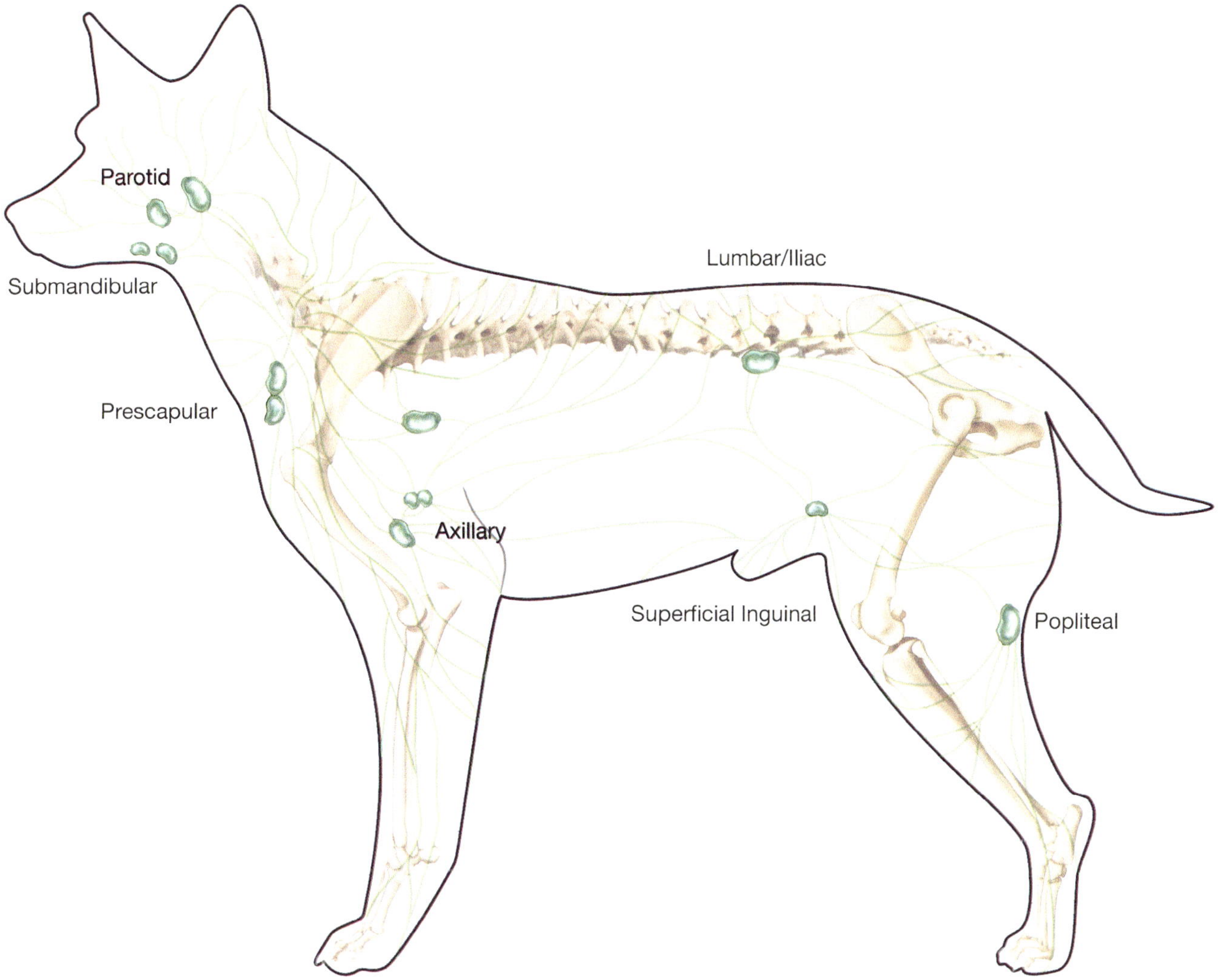

Figure 5.34: Lymph nodes.
© NCSU.

Overview of the Circulation

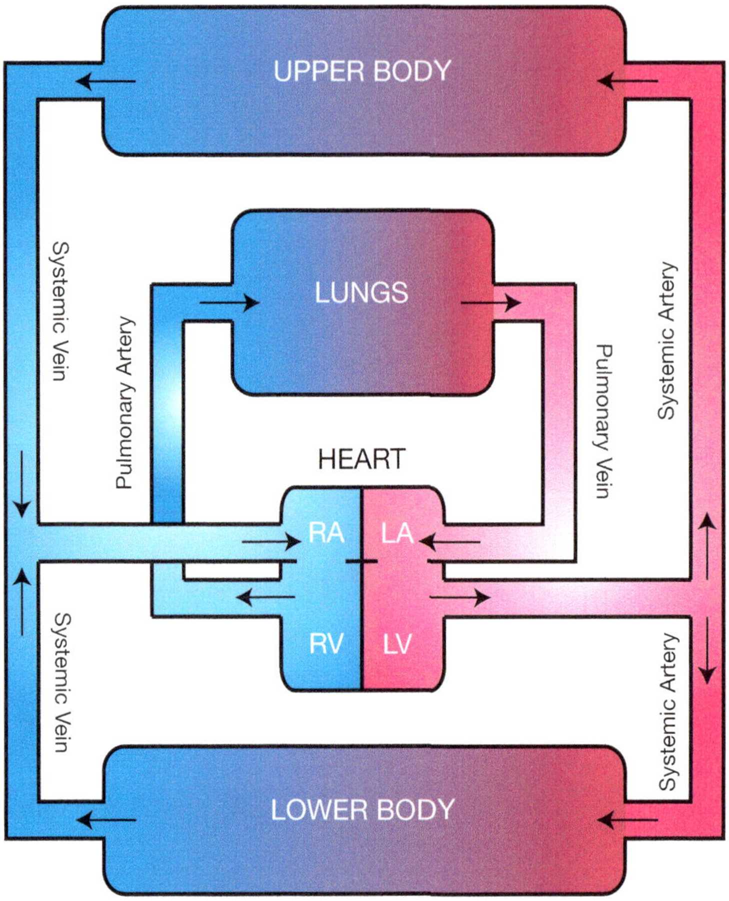

Figure 5.35: Circulation.

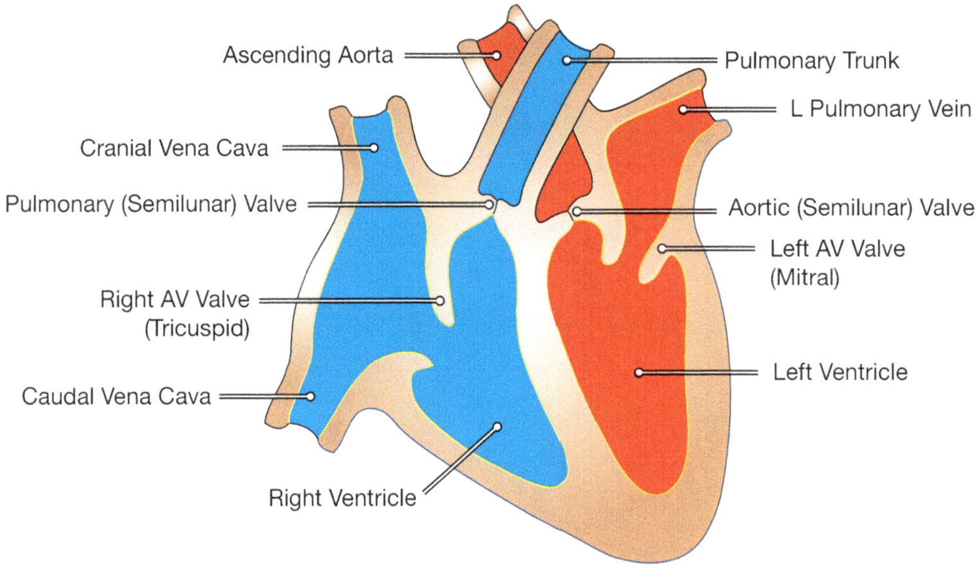

Figure 5.36: Organization of the heart.

Canine Heart

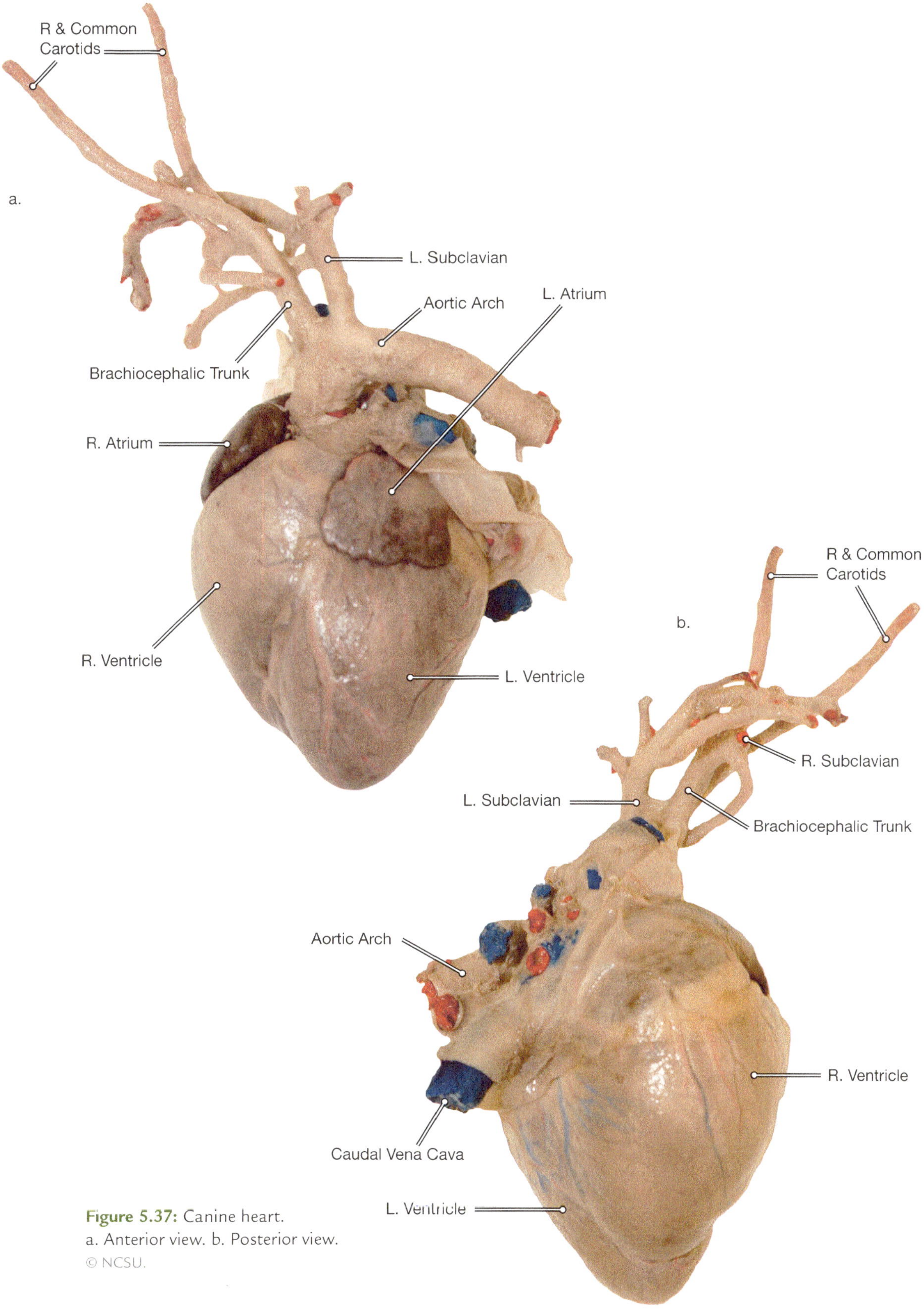

Figure 5.37: Canine heart.
a. Anterior view. b. Posterior view.

Canine Heart

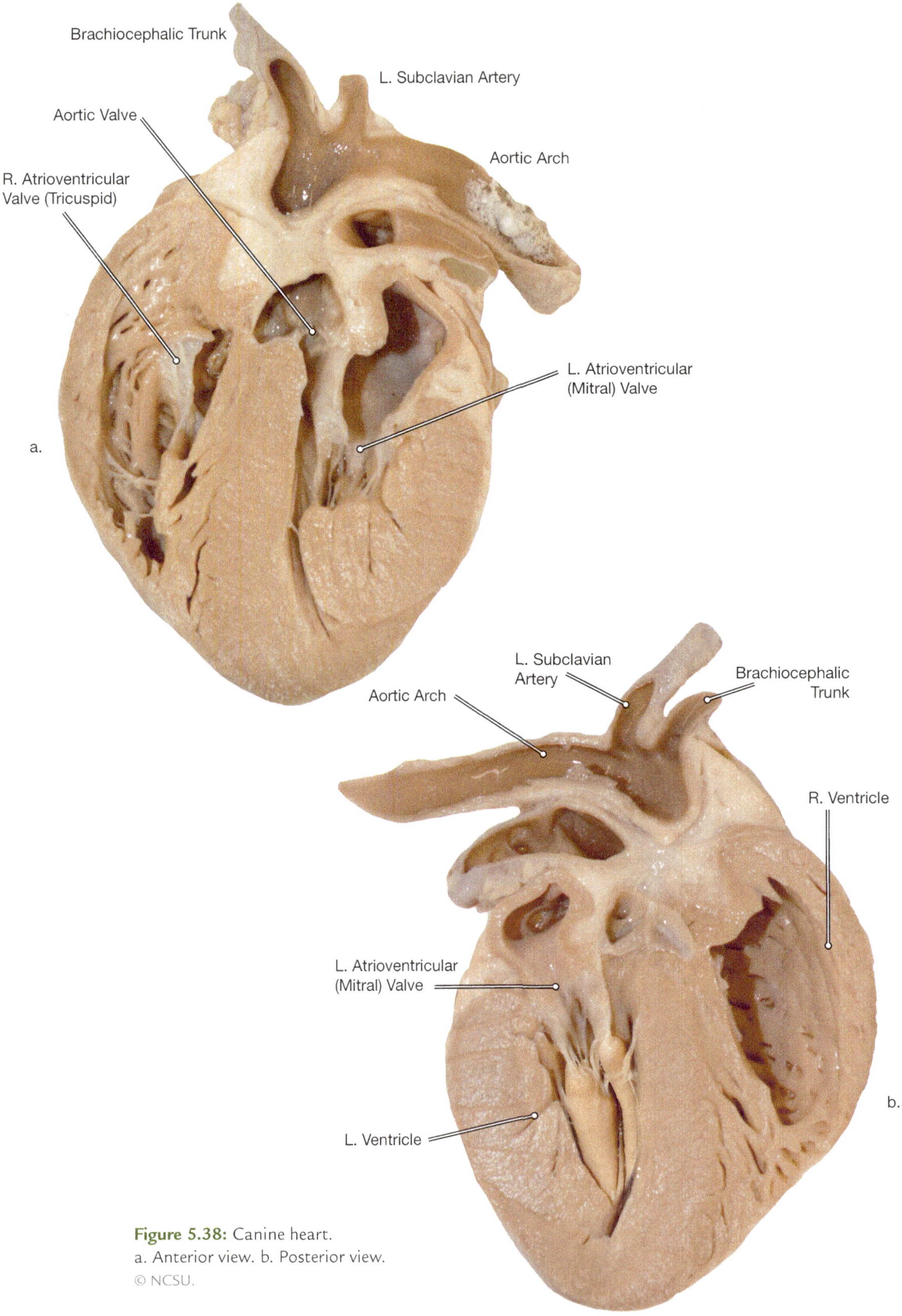

Figure 5.38: Canine heart.
a. Anterior view. b. Posterior view.

Heart: Interior

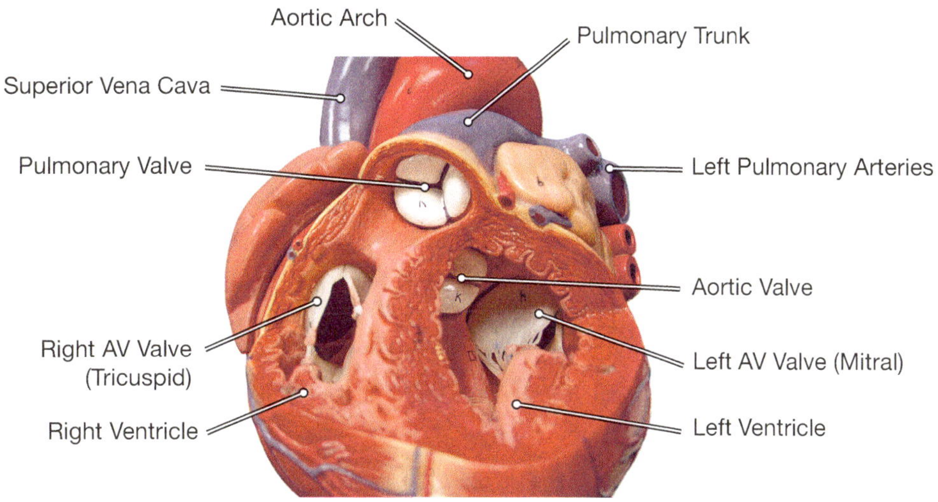

Figure 5.39: Ventral view.

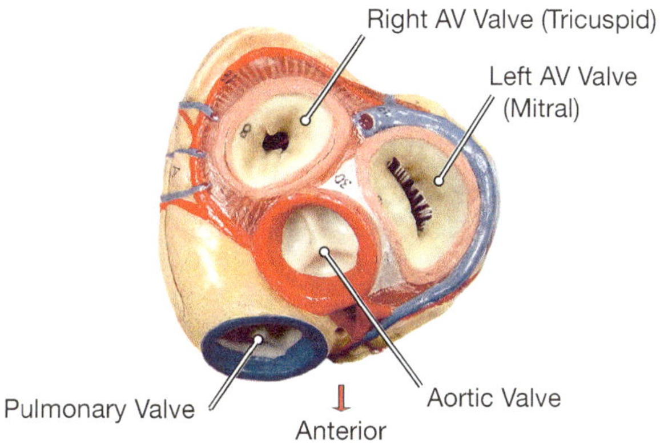

Figure 5.40: Valves, dorsal view.

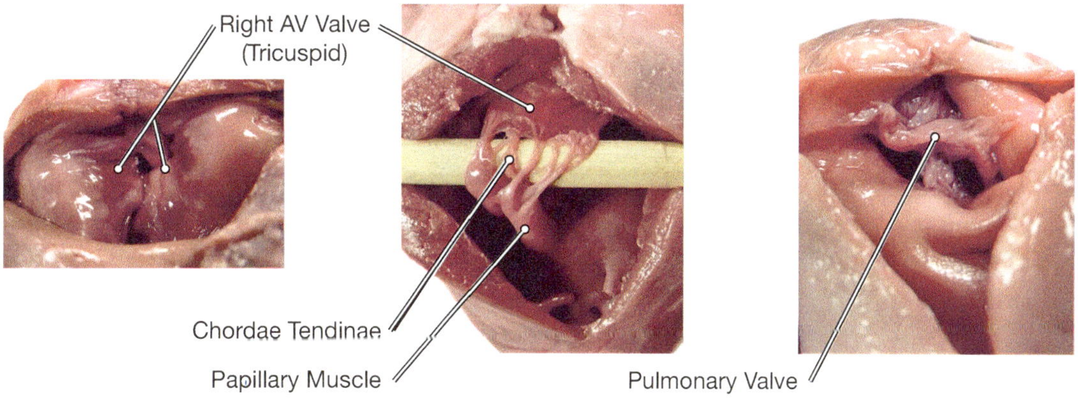

Figure 5.41: Heart. a. Right AV valve. b. Chordae tendinae. c. Pulmonary valve.

Figure 5.42: Intercostal ausculation for heart valves, PAM highlighted.
© NCSU.

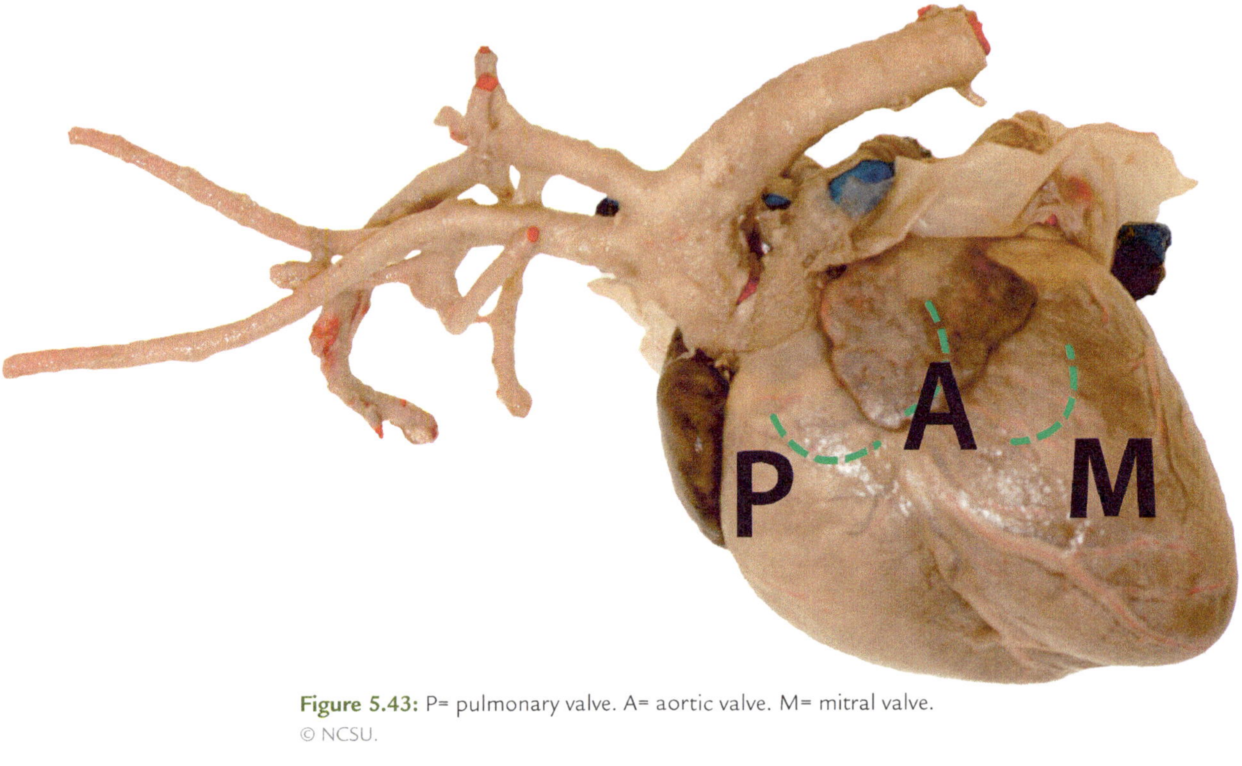

Figure 5.43: P= pulmonary valve. A= aortic valve. M= mitral valve.
© NCSU.

Intercostal Ausculation for Heart Valves: Right Side

Figure 5.44: Intercostal ausculation for heart valves, tricuspid valve highlighted.
© NCSU.

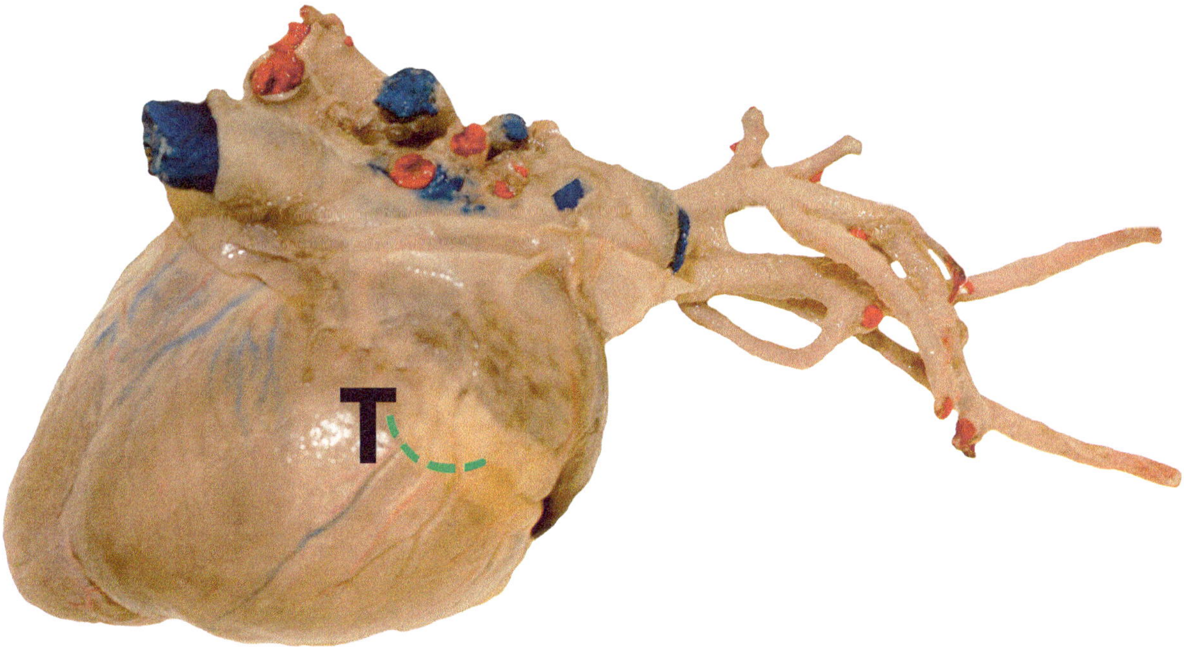

Figure 5.45: T= tricuspid valve.
© NCSU.

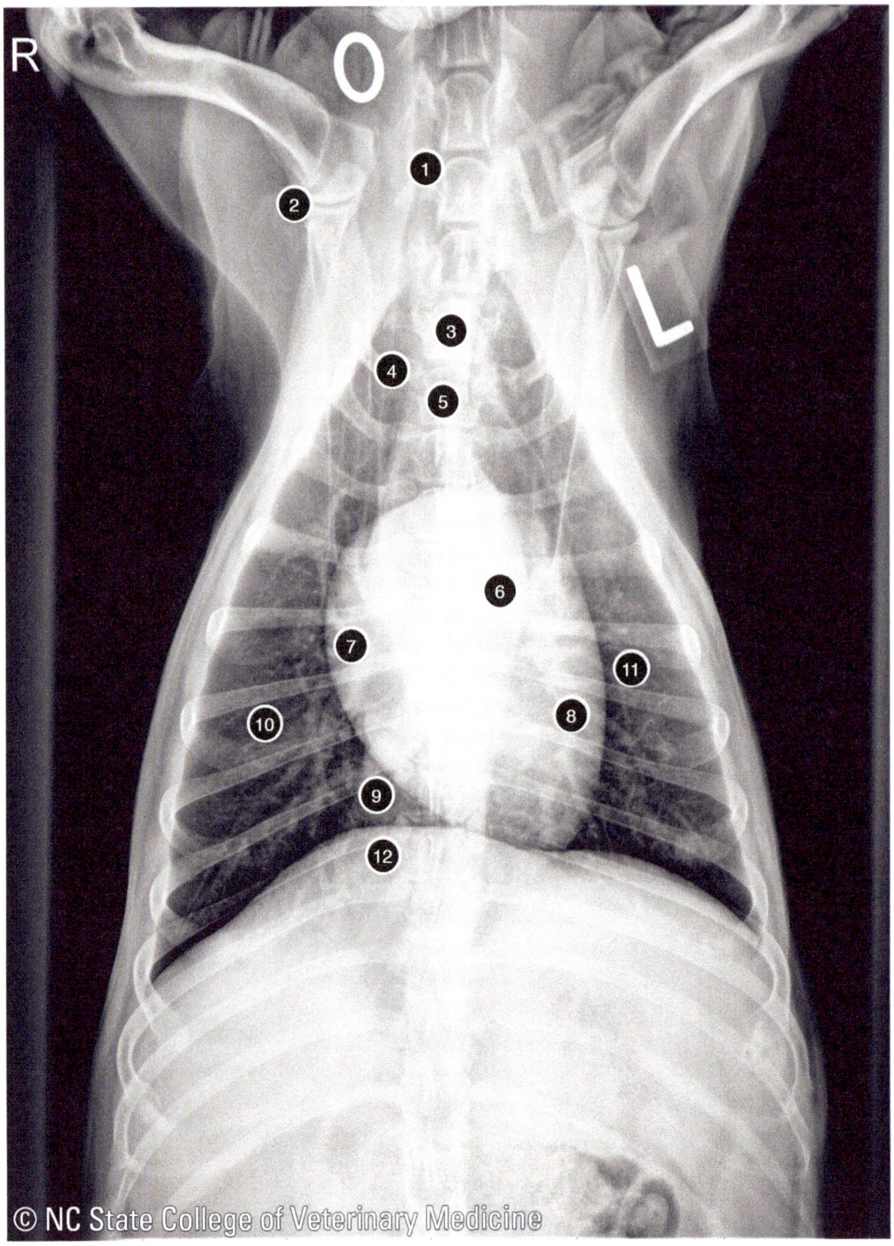

Figure 5.46: Vetral-dorsal (VD) view of thorax.

1. trachea
2. right shoulder joint
3. 7th cervical vertebra (C7)
4. 1st rib
5. 1st thoracic vertebra (T1)
6. descending aorta
7. right ventricle of heart
8. left ventricle of heart
9. caudal vena cava
10. right lung
11. left lung
12. diaphragm

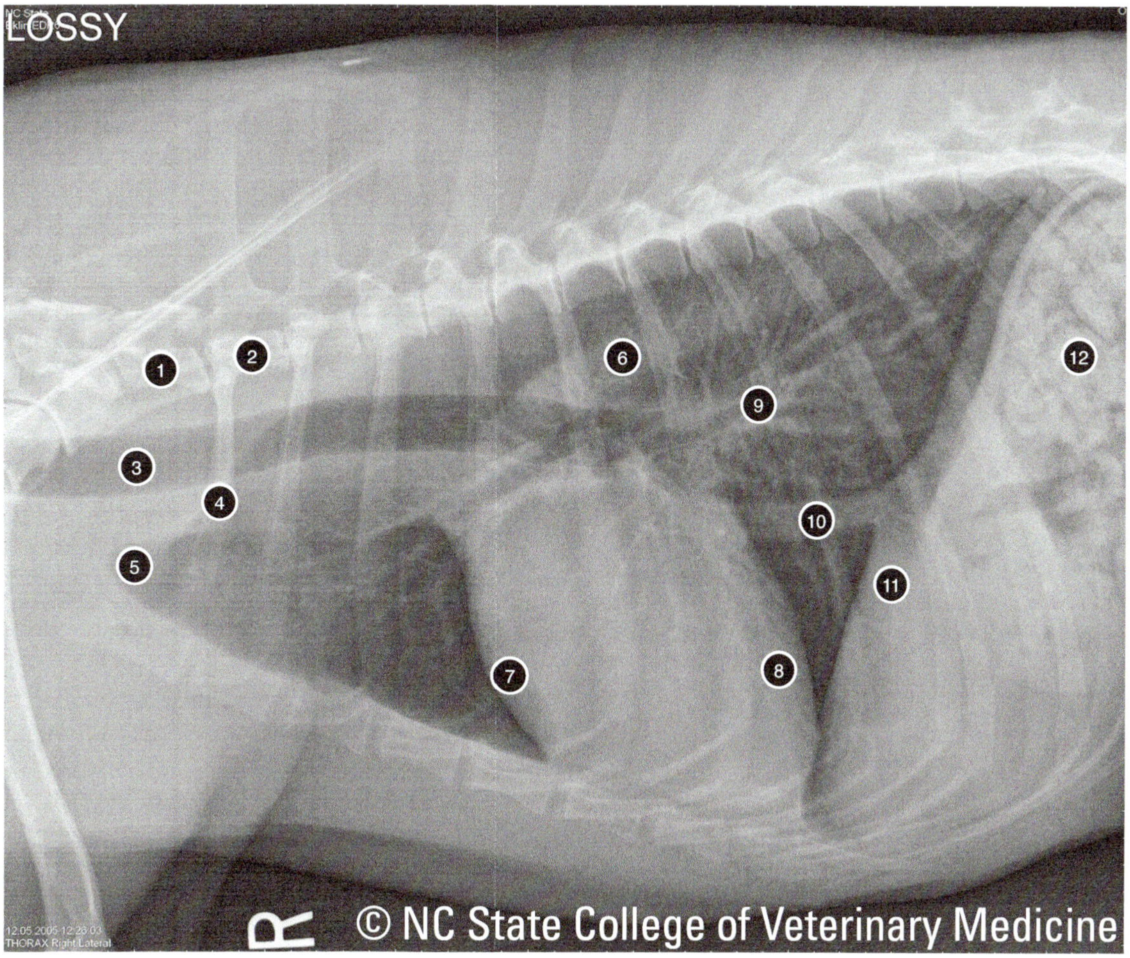

Figure 5.47: Right-left lateral view of thorax.

1. 7th cervical vertabra (C7)
2. 1st thoracic vertebra (T1)
3. trachea
4. 1st rib
5. apex of left lung
6. descending aorta
7. right ventricle of heart
8. left ventricle of heart
9. pulmonary vessels
10. caudal vena cava
11. diaphragm
12. gas in fundus of stomach

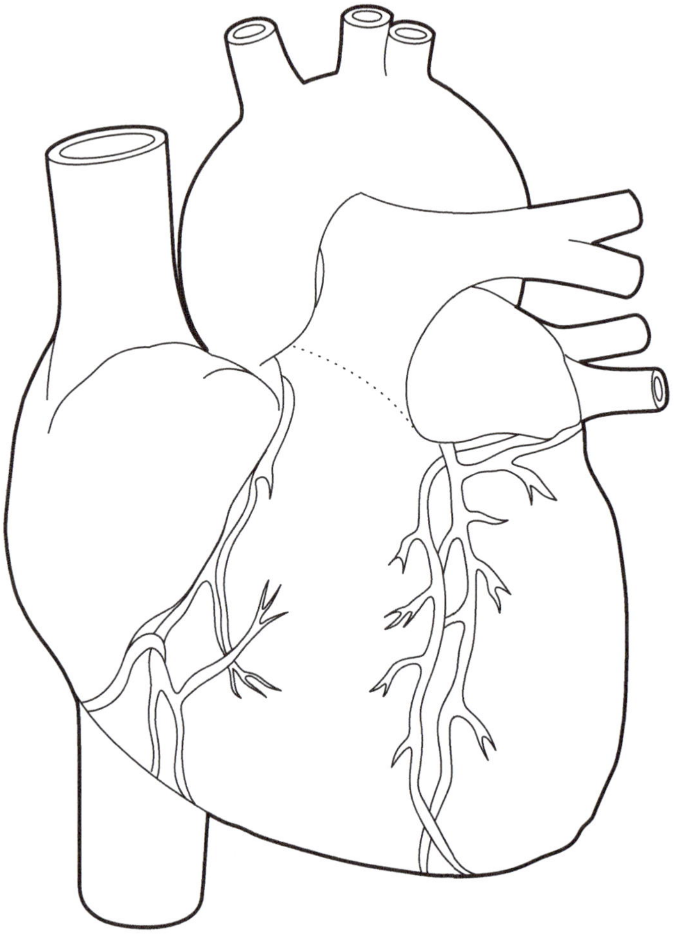

AORTA

INFERIOR VENA CAVA

LEFT ATRIUM

LEFT VENTRICLE

PULMONARY ARTERIES

PULMONARY TRUNK

PULMONARY VEINS

RIGHT ATRIUM

RIGHT VENTRICLE

SUPERIOR VENA CAVA

Figure 5.48: Heart coloring activity.

Review: Crossword Puzzle

Across

2. vessel that contains a thick tunica media layer
5. substance found in blood that aids in clotting
9. red blood cells
12. determines the amount of erythrocytes in a blood sample
14. target vein in the thoracic limb
15. most common leukocyte

Down

1. system that returns interstitial fluid back to the heart
3. white blood cells
4. target vein in the pelvic limb
6. least common leukocyte
7. target vein in the neck
8. muscular layer of the heart
10. binds oxygen
11. vessel that contains a thick tunica externa layer
13. muscle type that contains intercalated discs

Review: Questions

1. Put the following structures in order as blood flows through the heart

tricuspid valve	pulmonary valve
mital valve	lungs
ascending aorta	right atria
pulmonary vein	left atria
pulmonary trunk	right ventricle
aortic valve	left ventricle

1. vena cava
2. ______________________________
3. ______________________________
4. ______________________________
5. ______________________________
6. ______________________________
7. ______________________________
8. ______________________________
9. ______________________________
10. ______________________________
11. ______________________________
12. ______________________________
13. ______________________________

2. What side of the heart has a thicker muscular wall? Why?

3. Why do veins and lymphatic vessels contain valves?

4. Why are veins better vessels for blood withdrawal than arteries?

5. Define venipuncture and name three common veins used for this procedure.

6. What diseases could initially be diagnosed from listening to the heart rate, sounds and rhythm?

7. What are the functions of the lymphatic system?

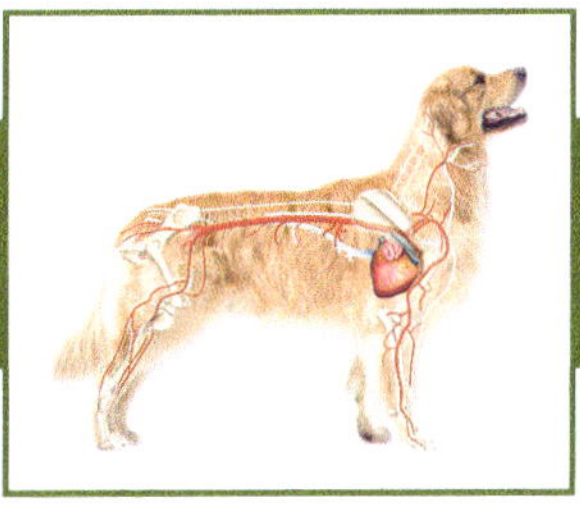

Chapter 6

Airways, Lungs and the Respiratory System

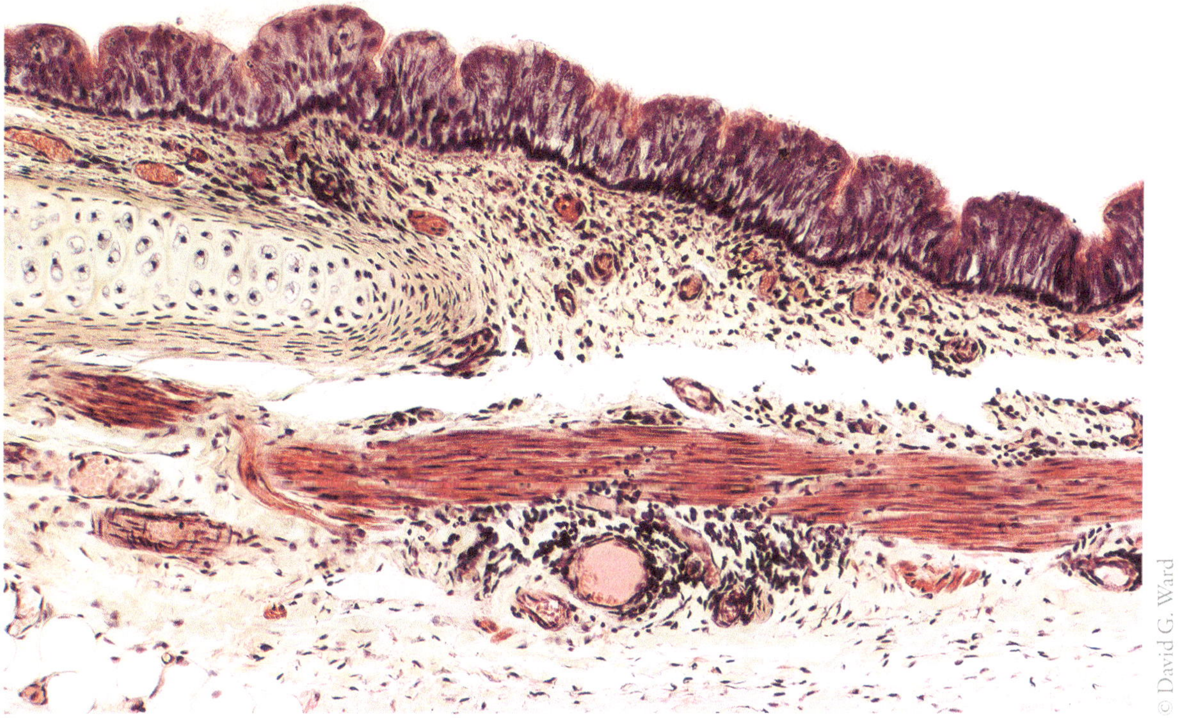

Trachea

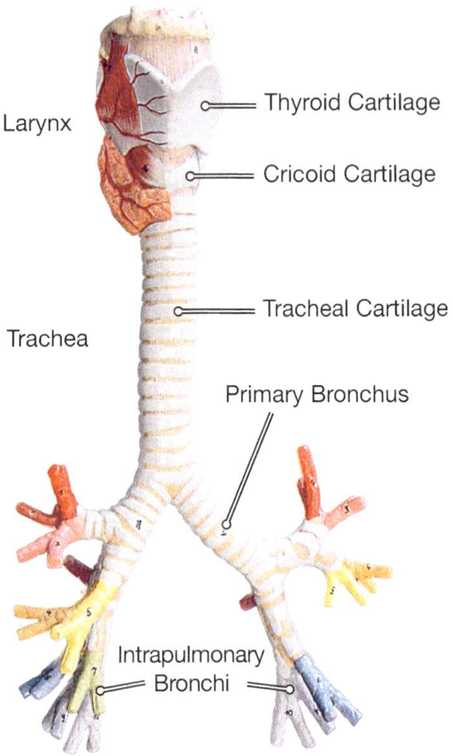

Figure 6.1: Larynx, trachea, and bronchi.

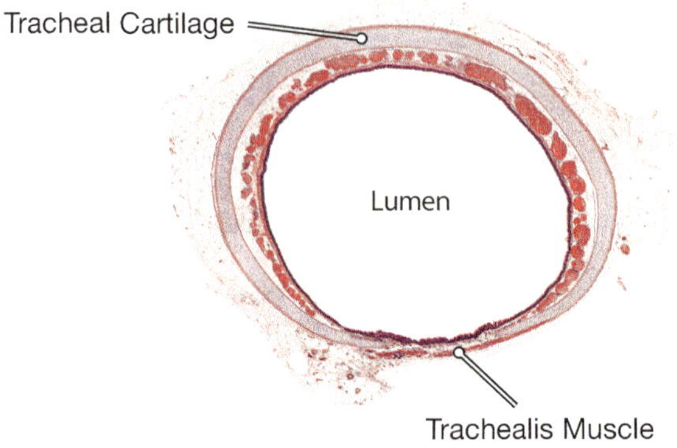

Figure 6.2: Trachea, histology cross section.

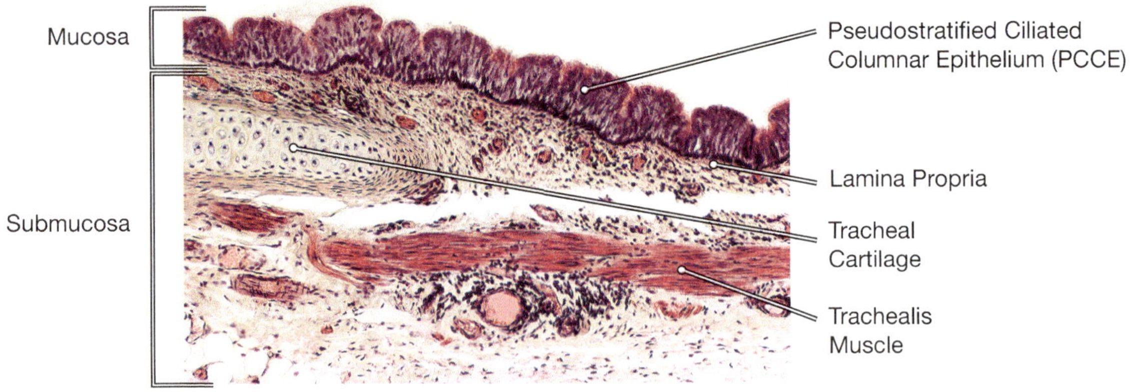

Figure 6.3: Trachea, histology.

Alveolar Sacs and Alveoli

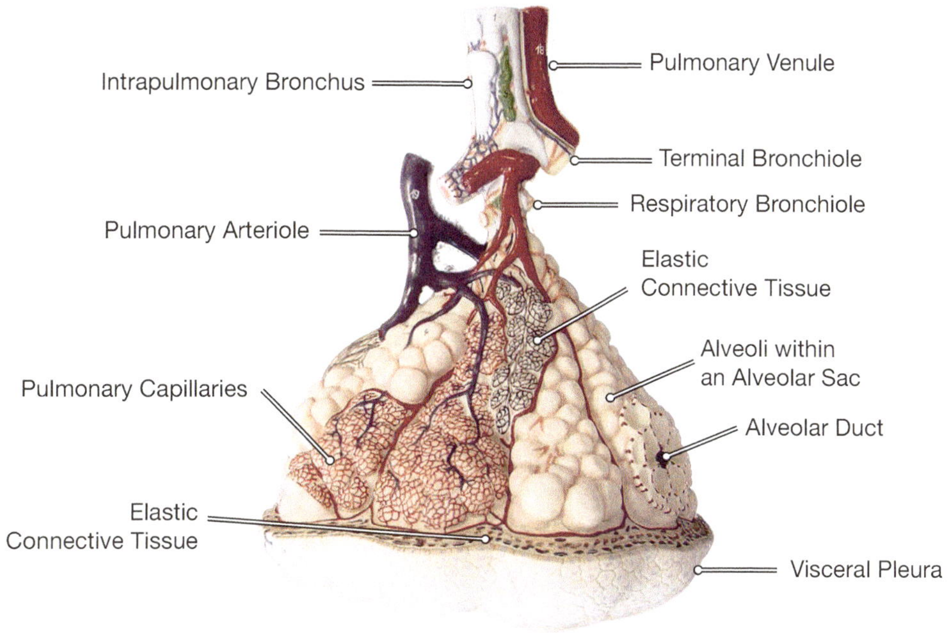

Figure 6.4: Alveolar sacs.
© David G. Ward.

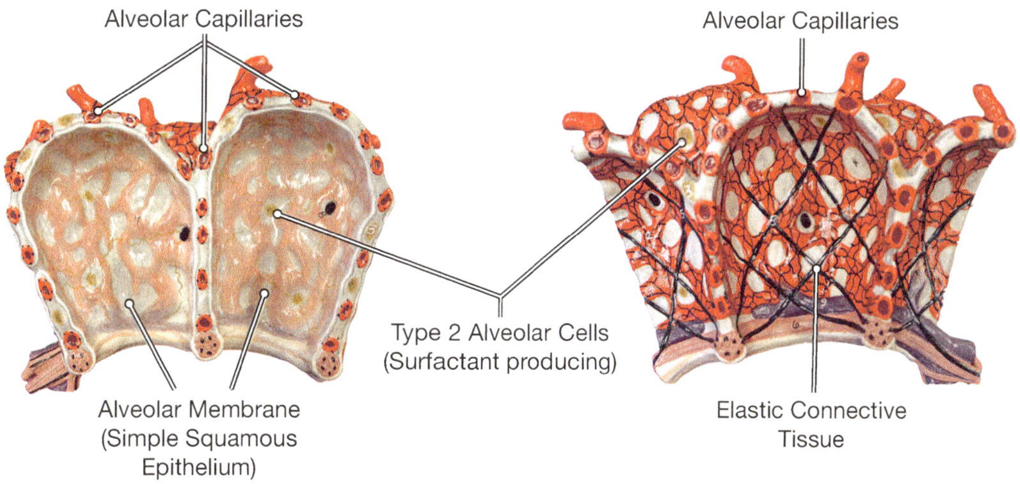

Figure 6.5: Alveoli.
© David G. Ward.

Intrapulmonary Bronchi

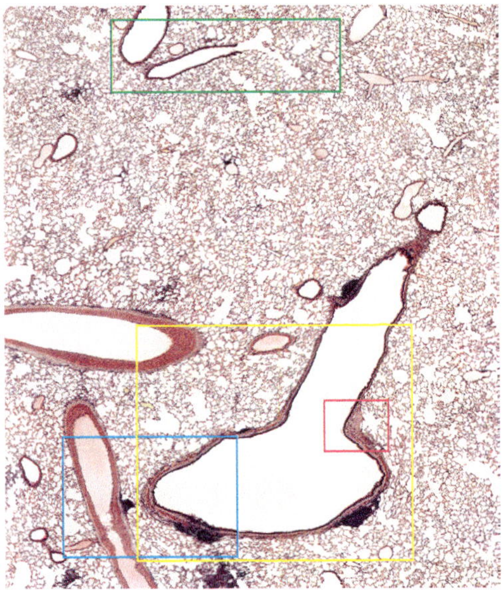

Figure 6.6: Lung, overview.

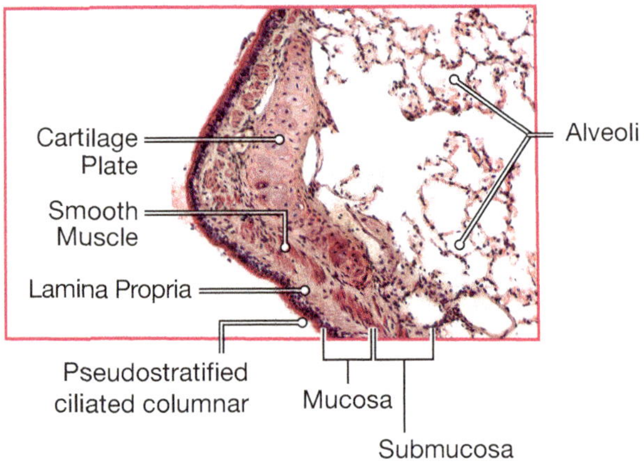

Figure 6.7: Artery and bronchus.

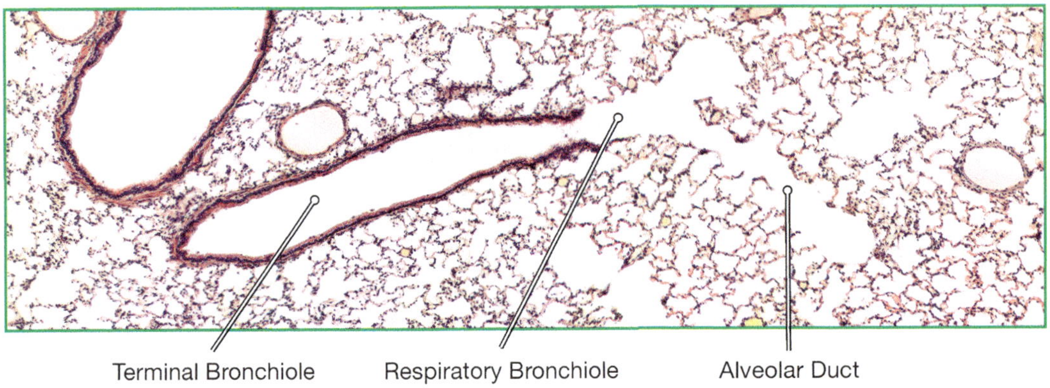

Figure 6.8: Terminal and respiratory bronchioles and alveolar ducts.

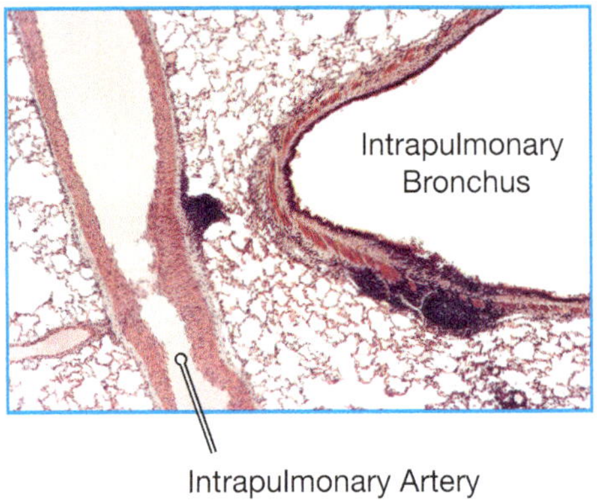

Figure 6.9: Mucosa and submucosa.

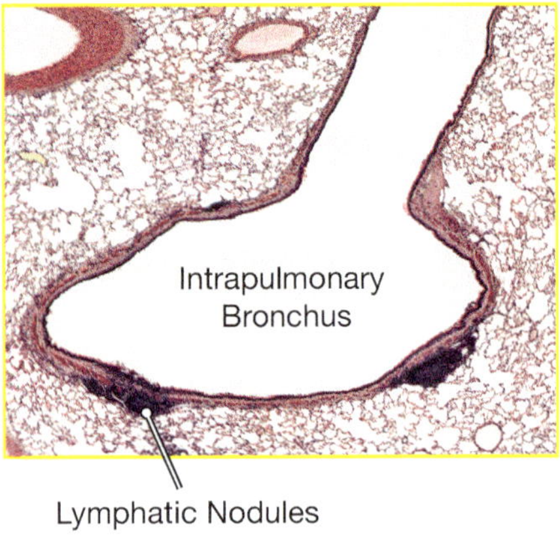

Figure 6.10: Intrapulmonary bronchus.

Bronchioles and Alveoli

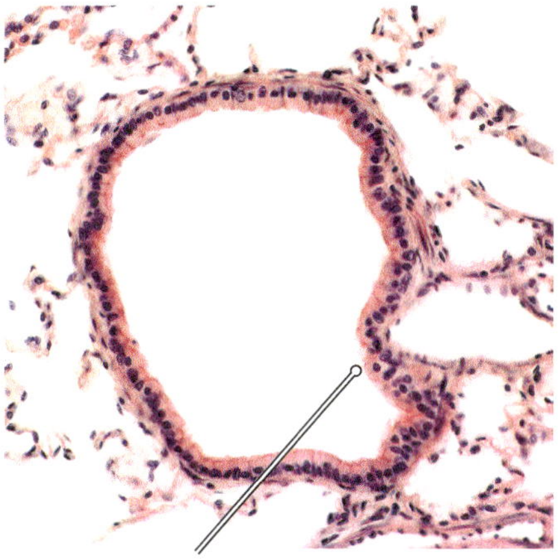

Figure 6.11: Terminal bronchiole.

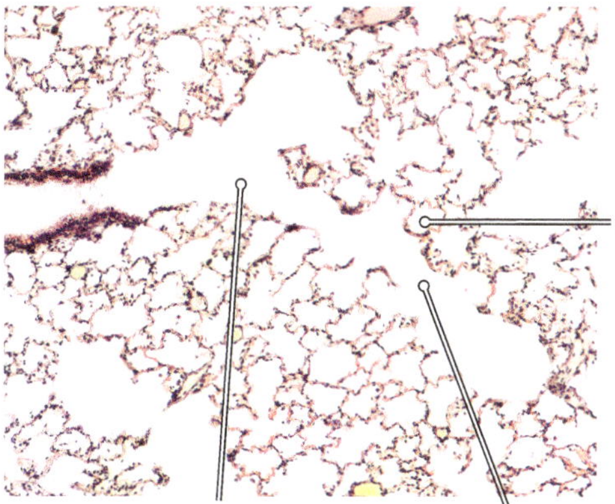

Figure 6.12: Respiratory brochiole and alveoli.

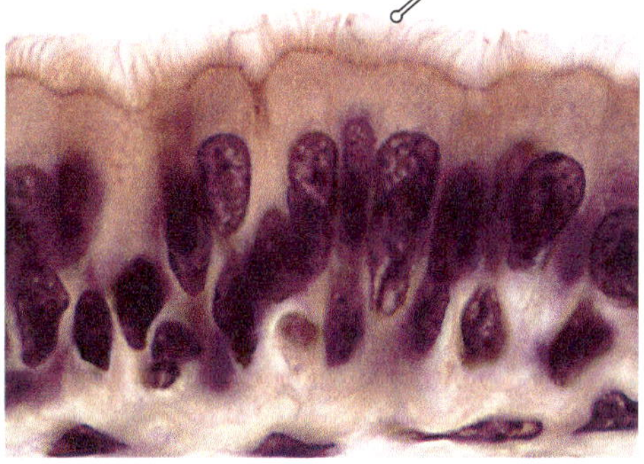

Figure 6.13: Pseudostratified Ciliated Columnar Epithelium.

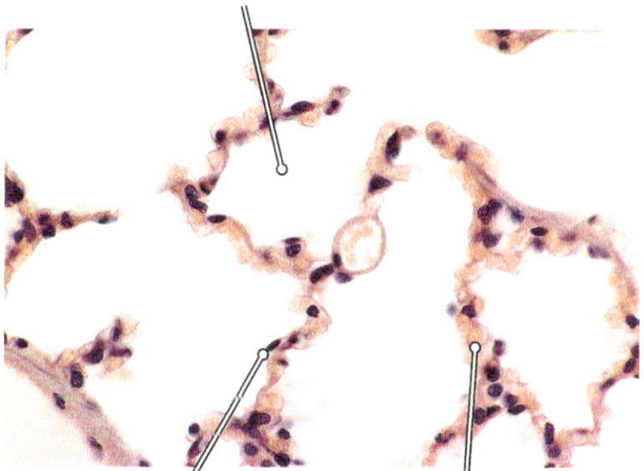

Figure 6.14: Alveoli.

Alveoli

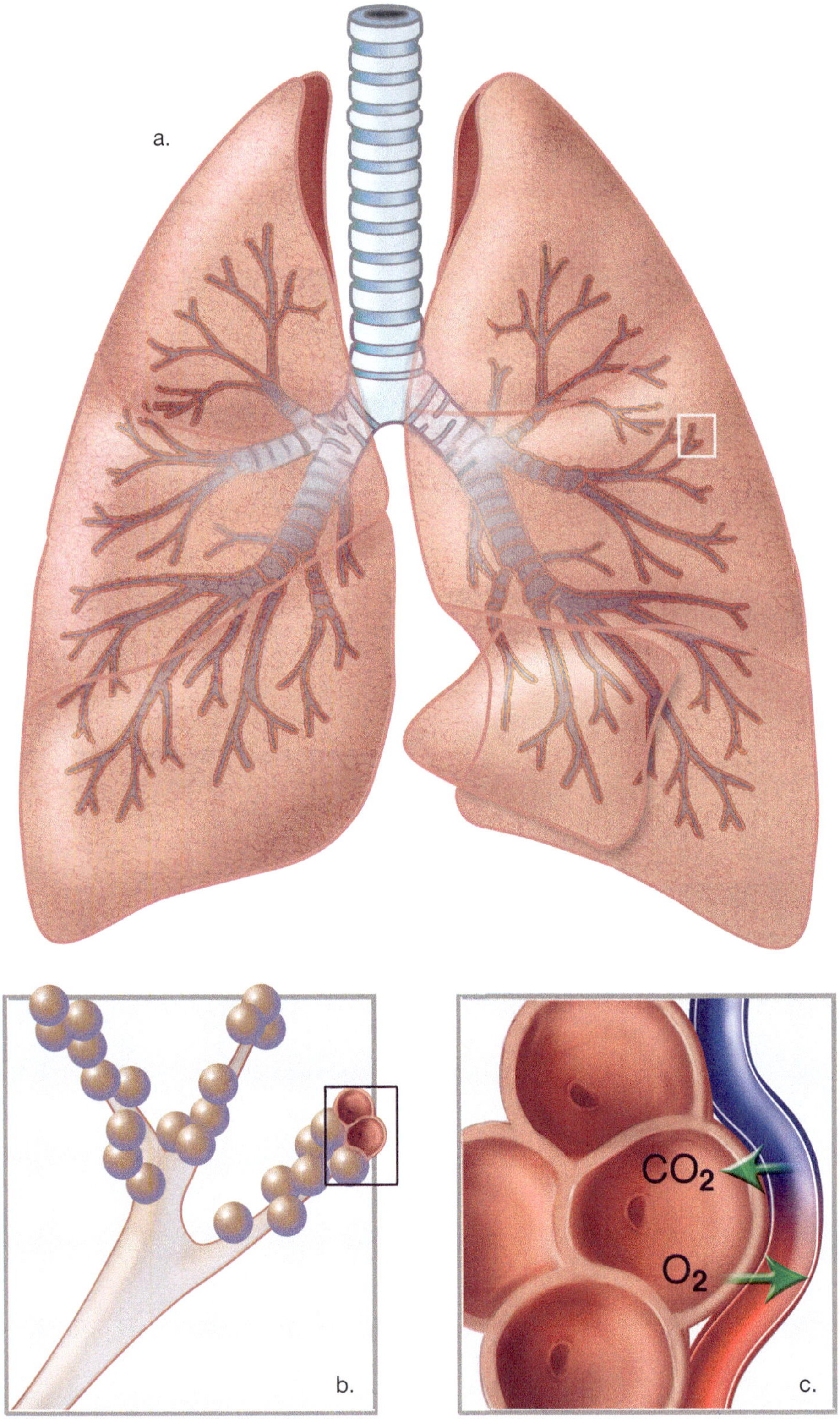

Figure 6.15: Alveoli is the site of oxgen exchange. a. Alveoli highlighted. b. Detail of alveoli. c. Site of oxygen exchange.

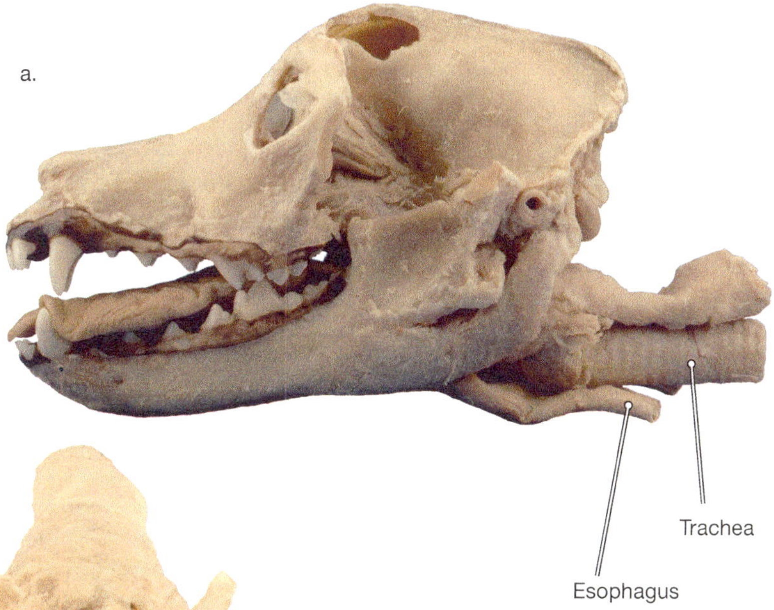

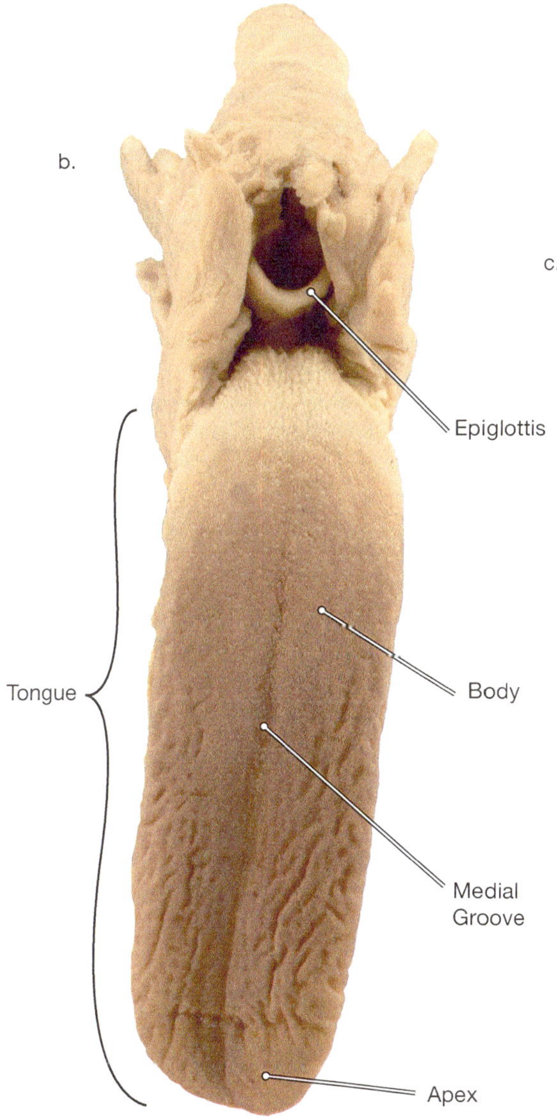

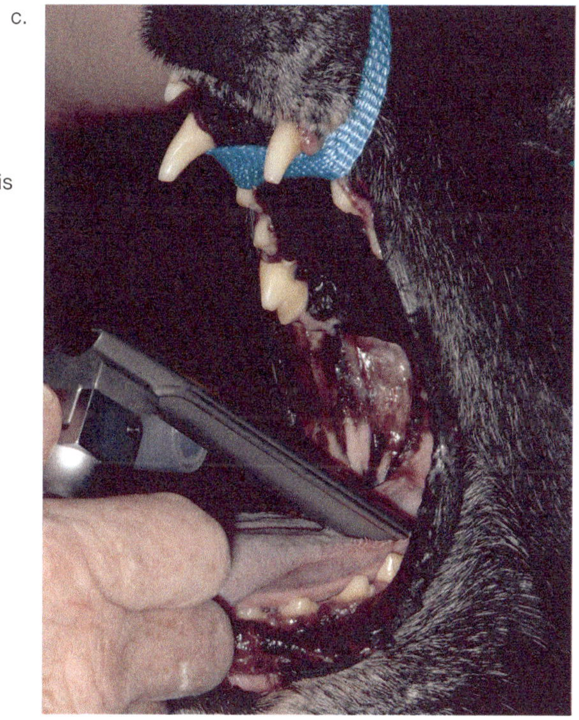

Figure 6.16: Upper airway. a. Upper airway, lateral view b. Upper airway, viewed from inside the mouth. c. Use of laryngoscope to visualize airway, lateral view.

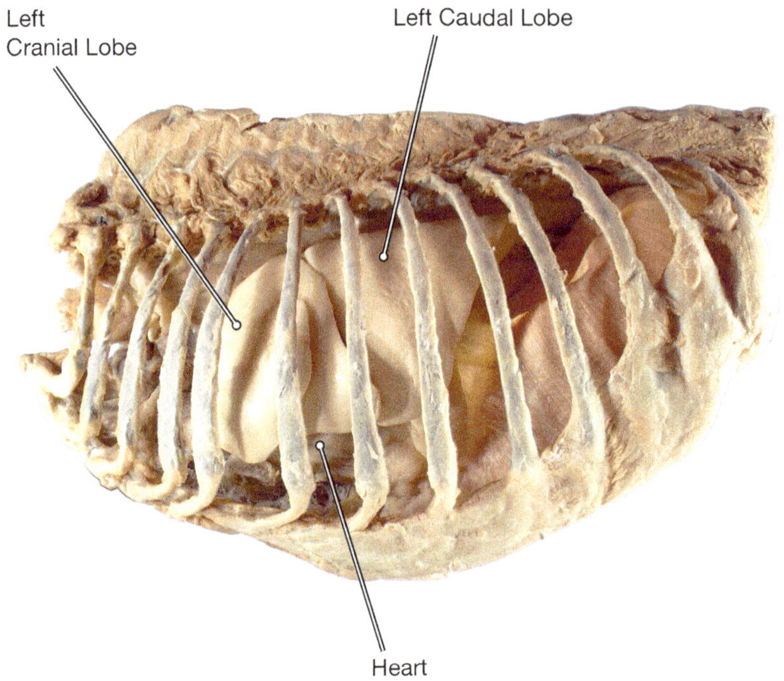

Figure 6.17: Thoracic cavity.

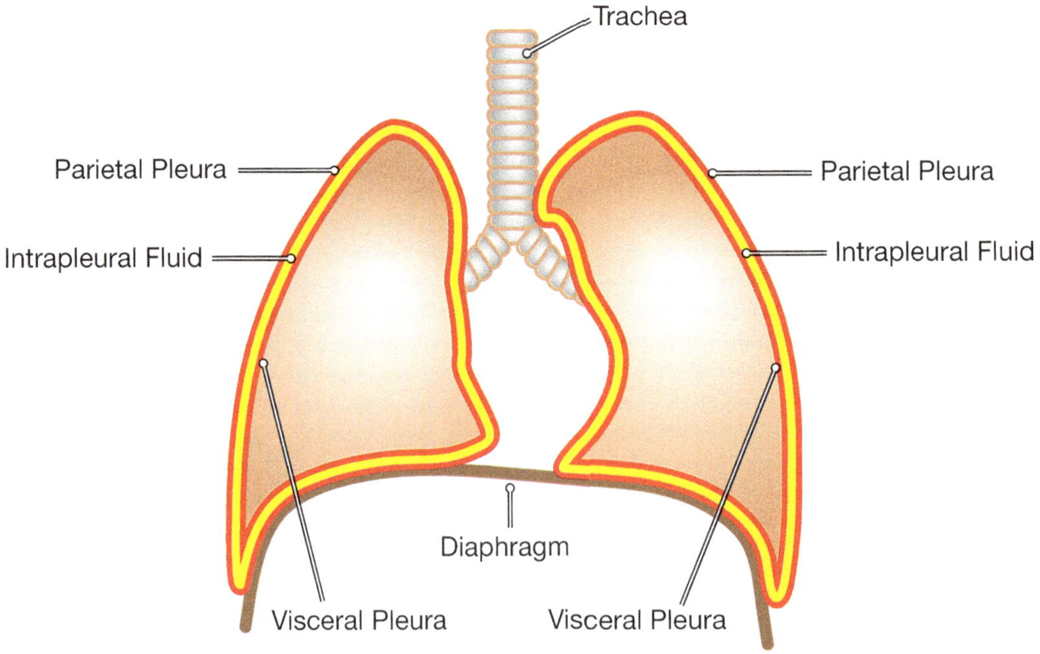

Figure 6.18: Pleural membranes.

Diaphragm Movement with Breathing

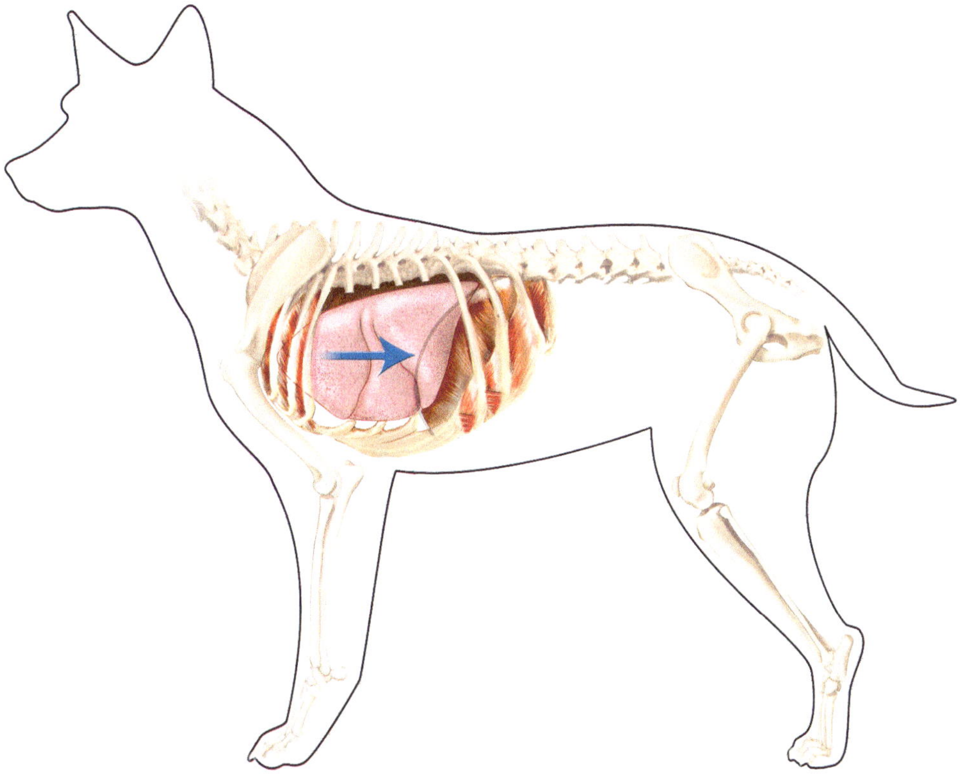

Figure 6.19: Inhalation, diaphragm moves caudally.
© NCSU.

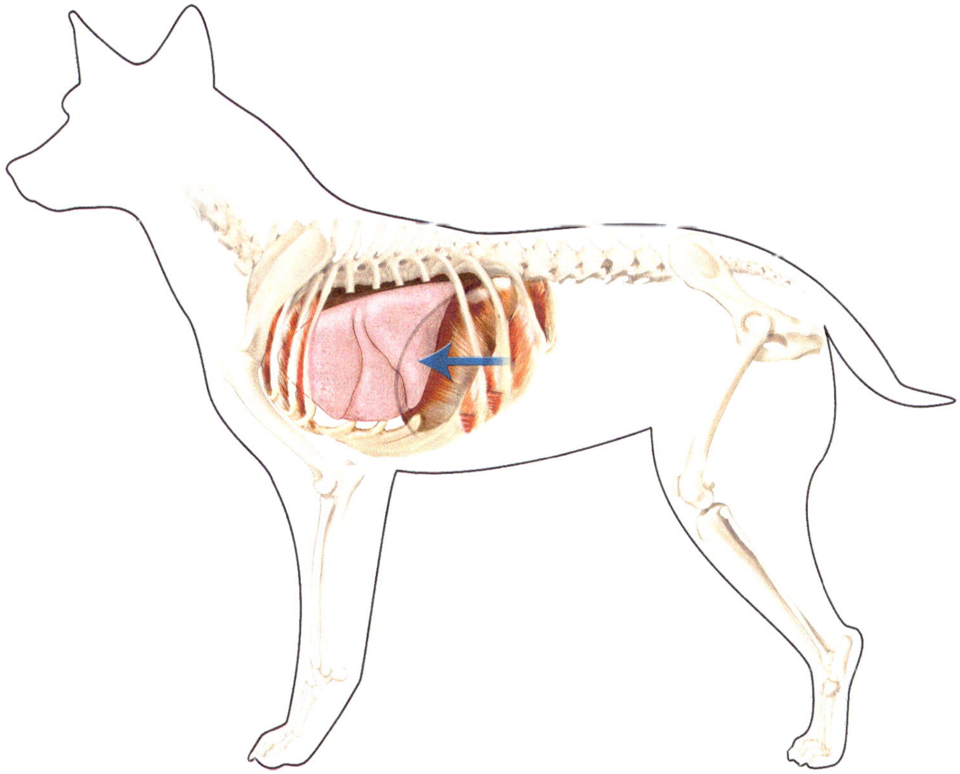

Figure 6.20: Exhalation, diaphragm moves cranially.
© NCSU.

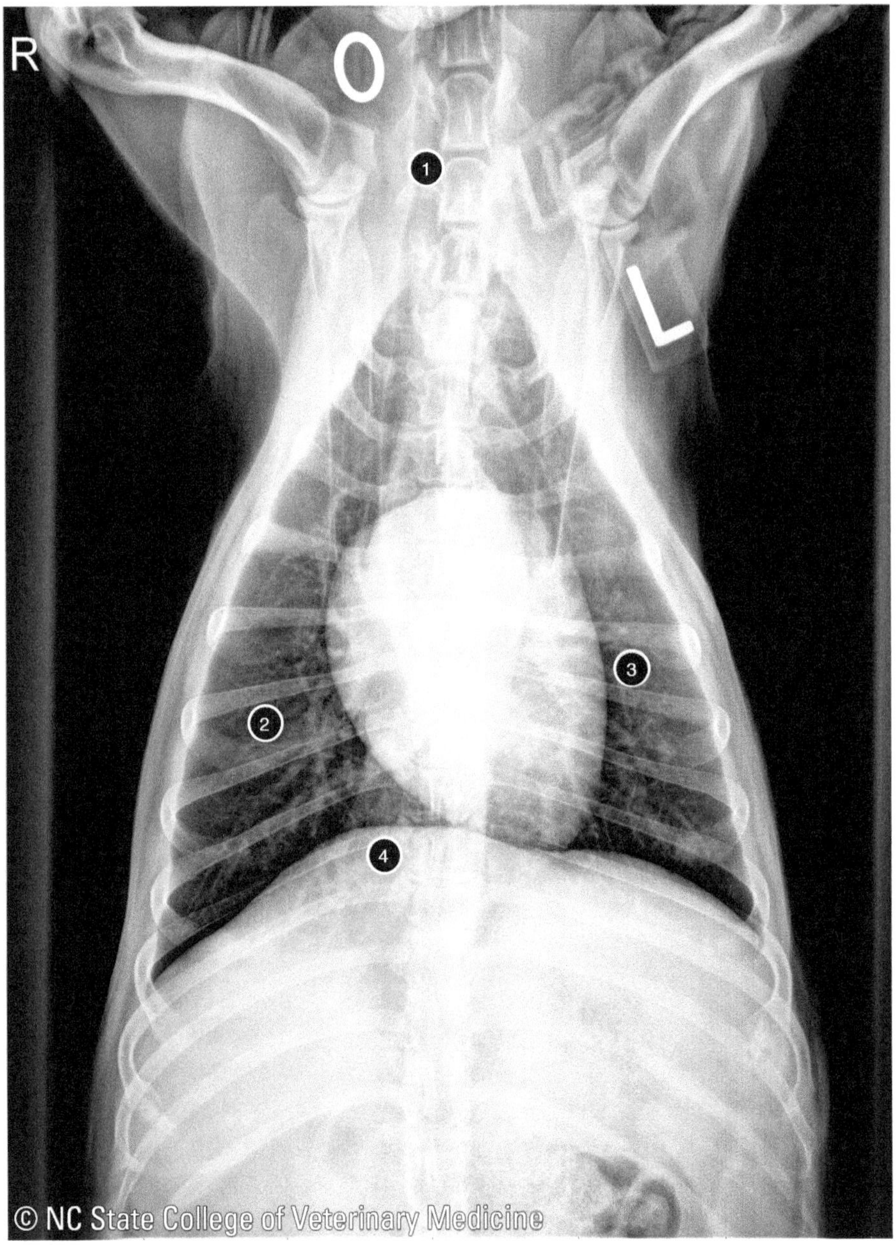

Figure 6.21: Ventral-dorsal (VD) view of thorax.

1. Trachea
2. Right lung
3. Left lung
4. Diaphragm

Review: Labeling Activity

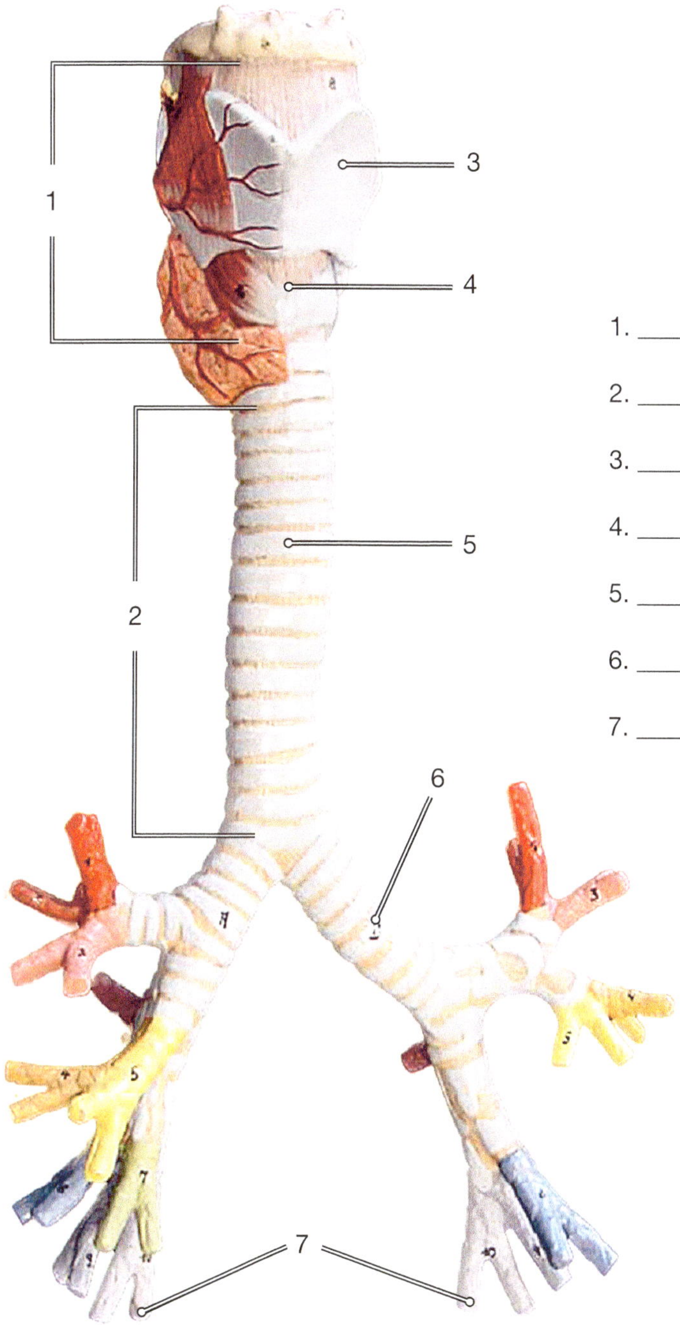

1. ______________________________

2. ______________________________

3. ______________________________

4. ______________________________

5. ______________________________

6. ______________________________

7. ______________________________

Figure 6.22: Trachea labeling activity.

Review: Labeling Activity

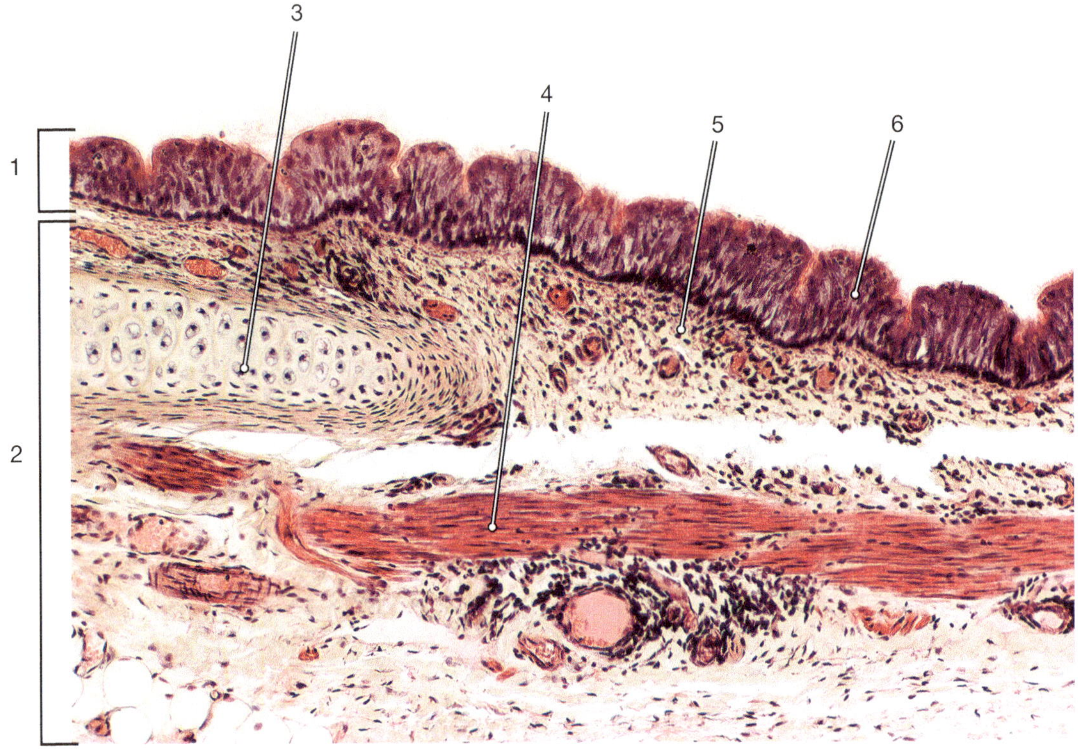

Figure 6.23: Trachea cross section labeling activity.
© David G. Ward.

1. ______________________________

2. ______________________________

3. ______________________________

4. ______________________________

5. ______________________________

6. ______________________________

Review: Labeling Activity

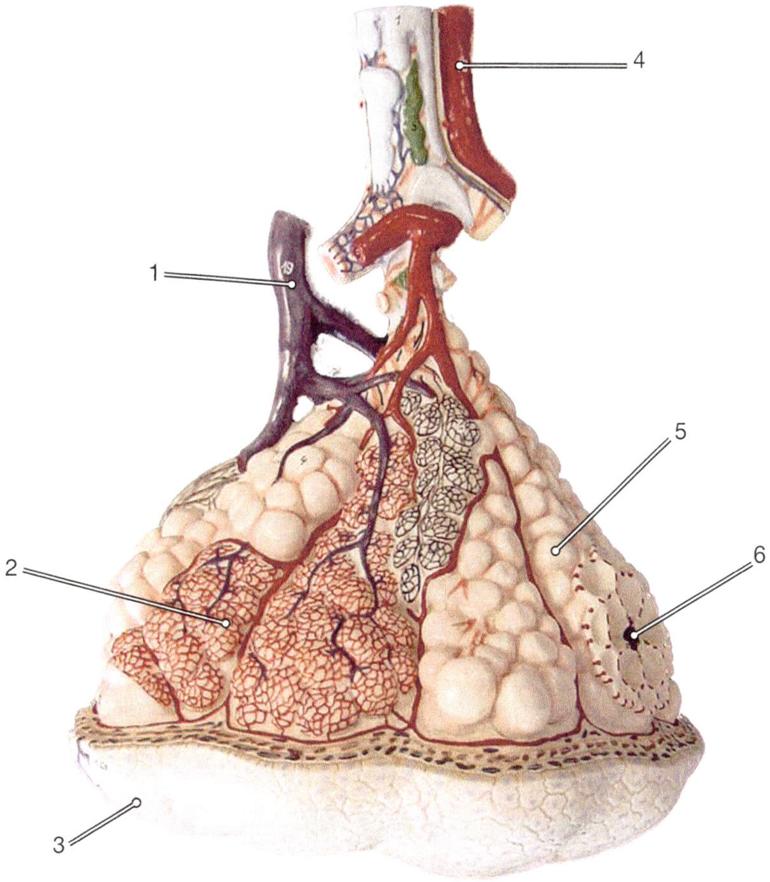

Figure 6.24: Alveolar sacs labeling activity.

1. ______________________________

2. ______________________________

3. ______________________________

4. ______________________________

5. ______________________________

6. ______________________________

Rewiew: Labeling Activity

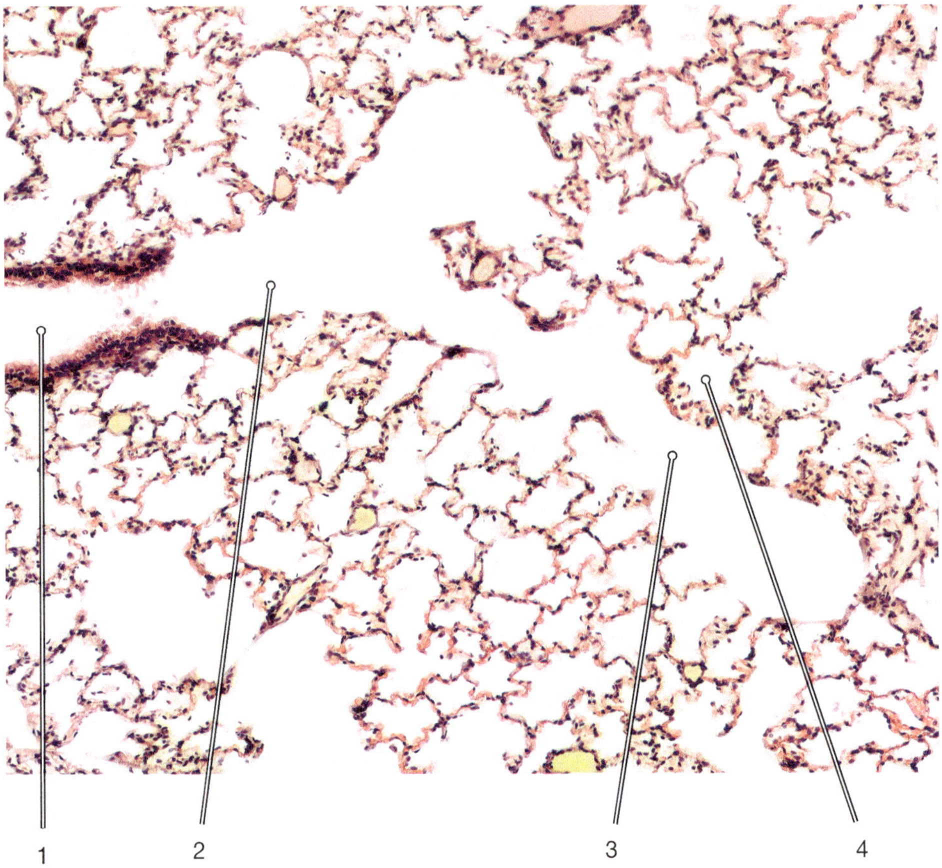

Figure 6.25: Respiratory bronchiole labeling activity.

1. ______________________________

2. ______________________________

3. ______________________________

4. ______________________________

Review: Crossword Puzzle

Across

1. muscle located in the submucosa of the trachea
5. structure located between the epiglottis and the trachea
6. structure located between the trachea and the alveolar sac
9. fluid located between the visceral and parietal pleural
11. layer of trachea surrounding the lumen
12. muscular structure inferior to the lungs that aids in respiration
13. number of lobes in the right lung
14. number of lobes in the left lung
15. the insertion of a tube towards the trachea to aid in breathing during a medical procedure

Down

2. muscles located between ribs to aid in respiration
3. the process of drawing air into the lungs
4. cluster of alveoli
7. the process of pushing air out of the lungs
8. pseudostratified ciliated columnar epithelium
10. site of oxygen exchange

Review: Questions

1. Put the following structures in order as oxygen flows through the respiratory system.

intrapulmonary bronchi	alveolar sac
primary bronchi	trachea
terminal bronchiole	epiglottis
blood	larynx
alveoli	

1. nasal cavity
2. ______________________________
3. ______________________________
4. ______________________________
5. ______________________________
6. ______________________________
7. ______________________________
8. ______________________________
9. ______________________________
10. ______________________________

2. What vessel carries oxygen-rich blood to the heart?

3. What vessel carries oxygen-poor blood to the lungs?

4. What is the function of cartilage rings in the trachea?

5. How does the diaphragm aid in breathing?

6. How do the intercostals aid in breathing?

7. If the endotracheal tube was inserted too far (into a primary bronchi) what effect would this have?

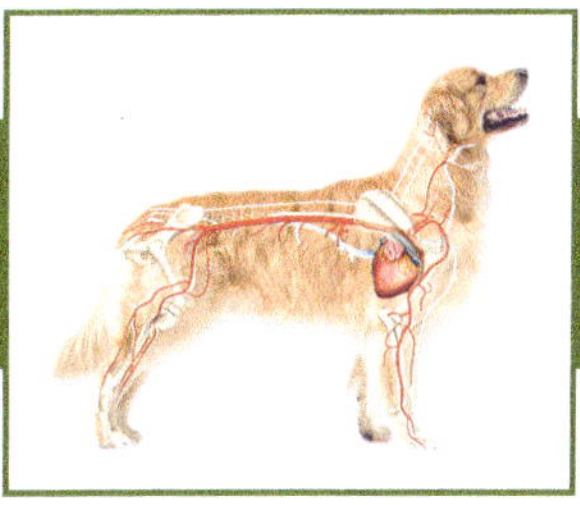

Chapter 7

GI Tract and the Digestive System

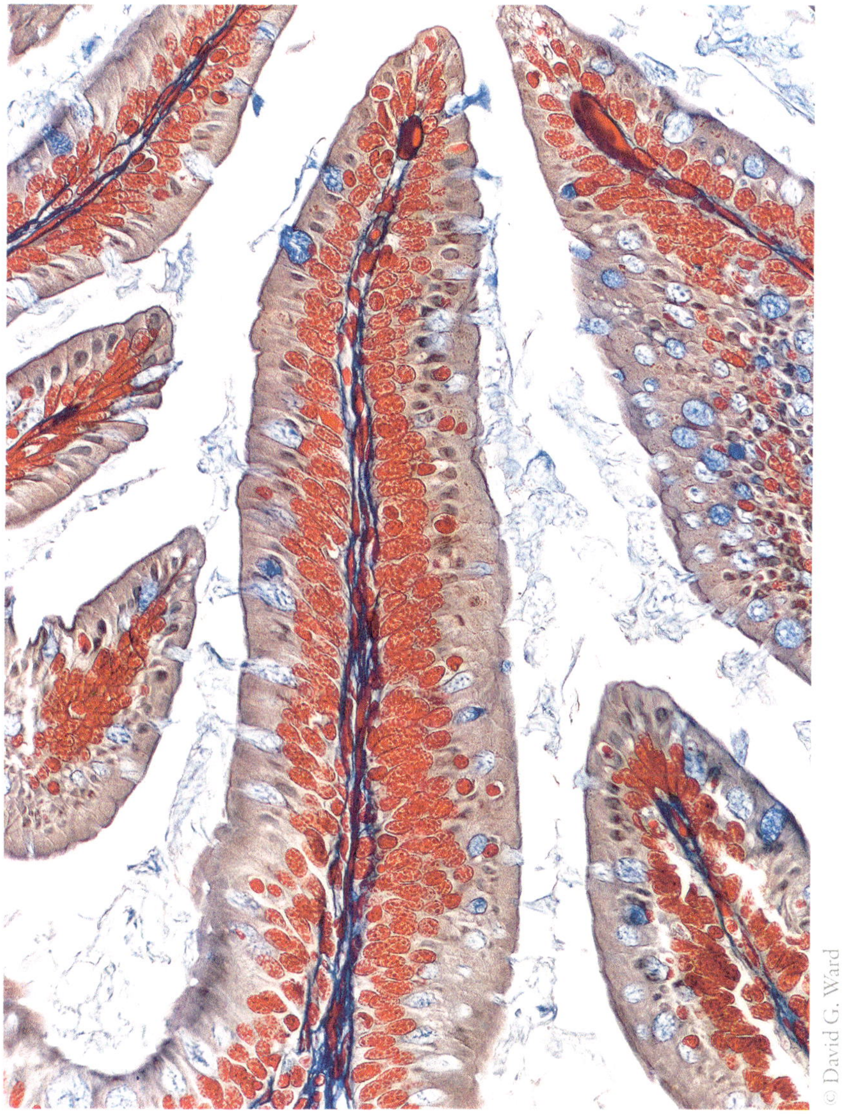

Intestinal Villi Showing Goblet Cells (Brightfield, Trichrome; x350)

Gastrointestinal Tract Overview

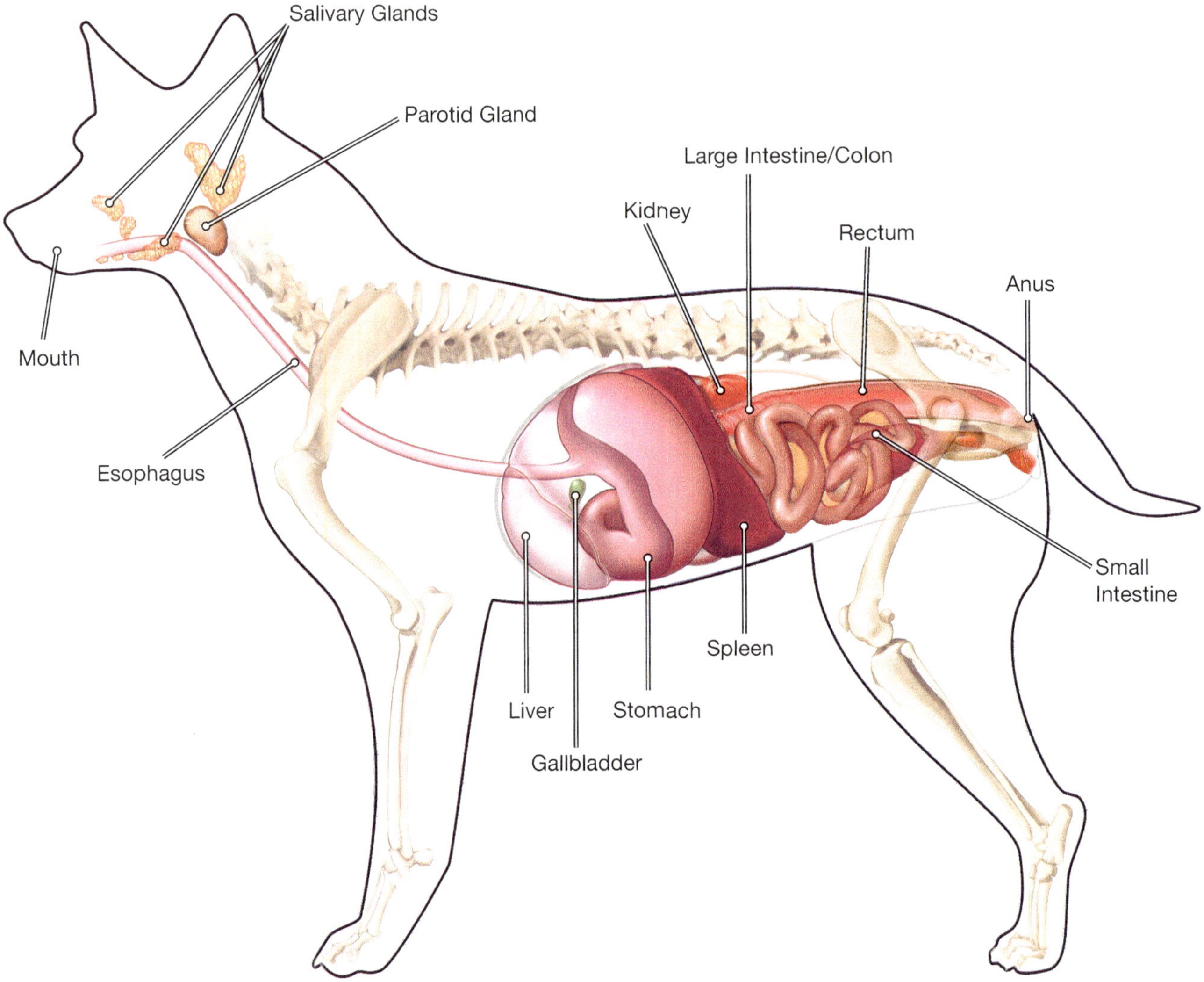

Figure 7.1: Gastrointestinal tract and related structures.
© NCSU.

Esophagus

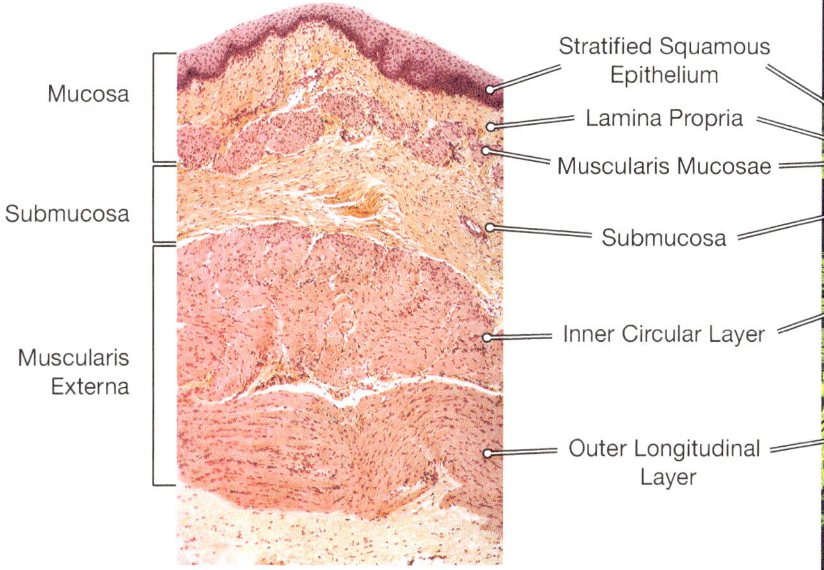

Figure 7.2: Esophagus.

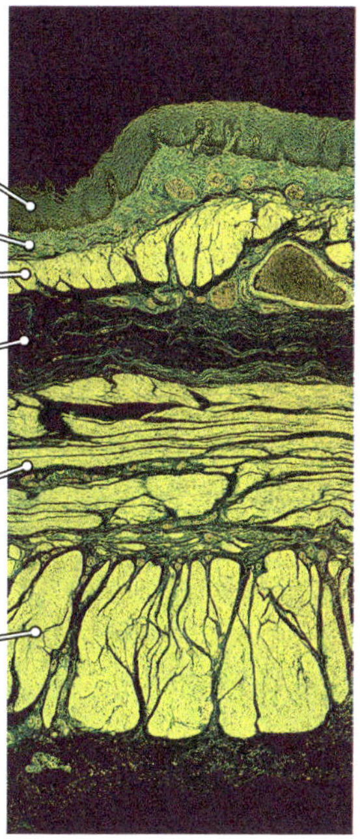

Figure 7.3: Esophagus. (autofluorescence)

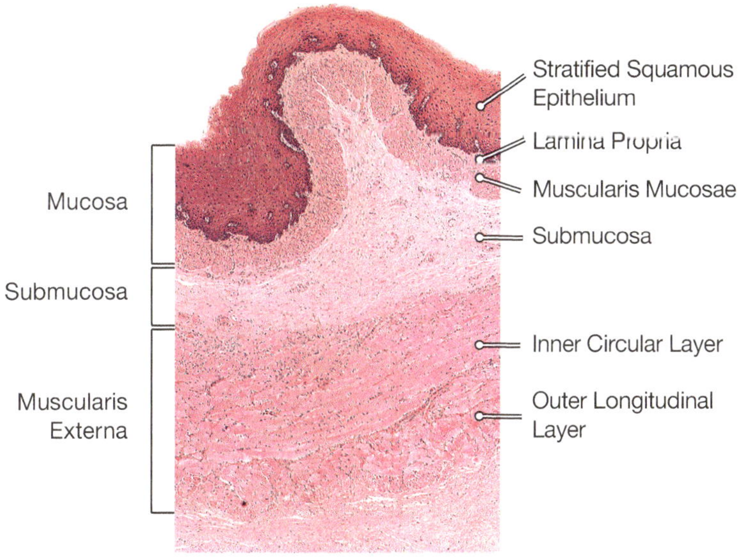

Figure 7.4: Esophagus.

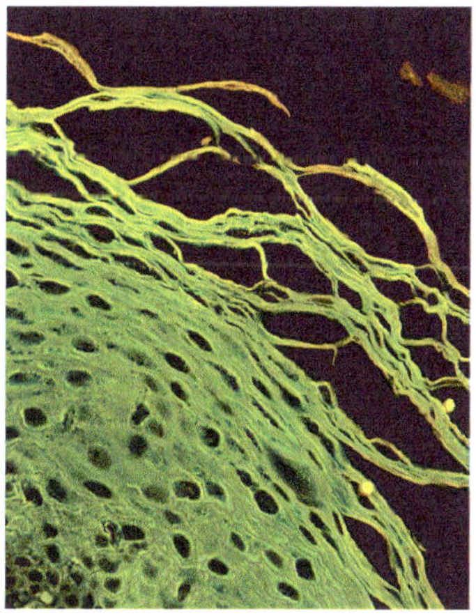

Figure 7.5: Esophagus. (autofluorescence)

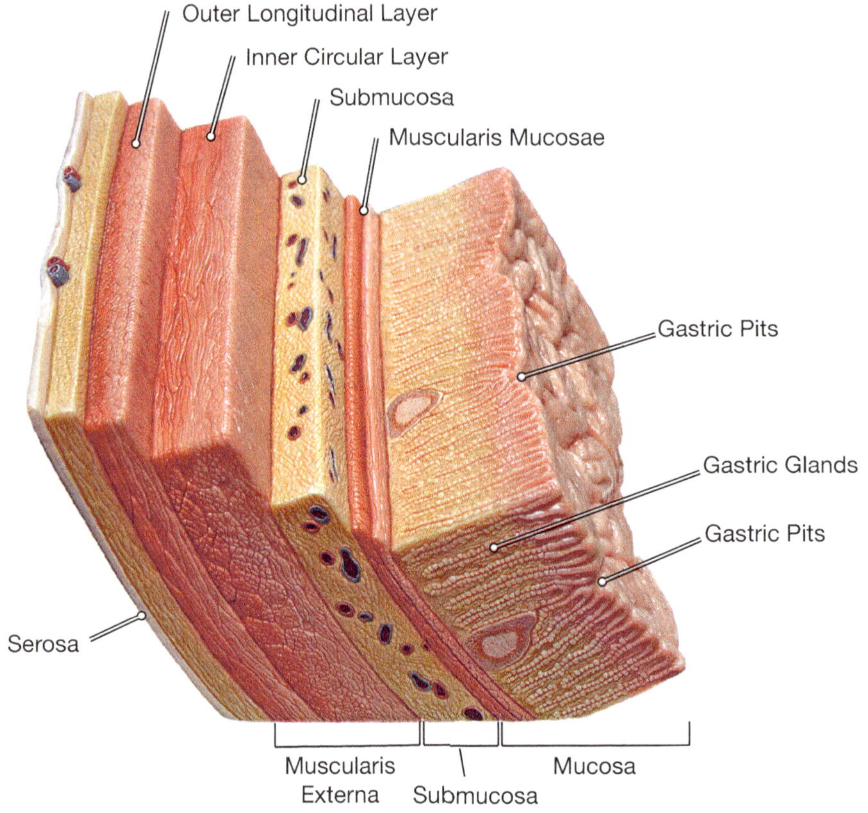

Figure 7.6: Wall of stomach.
© David G. Ward.

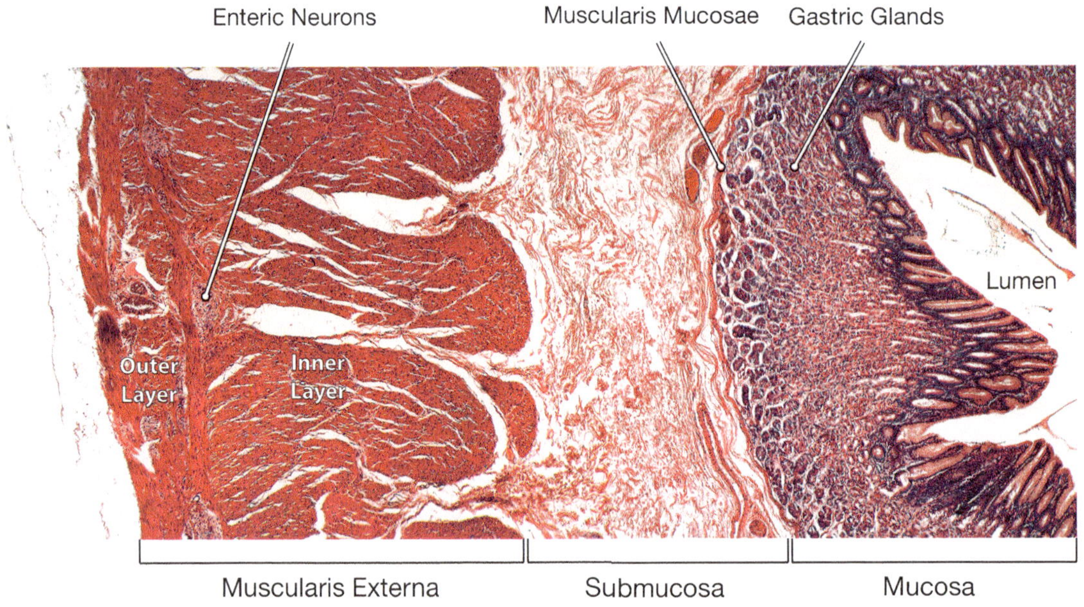

Figure 7.7: Wall of stomach, histology.
© David G. Ward.

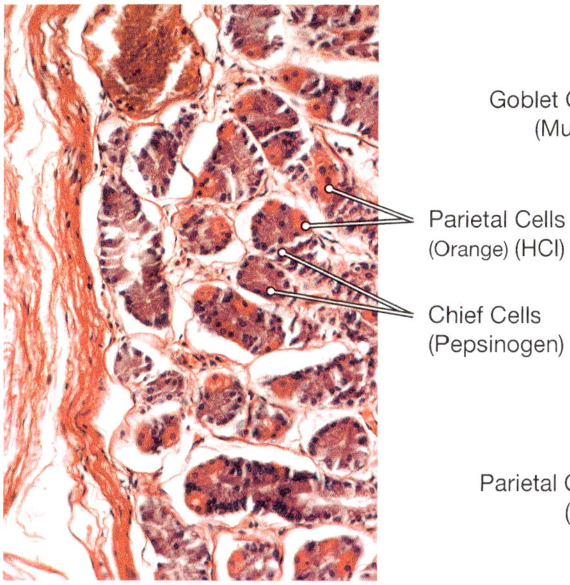

Figure 7.8: Mucosa of the stomach, histology, x400.

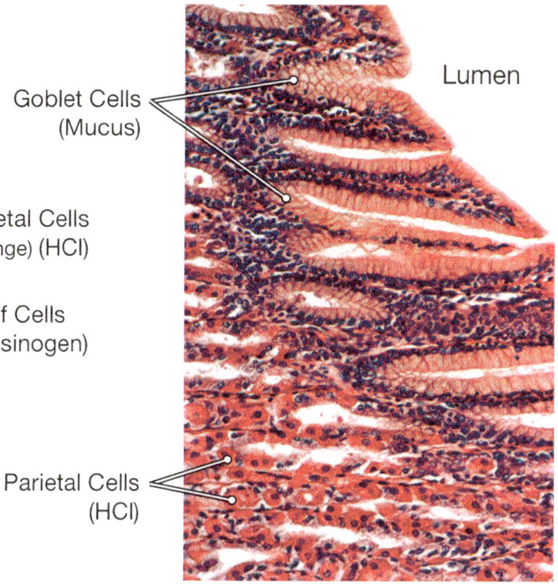

Figure 7.9: Mucosa of the stomach, histology, x200.

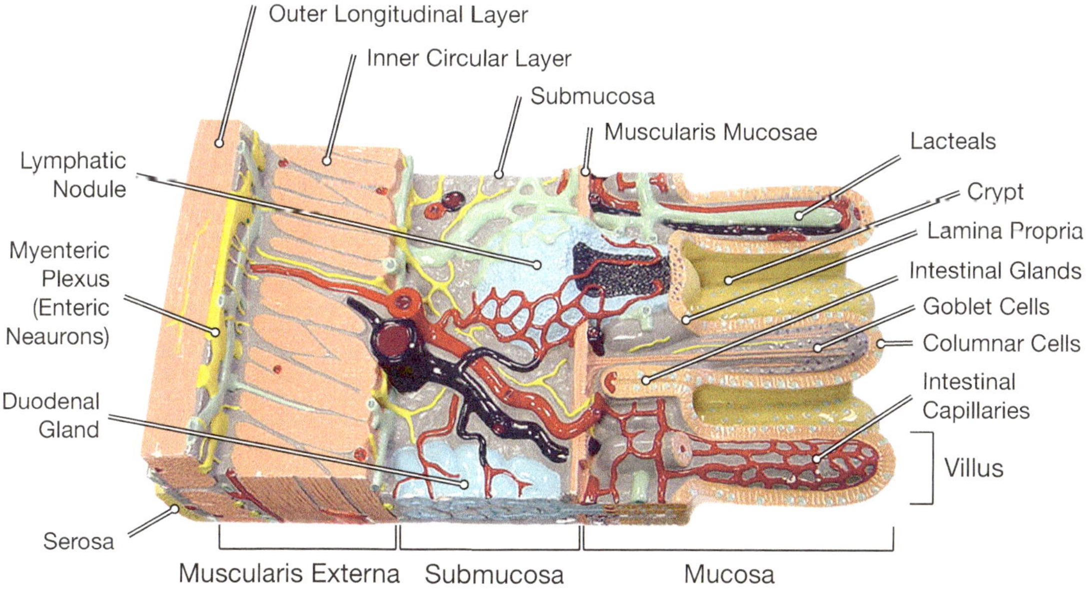

Figure 7.10: Small intesine.

Duodenum

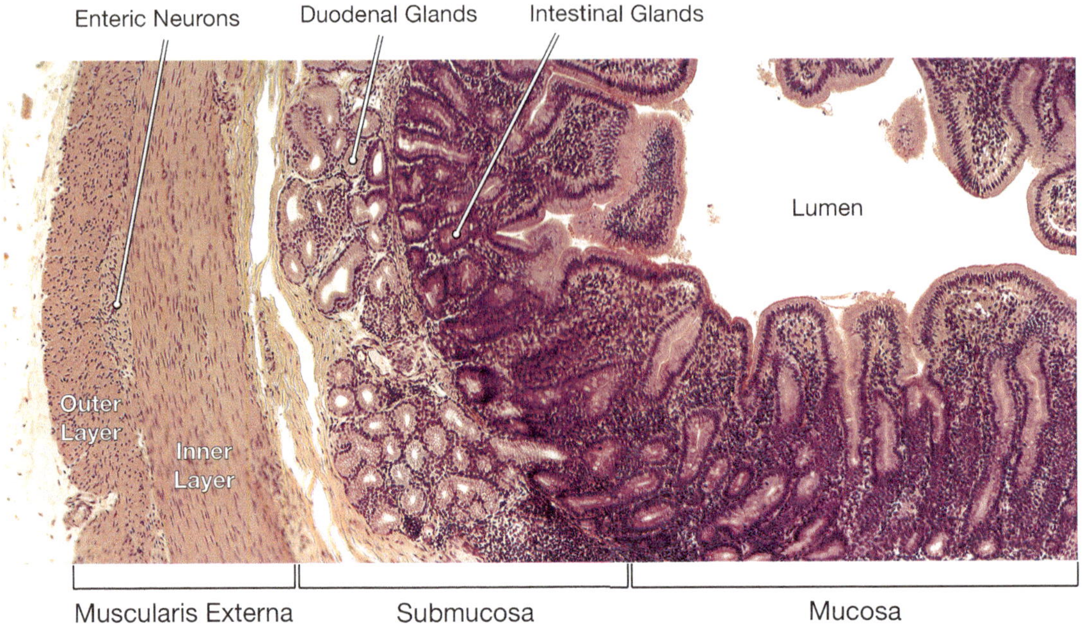

Figure 7.11: Wall of the duodenum.
© David G. Ward.

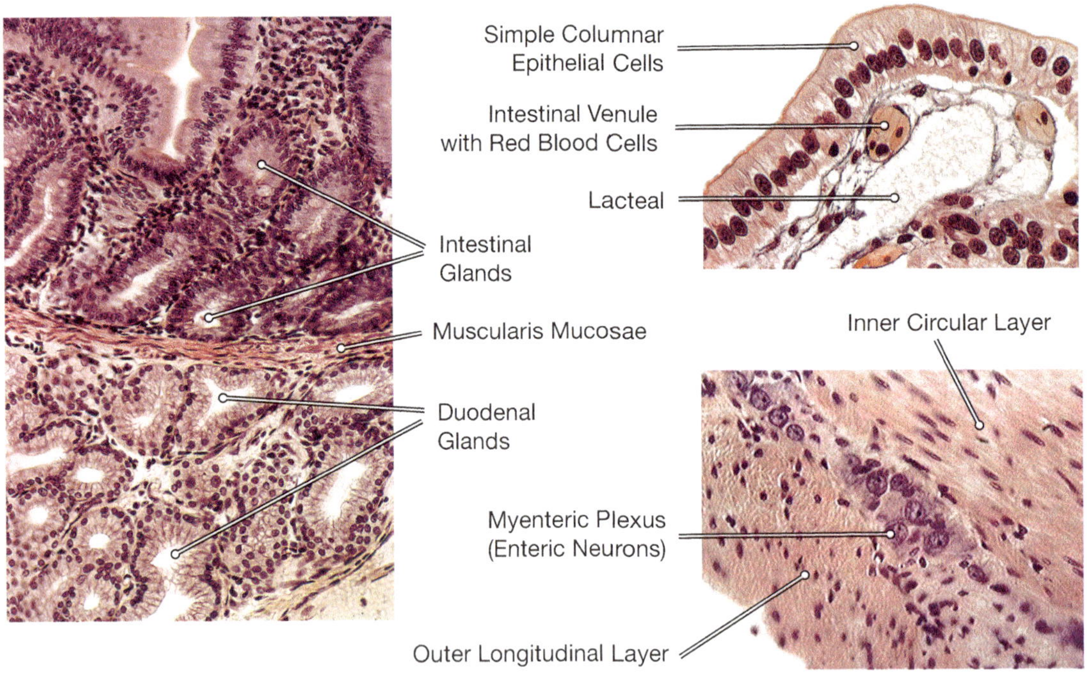

Figure 7.12: Duodenum.
© David G. Ward.

Liver and Pancreas

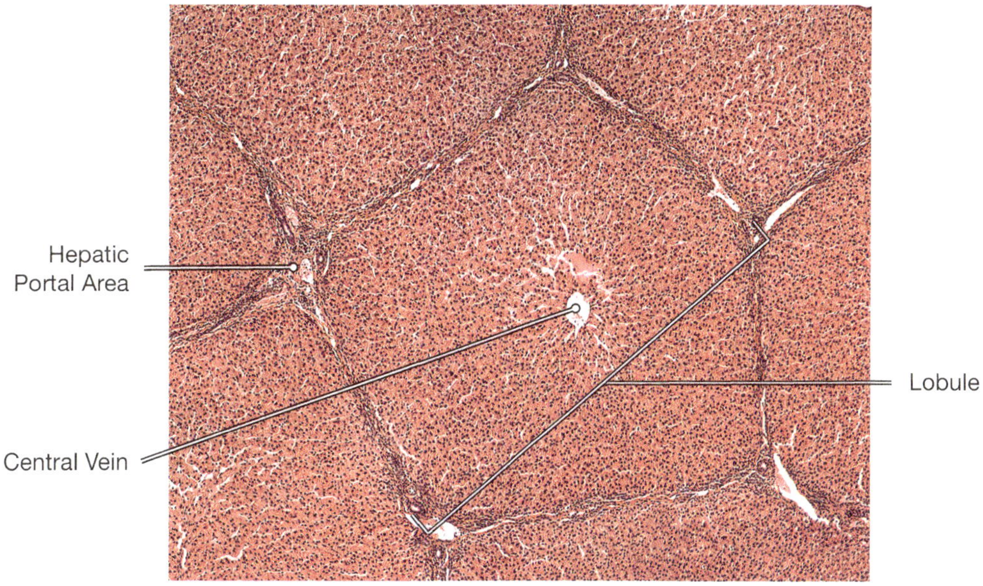

Figure 7.13: Liver lobules.

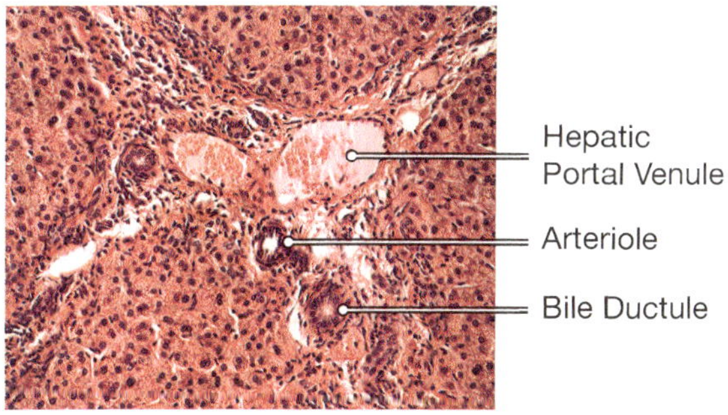

Figure 7.14: Hepatic portal area.

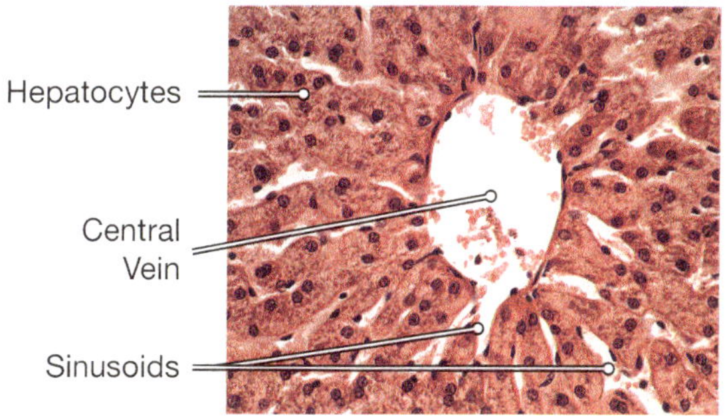

Figure 7.15: Central canal in liver.

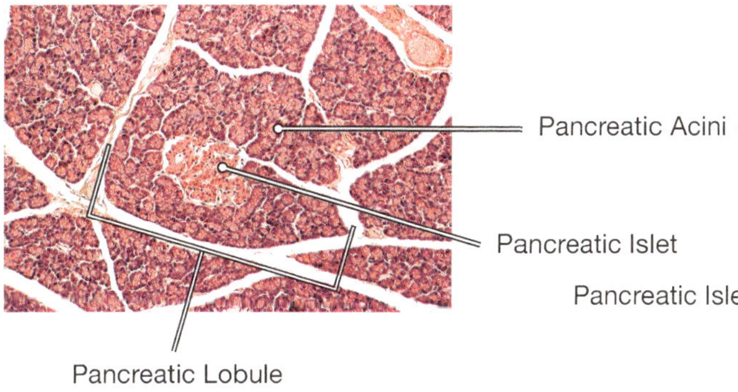

Figure 7.16: Pancreatic lobules.

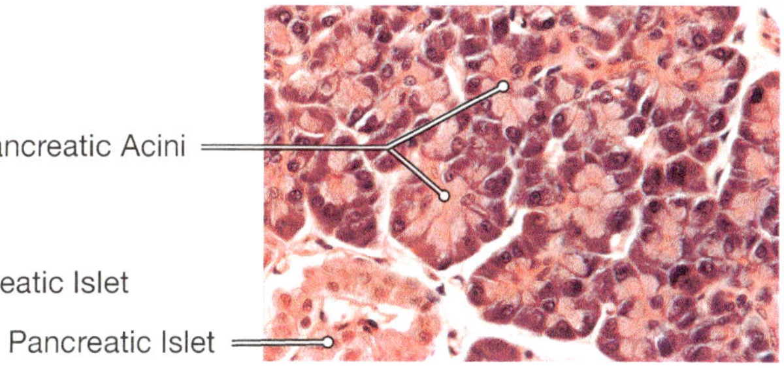

Figure 7.17: Pancreatic Acini.

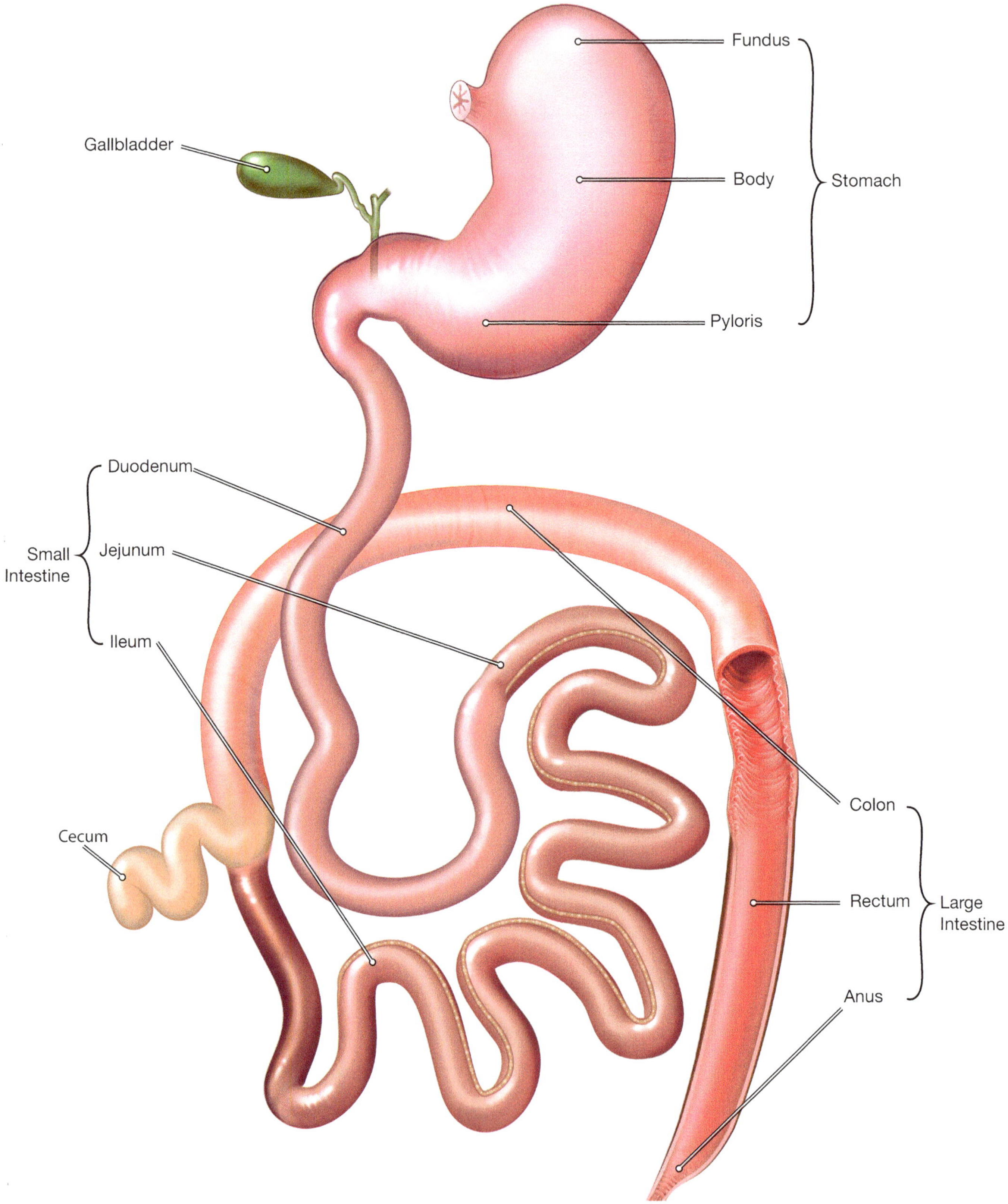

Figure 7.18: Digestive organs.
© NCSU.

Digestive Organs

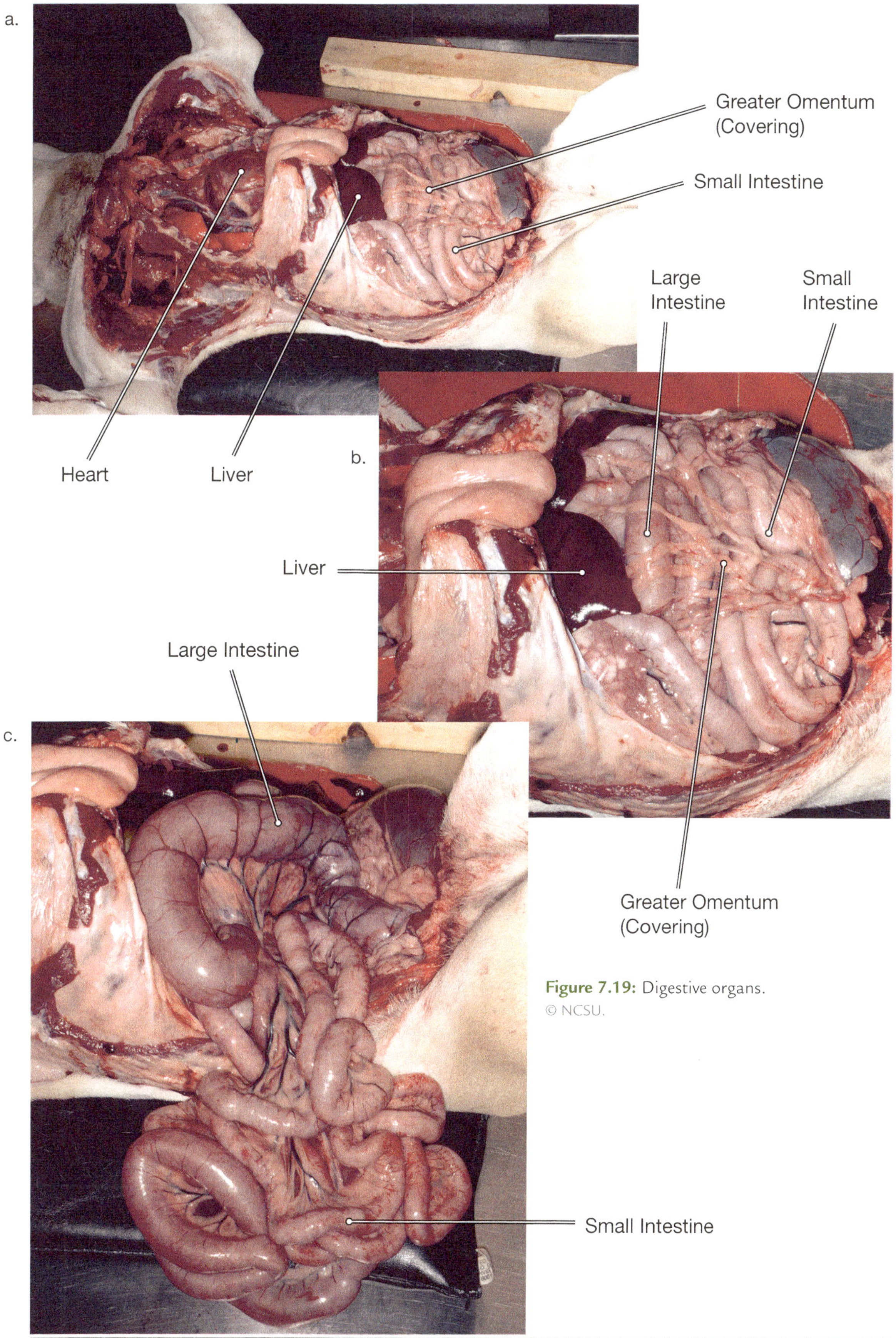

Figure 7.19: Digestive organs.
© NCSU.

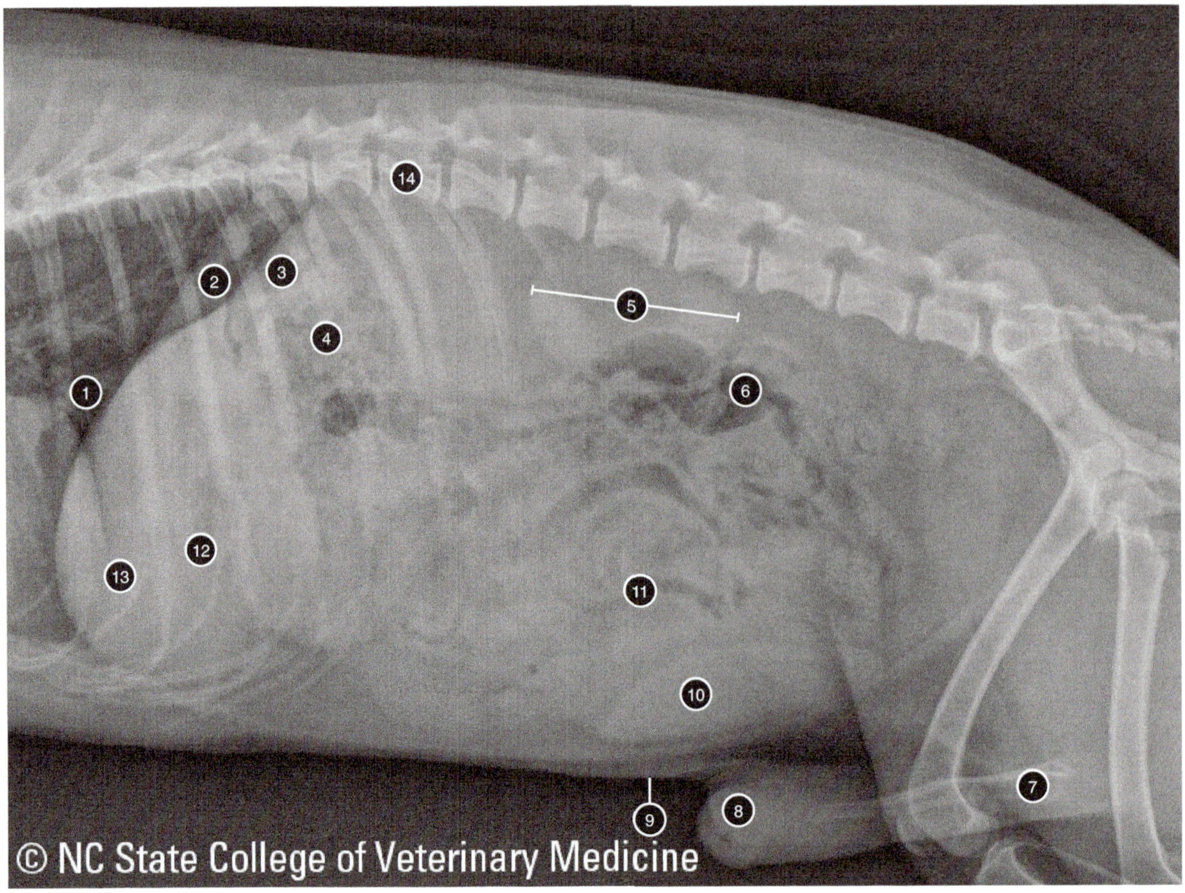

Figure 7.20: Right-left lateral view of abdomen (male).

1. caudaul vena cava
2. left side ("crus") of diaphragm
3. right side ("crus") of diaphragm
4. gas in fundus of stomach
5. left kidney
6. gas in descending colon
7. os penis
8. prepuce
9. teat or nipple
10. ventral region of spleen
11. gas in small bowel
12. body of stomach
13. liver
14. 13th thoracic vertebra (T13)

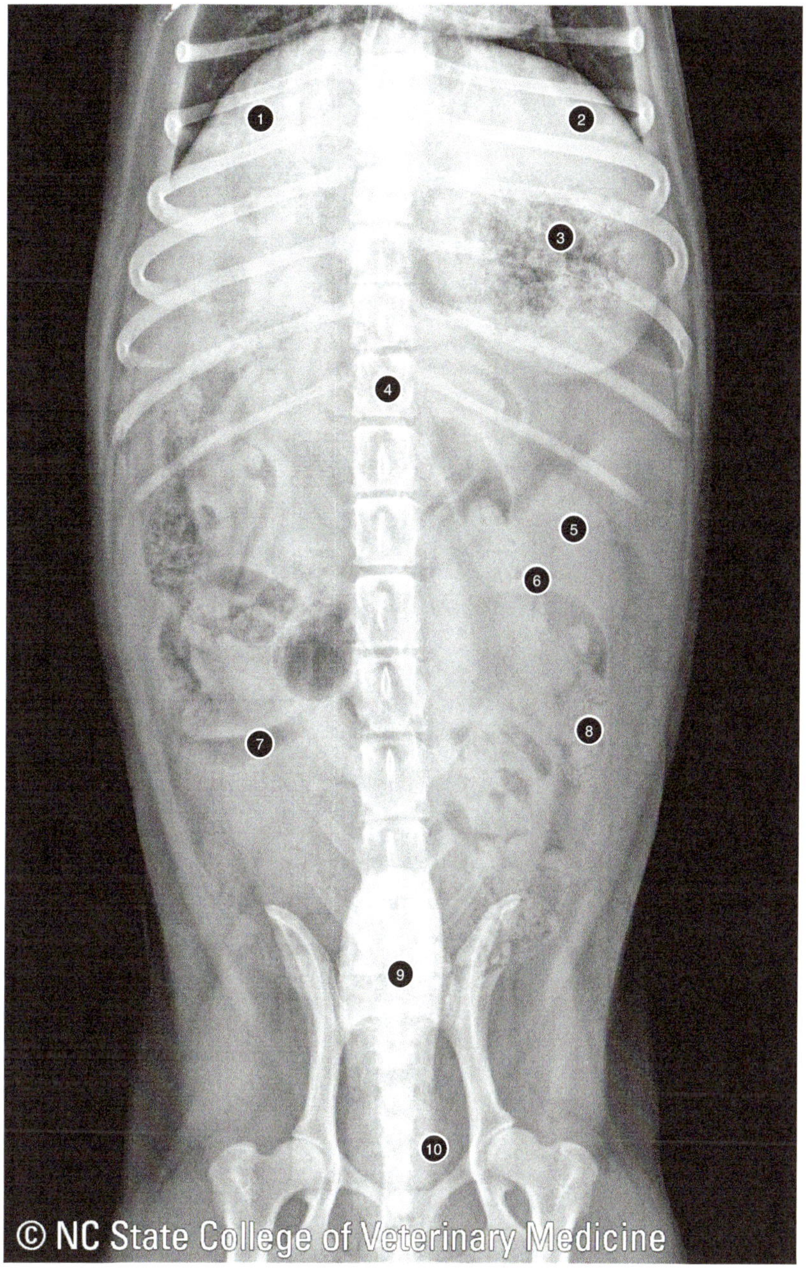

Figure 7.21: Vetral-dorsal (VD) view of abdomen (male).

1. left lobe
2. right lobe of liver
3. gas and food in fundus of stomach
4. 13th thoracic vertebra (T13)
5. spleen
6. left kidney
7. gas in small bowel
8. feces in descending colon
9. prepuce
10. prostate

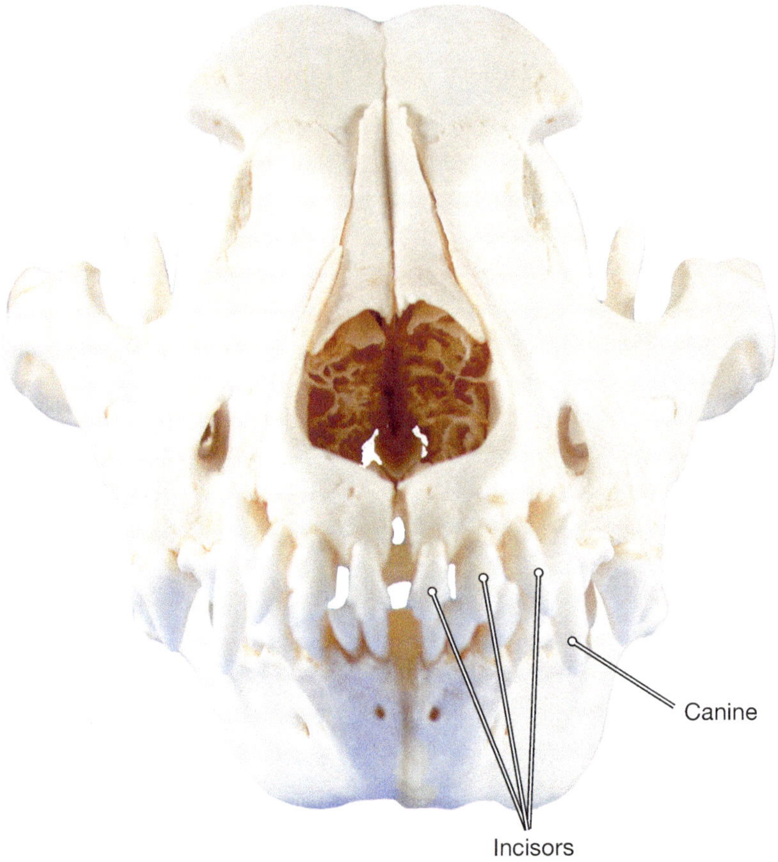

Figure 7.22: Rostral view.

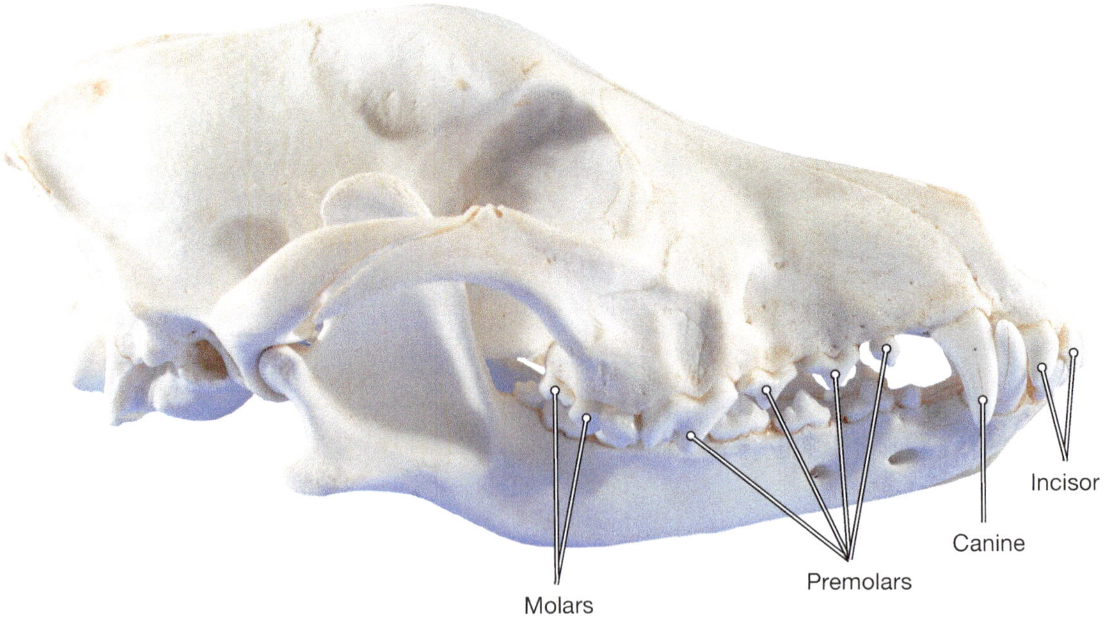

Figure 7.23: Lateral view.

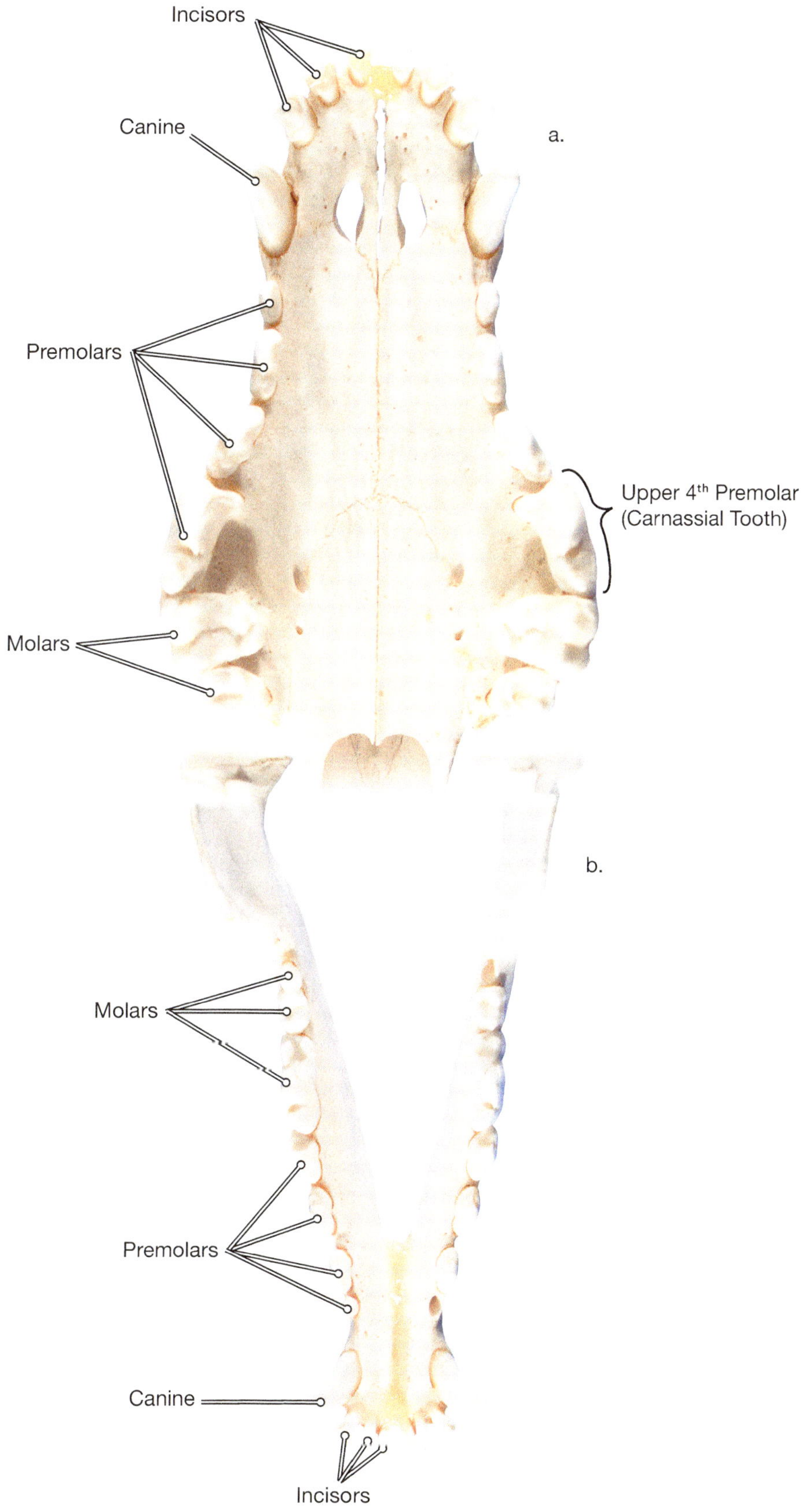

Figure 7.24: Teeth. a. Upper jaw. b. Lower jaw.

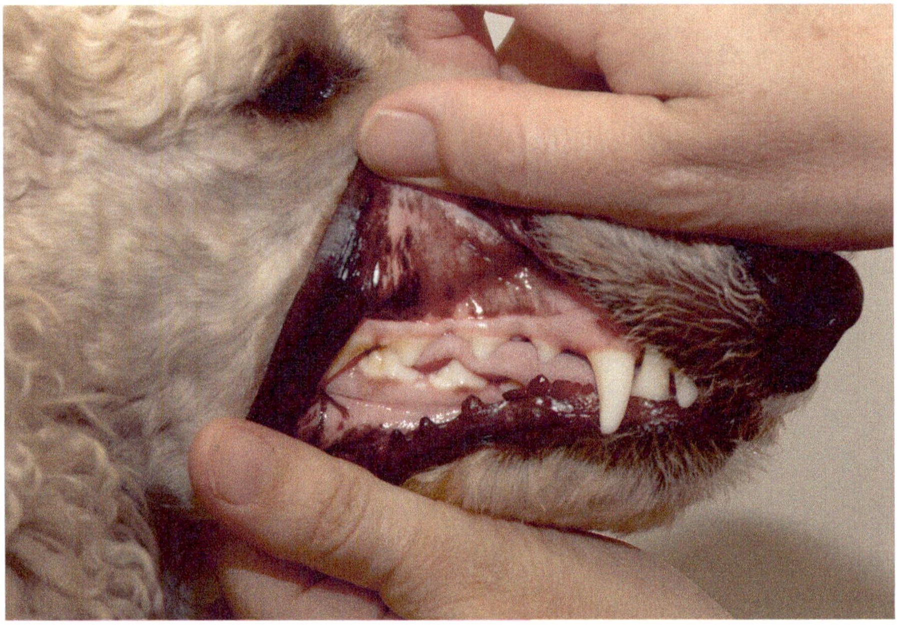

Figure 7.25: Heathly Teeth.

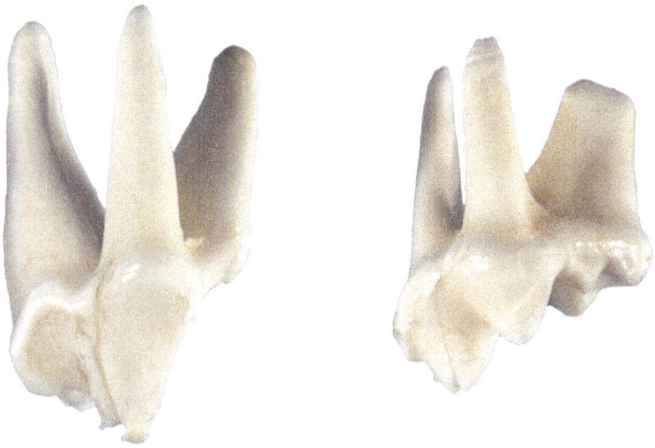

Figure 7.26: Upper teeth (premolars).

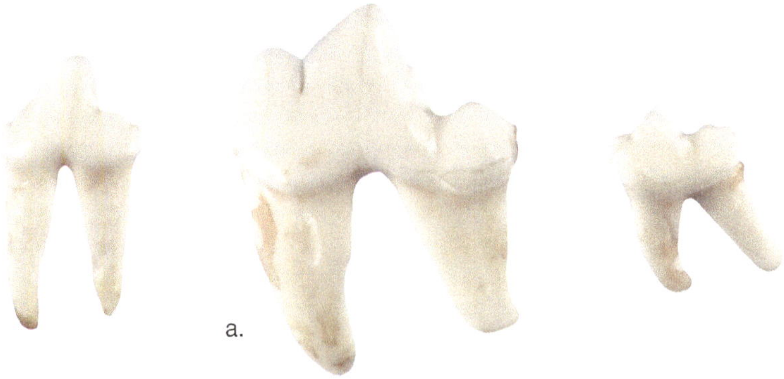

Figure 7.27: Lower teeth. a. First molar.

Stomach

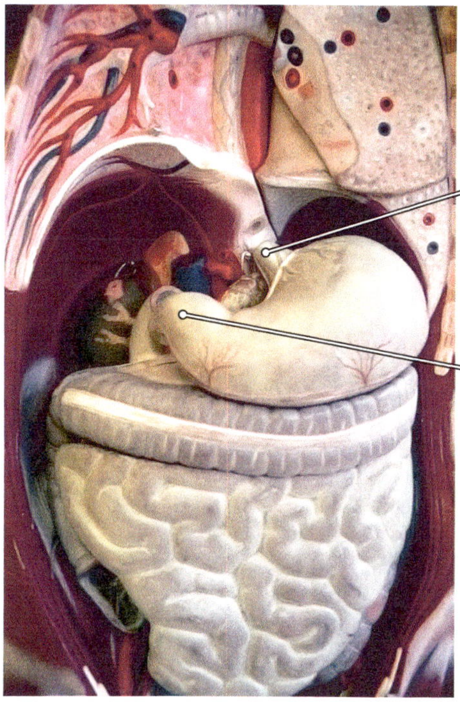

Figure 7.28: Abdominal cavity.
© David G. Ward.

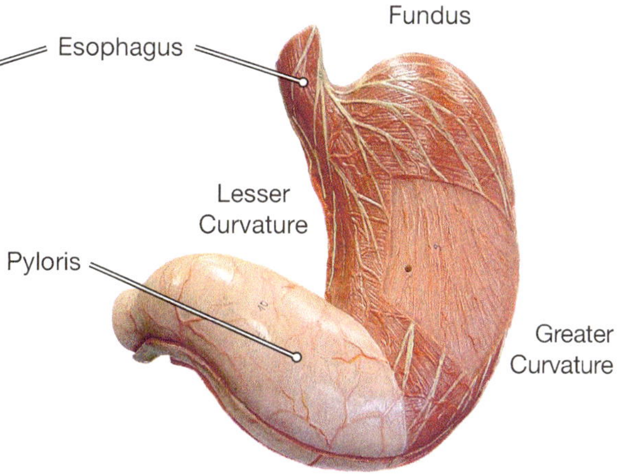

Figure 7.29: Stomach.
© David G. Ward.

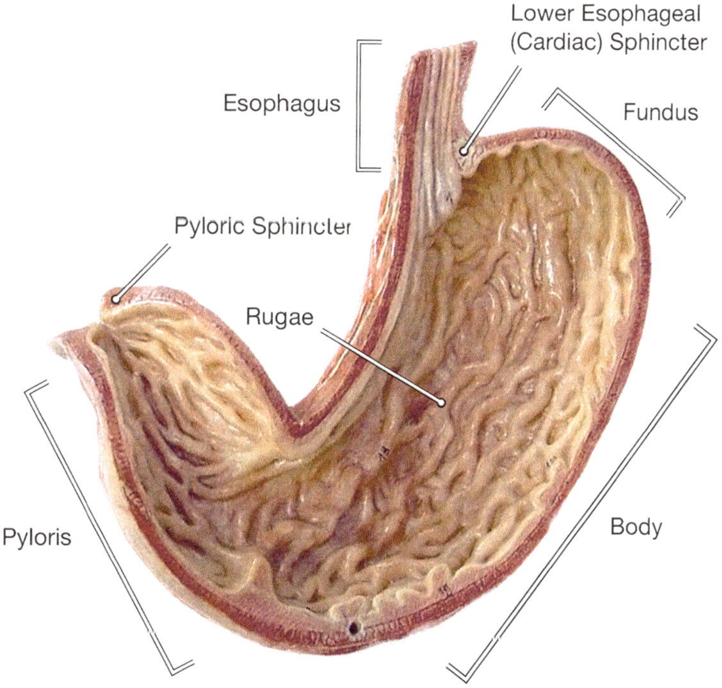

Figure 7.30: Stomach.
© David G. Ward.

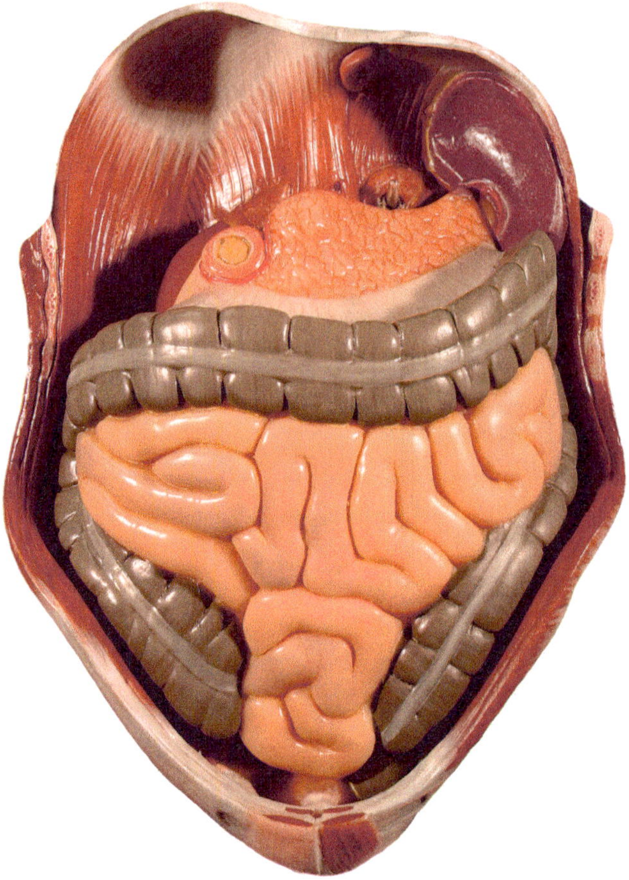

Figure 7.31: Small intestine.
© blueboor, LLC.

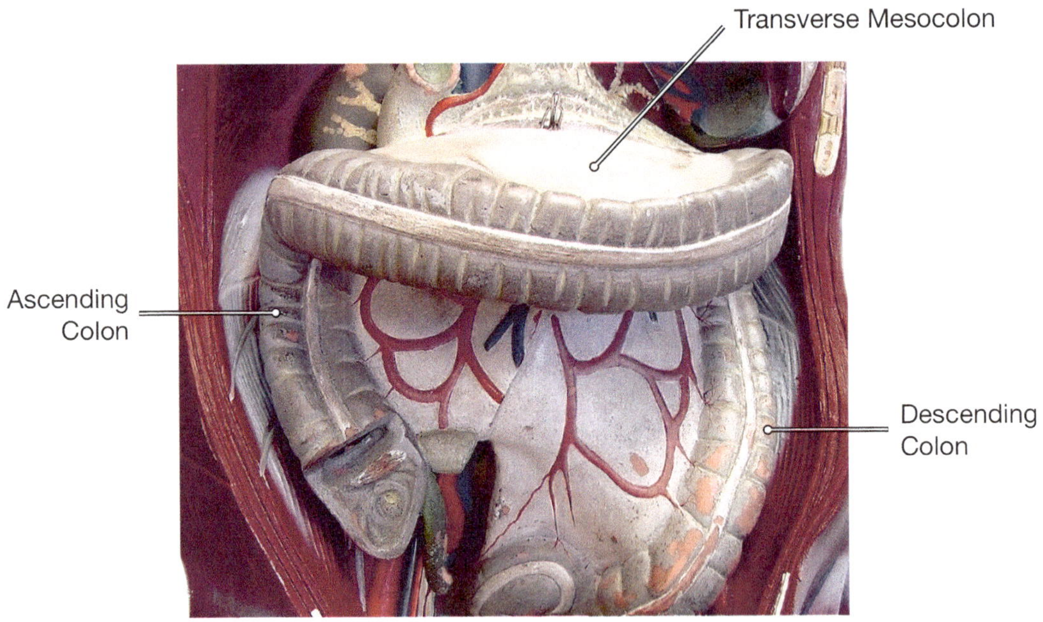

Figure 7.32: Large intestine (colon).
© Davis G. Ward.

Liver and Pancreas

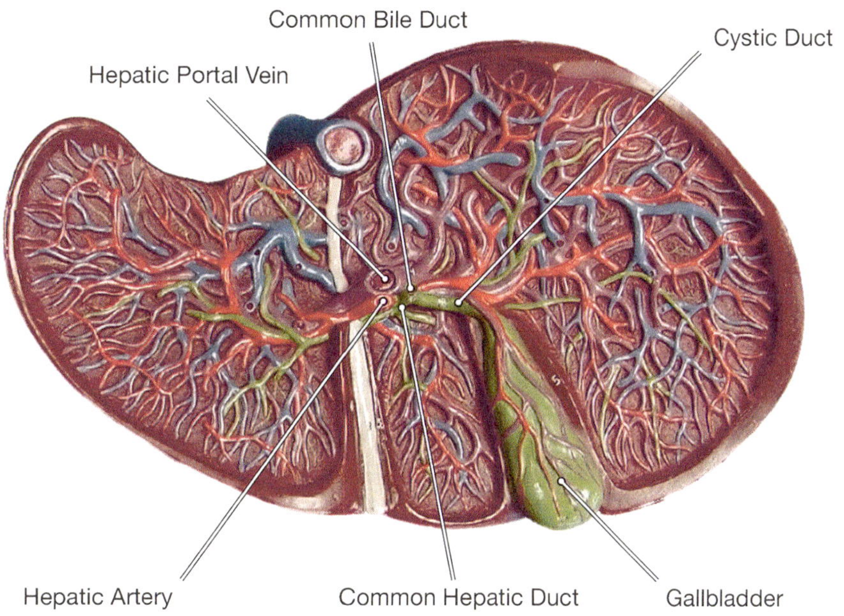

Figure 7.33: Liver.

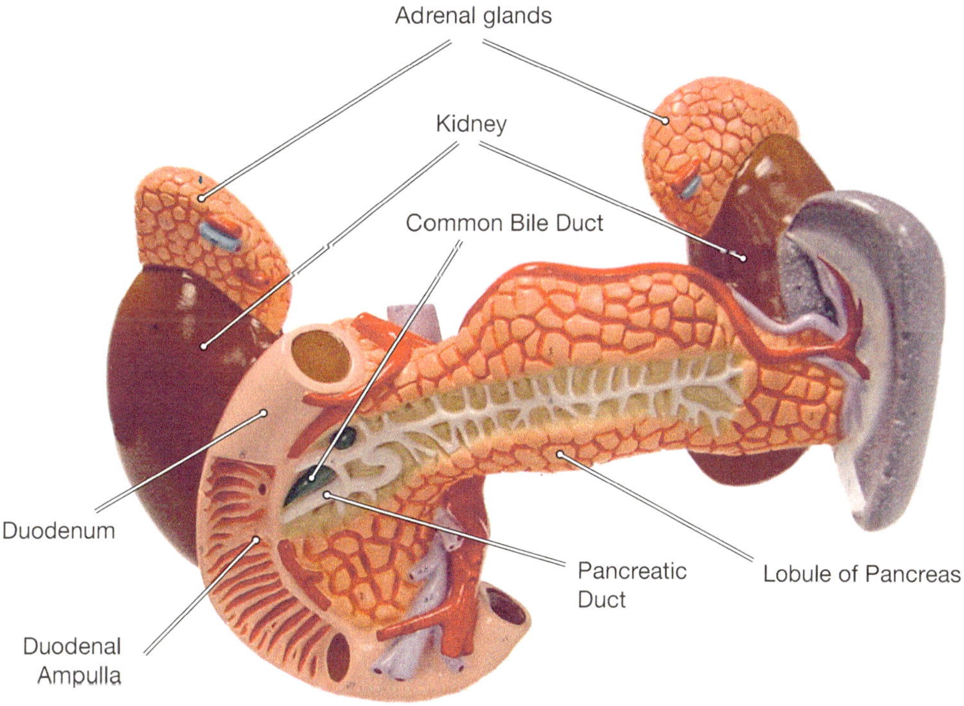

Figure 7.34: Pancreas.

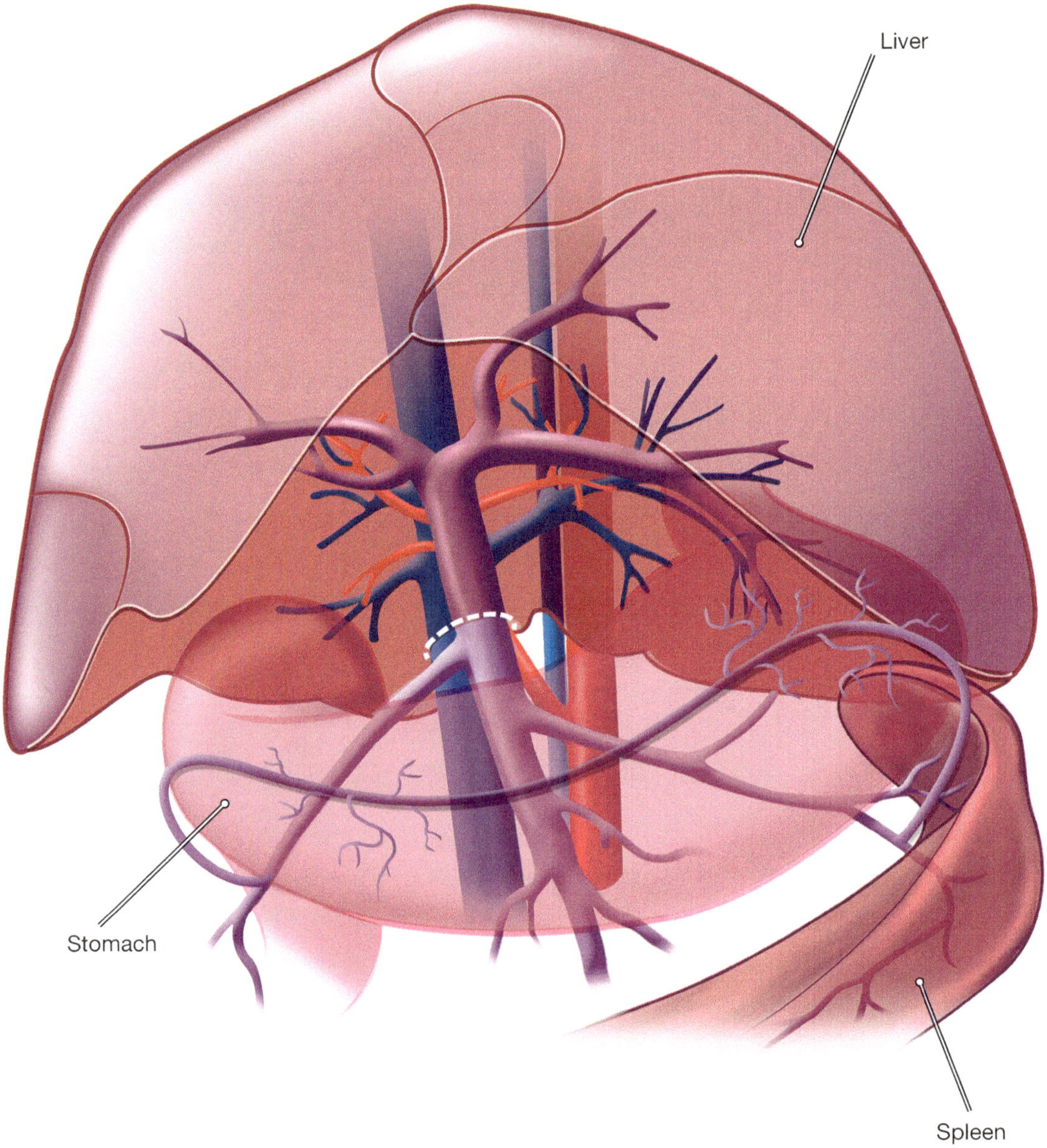

Figure 7.35: Hepatic circulation.

Rectum and Anal Sacs

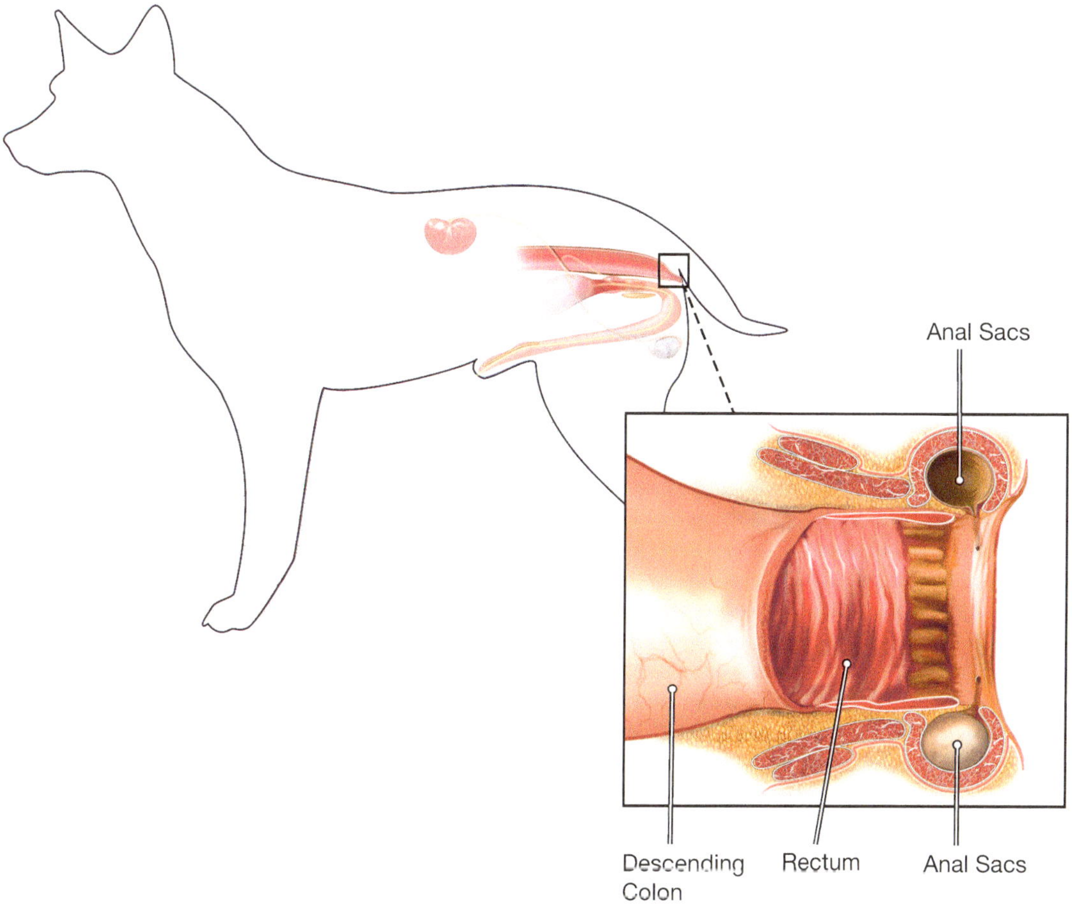

Figure 7.36: Rectum and anal sacs, dorsal view.

Review: Stomach

The stomach is responsible for receiving the food you have swallowed. It then churns and mixes food with stomach juices to begin the digestion process. To accomplish this function, the stomach needs the following types of tissues.

1. This hollow organ needs a protective inner lining that is also able to secrete digestive juices. A protective tissue that also secretes is ______________________ tissue.
2. In order to mix food and digestive juices the stomach contracts to squeeze its contents. The tissue that produces movement of an internal organ is __________________ tissue.
3. The stomach is covered by a protective membrane. A protective tissue is ________________ tissue.
4. The stomach is anchored in the abdominal cavity by fibrous tissue. Tissue that binds and has strong fibers is __________________ tissue.
5. The stomach receives information regarding the digestive process. Tissue that is able to carry information throughout the body is ____________________ tissue.
6. Put the following structures in order as food passes through the digestive system.

descending colon	esophagus	anus
ascending colon	lower esophageal sphincter	stomach
transverse colon	pyloric sphincter	rectum

   1. mouth
   2.
   3.
   4.
   5.
   6.
   7.
   8.
   9.
   10.

7. List the three sections of the small intestine.
   1.
   2.
   3.

Review: Labeling Activity

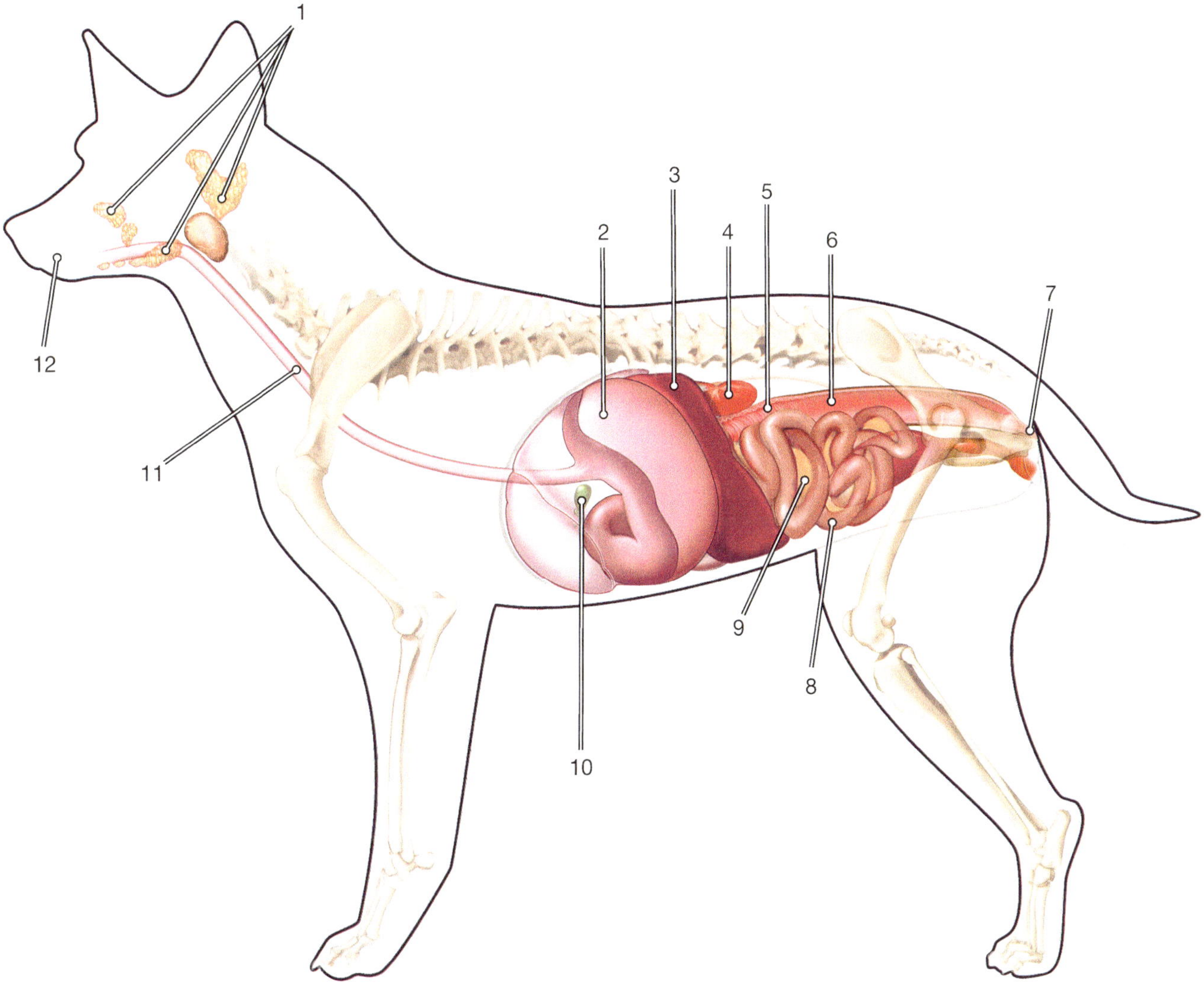

Figure 7.37: Gastrointestinal tract labeling activity.

1. ______________________ 7. ______________________

2. ______________________ 8. ______________________

3. ______________________ 9. ______________________

4. ______________________ 10. ______________________

5. ______________________ 11. ______________________

6. ______________________ 12. ______________________

Review: Crossword Puzzle

Across

3. the lower region of the stomach
4. cell type located in the liver
6. the center region of the small intestine
7. the center region of the stomach
10. the outermost layer of the muscularis externa
12. folds located in the stomach that increase surface area
13. another name for mechanical digestion
14. this structure joins the descending colon and the anal area
15. the final region of the small intestine that connects to the colon

Down

1. the upper region of the stomach
2. this acid starts the hydrolysis of proteins
5. the inactive precursor to gastric pepsin
8. the first region of the small intestine
9. where the digestive process begins
11. the first section of the colon

Review: Word Search

V	A	A	L	Q	I	I	E	S	C	R	F	R	M	M	I	M	D
Z	Z	P	R	A	J	E	U	P	A	S	D	N	O	Z	U	K	I
O	M	D	A	S	N	B	W	L	N	N	O	L	W	S	J	O	C
E	E	U	E	N	M	E	O	C	I	R	A	I	C	W	X	S	A
A	L	Q	C	U	C	M	D	P	N	R	H	U	V	C	P	A	C
H	Z	G	C	O	E	R	H	O	E	Y	L	Z	P	K	F	E	I
U	T	O	P	R	S	B	E	E	U	A	X	K	Y	R	P	R	R
A	S	K	P	H	Y	A	W	A	R	D	O	H	A	N	E	C	O
A	U	B	M	D	B	X	D	I	T	W	I	A	I	Z	P	N	L
S	U	G	A	H	P	O	S	E	B	I	C	R	J	P	S	A	H
P	J	X	C	L	L	E	G	P	R	R	C	W	R	M	I	P	C
C	E	K	O	H	X	O	S	Z	O	N	V	I	P	X	N	H	O
L	M	Q	F	T	J	S	B	V	F	L	I	W	S	T	O	B	R
U	S	W	E	V	G	C	P	O	X	V	I	Q	D	L	G	D	D
A	E	R	A	L	A	T	R	O	P	C	I	T	A	P	E	H	Y
Q	N	H	T	X	X	D	L	I	M	Z	L	Y	Z	Q	N	T	H
A	O	Y	R	N	V	N	C	X	T	R	I	N	C	I	S	O	R
P	Y	L	O	R	I	C	S	P	H	I	N	C	T	E	R	A	L

1. produced by parietal cells
2. produced by chief cells
3. esophagus, stomach and small intestinal layer lining the lumen
4. layer between the mucosa and muscularis externa in the digestive system
5. muscular layer deep to the submucosa
6. a gland located in the small intestine submucosa
7. contains hepatic portal venule, arteriole and bile ductule
8. located in a pancreatic lobule
9. connects the stomach to the duodenum of the small intestine
10. muscular structure connecting the mouth to the stomach
11. organ that produces digestive enzymes released into the small intestine
12. small teeth in the front of the jaw
13. relatively long tooth also called a cuspid or fang
14. crushing teeth in the back of the jaw
15. four teeth on each side of the jaw located behind the canine

Chapter 8

Kidneys, Bladder and the Urinary System

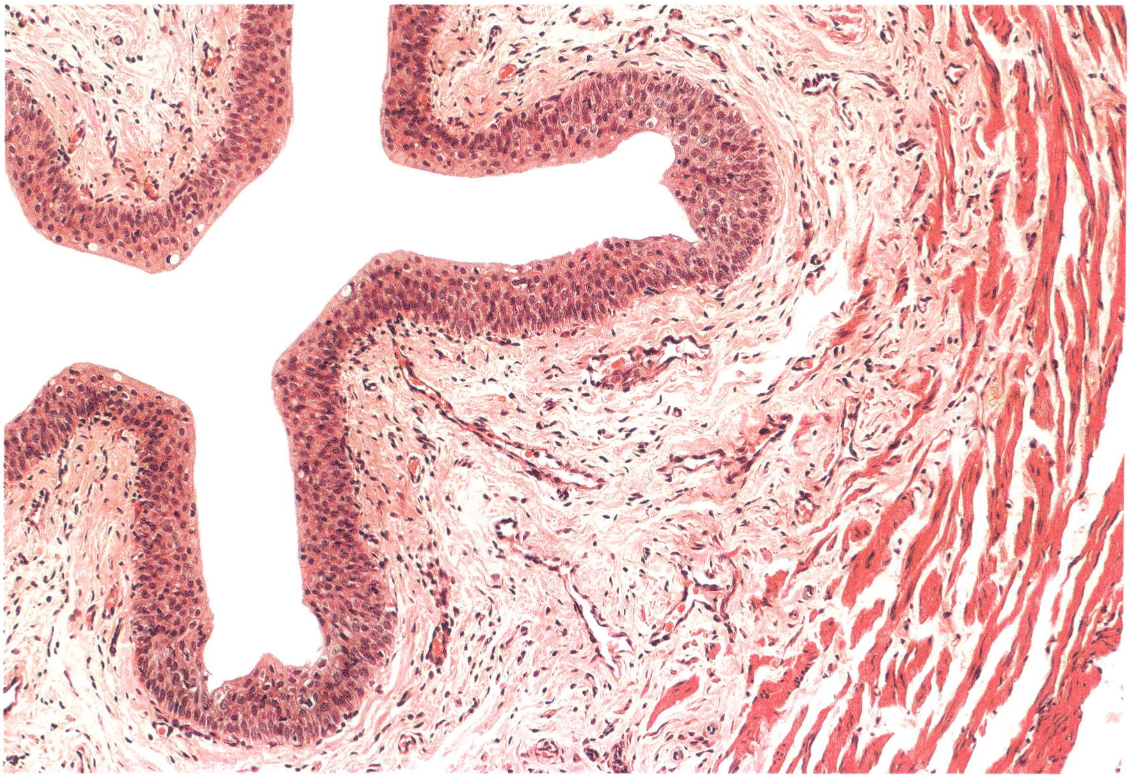

Ureter

Urinary System Overview: Female

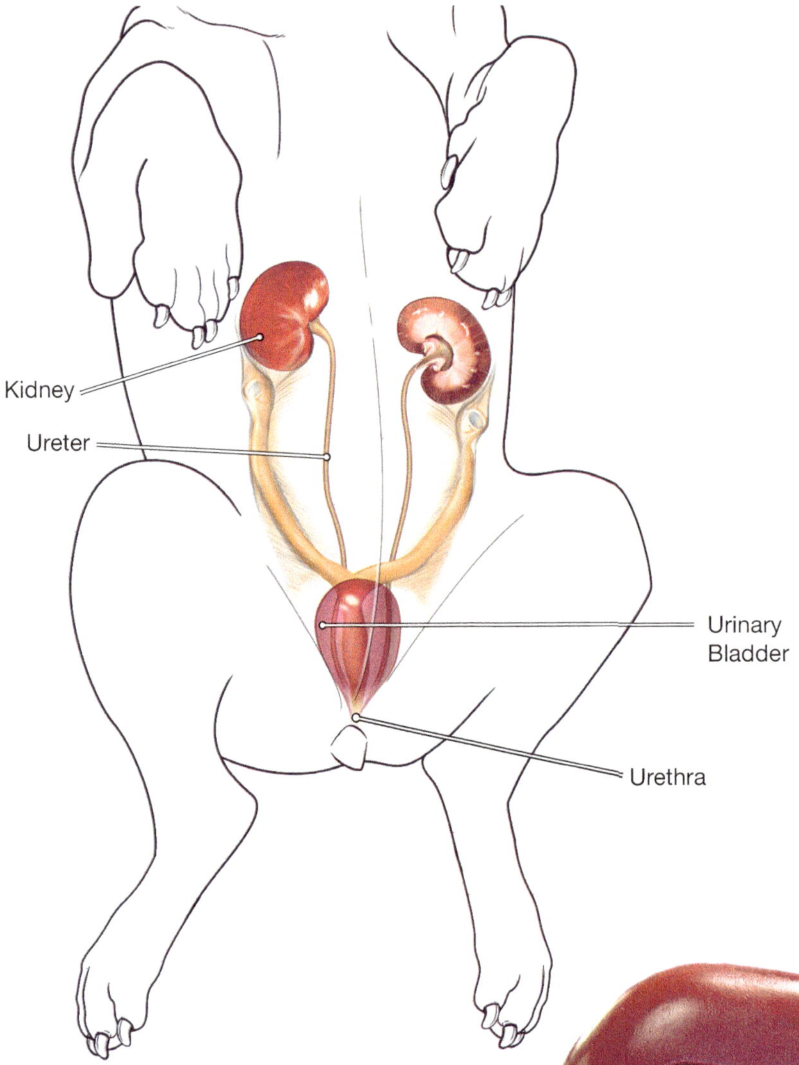

Figure 8.1: Urinary system (female).
© NCSU.

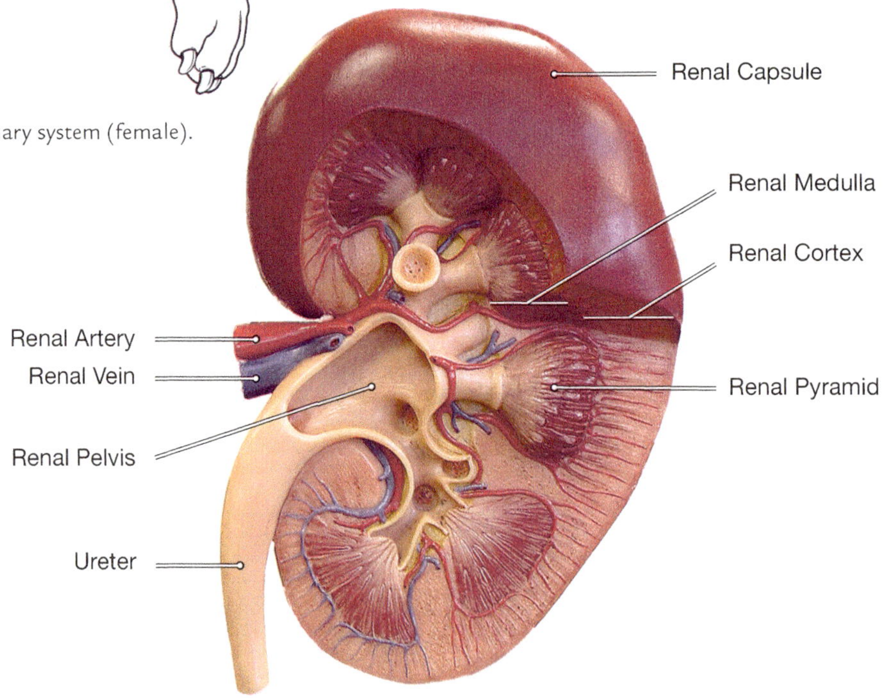

Figure 8.2: Kidney structure.
© David G. Ward.

Renal Corpuscles and Ureter

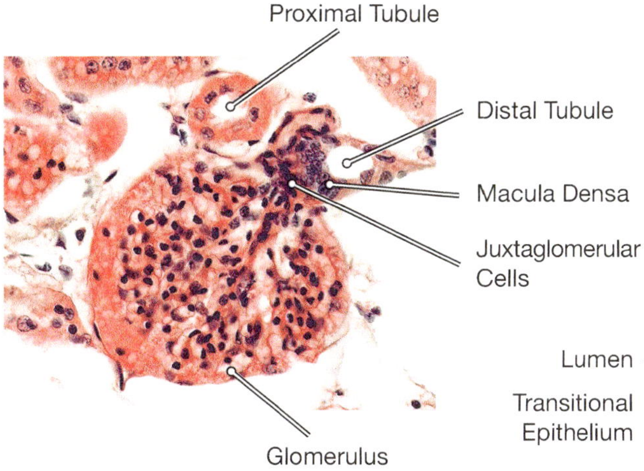

Figure 8.3: Renal corpuscle.

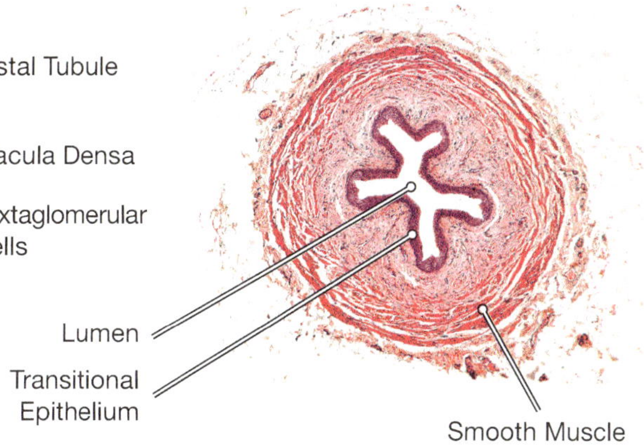

Figure 8.4: Ureter, cross-section.

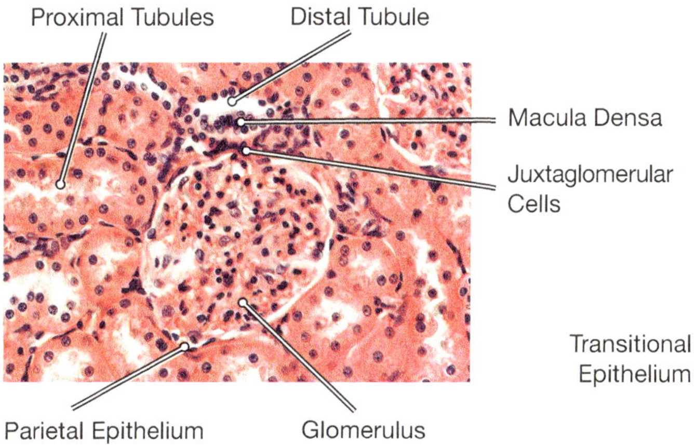

Figure 8.5: Renal corpuscle.

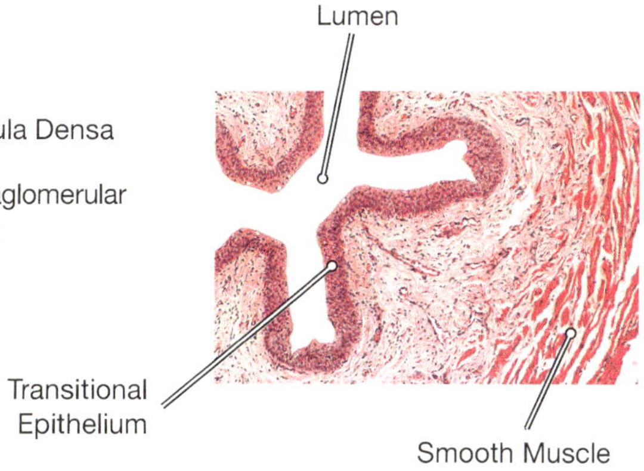

Figure 8.6: Ureter, cross-section

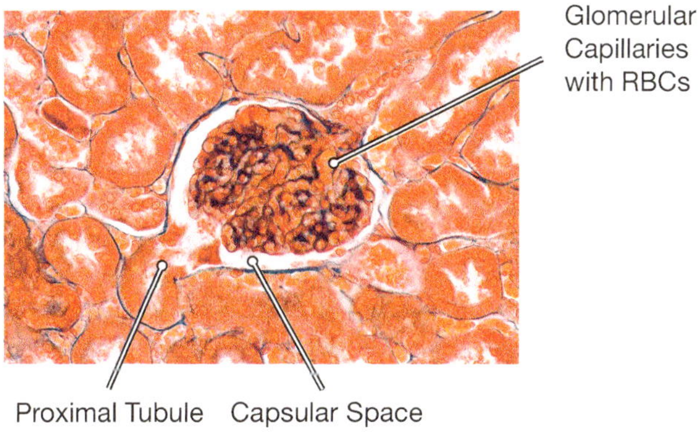

Figure 8.7: Renal corpuscle.

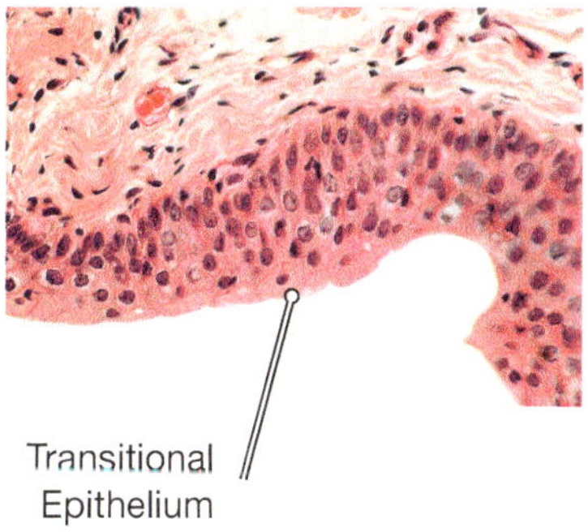

Figure 8.8: Ureter.

Male Urinary System

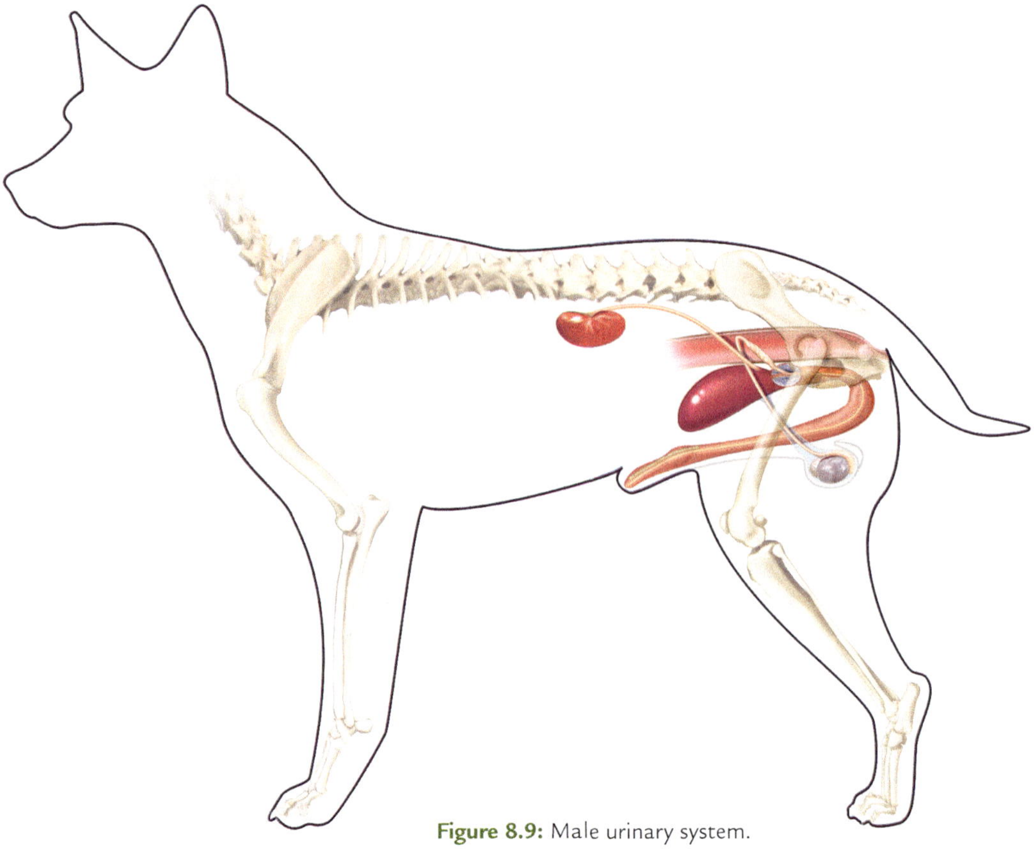

Figure 8.9: Male urinary system.
© NCSU.

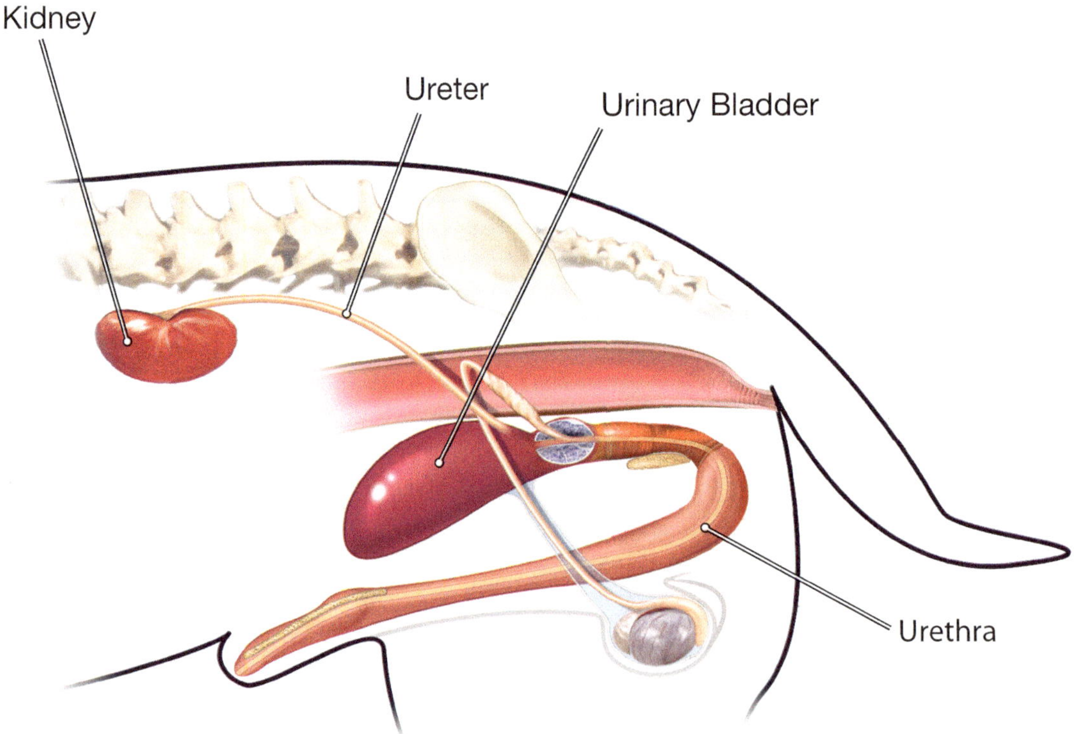

Figure 8.10: Male urinary system.
© NCSU.

Female Urinary System

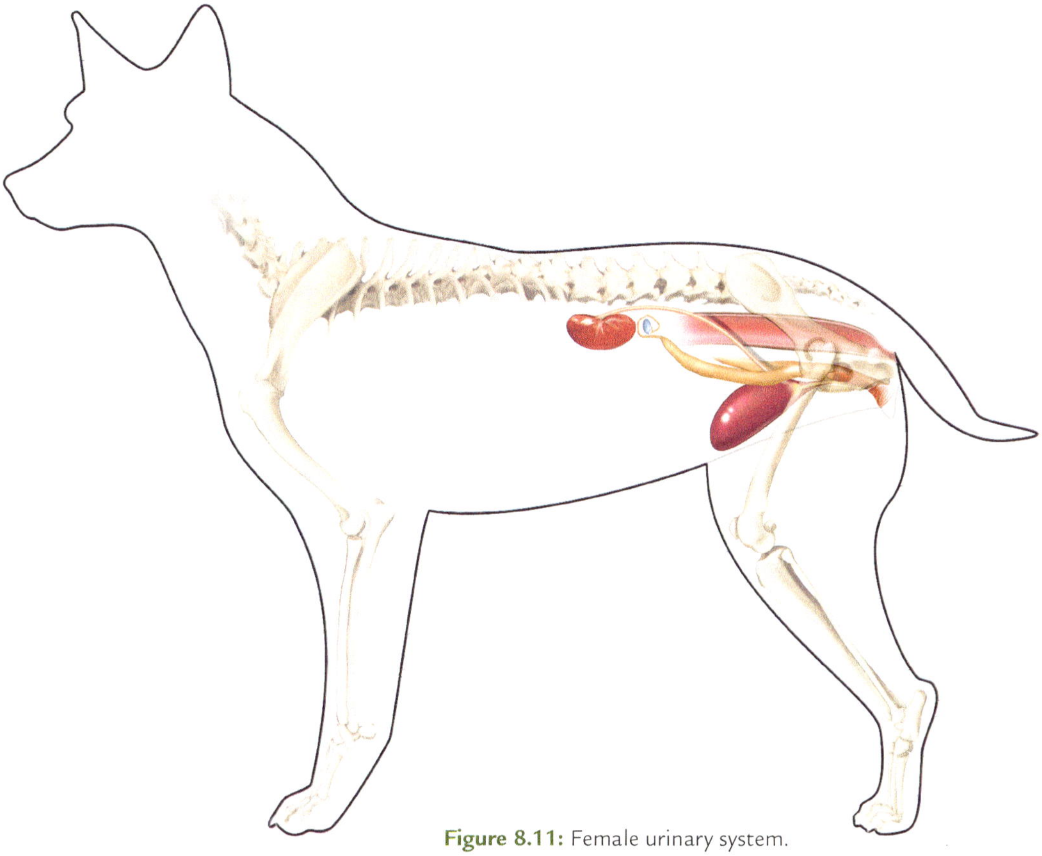

Figure 8.11: Female urinary system.
© NCSU.

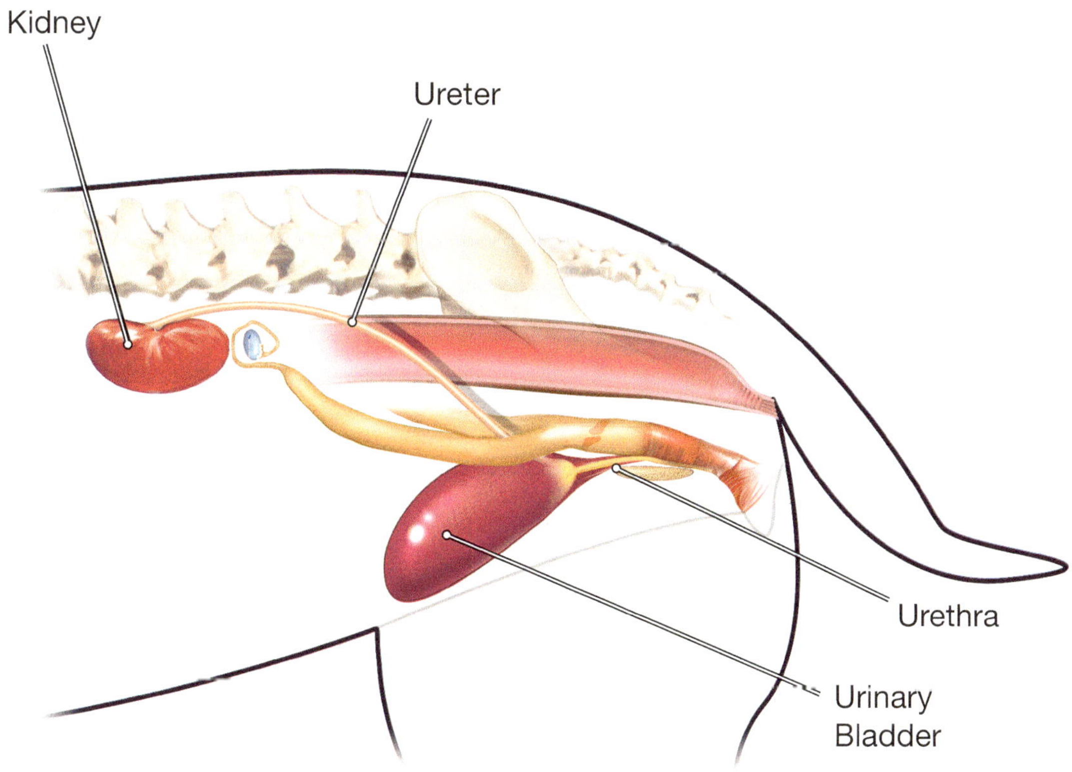

Figure 8.12: Female urinary system.
© NCSU.

Nephron Structure

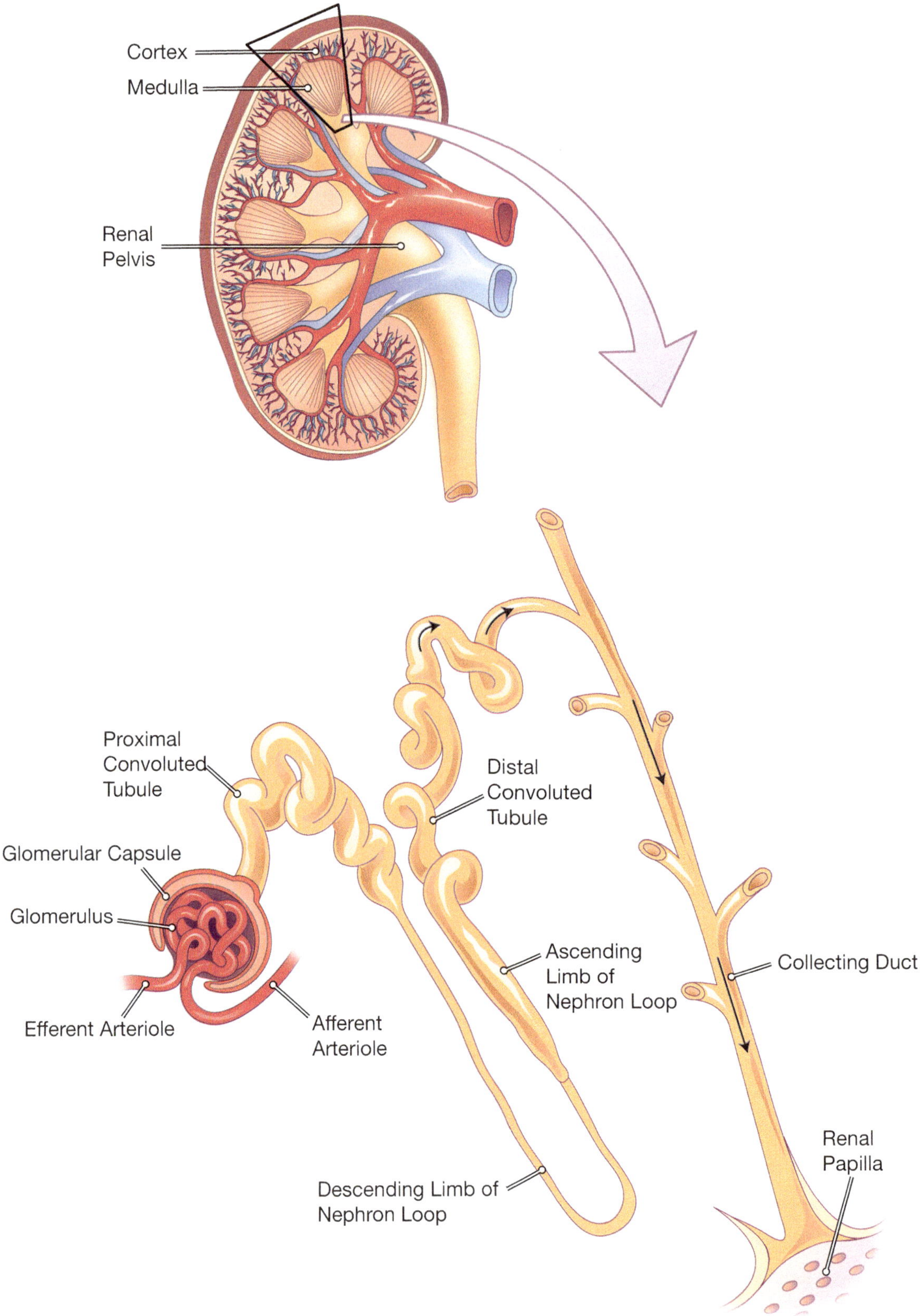

Figure 8.13: Nephron structure.

Nephron and Renal Corpuscle

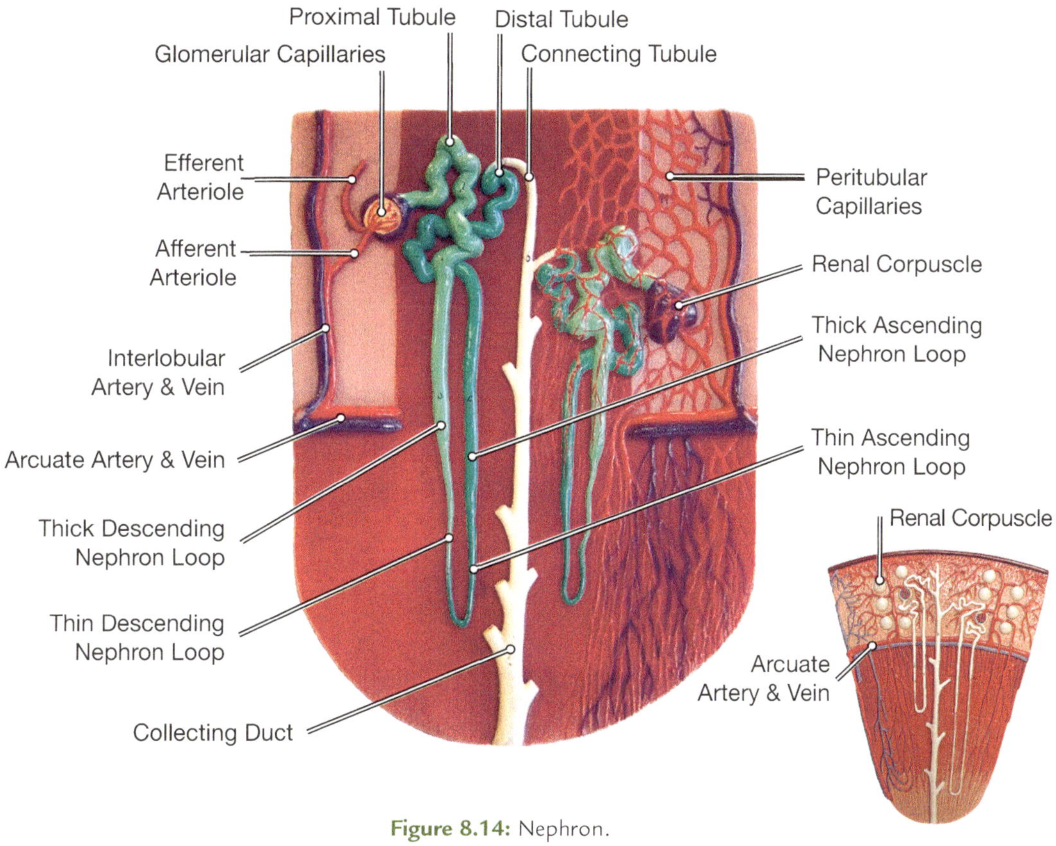

Figure 8.14: Nephron.

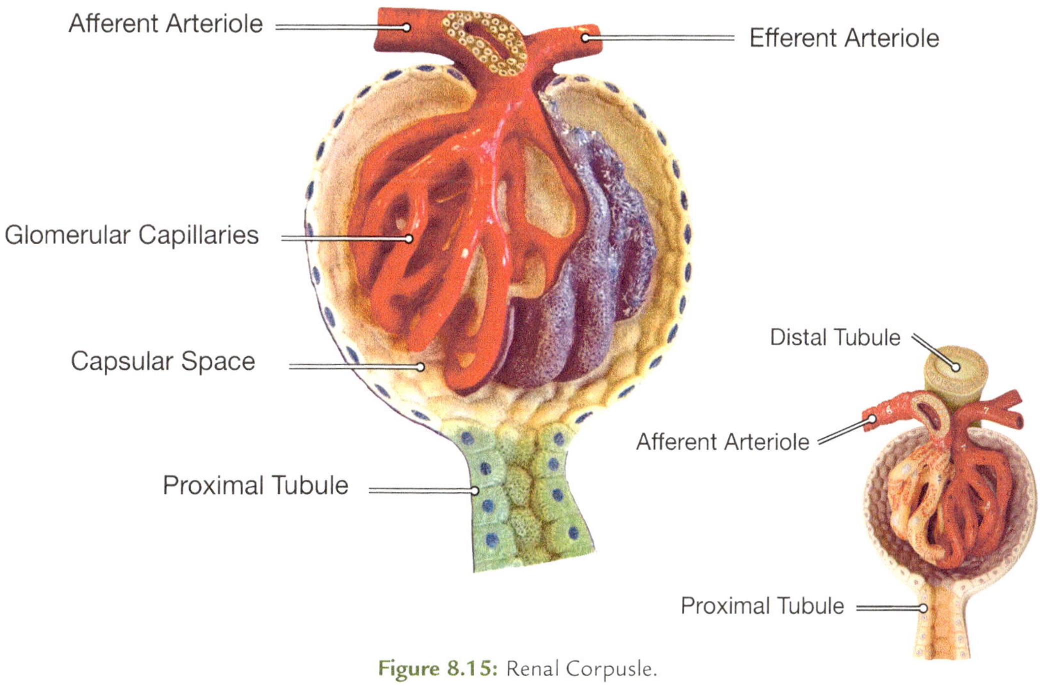

Figure 8.15: Renal Corpusle.

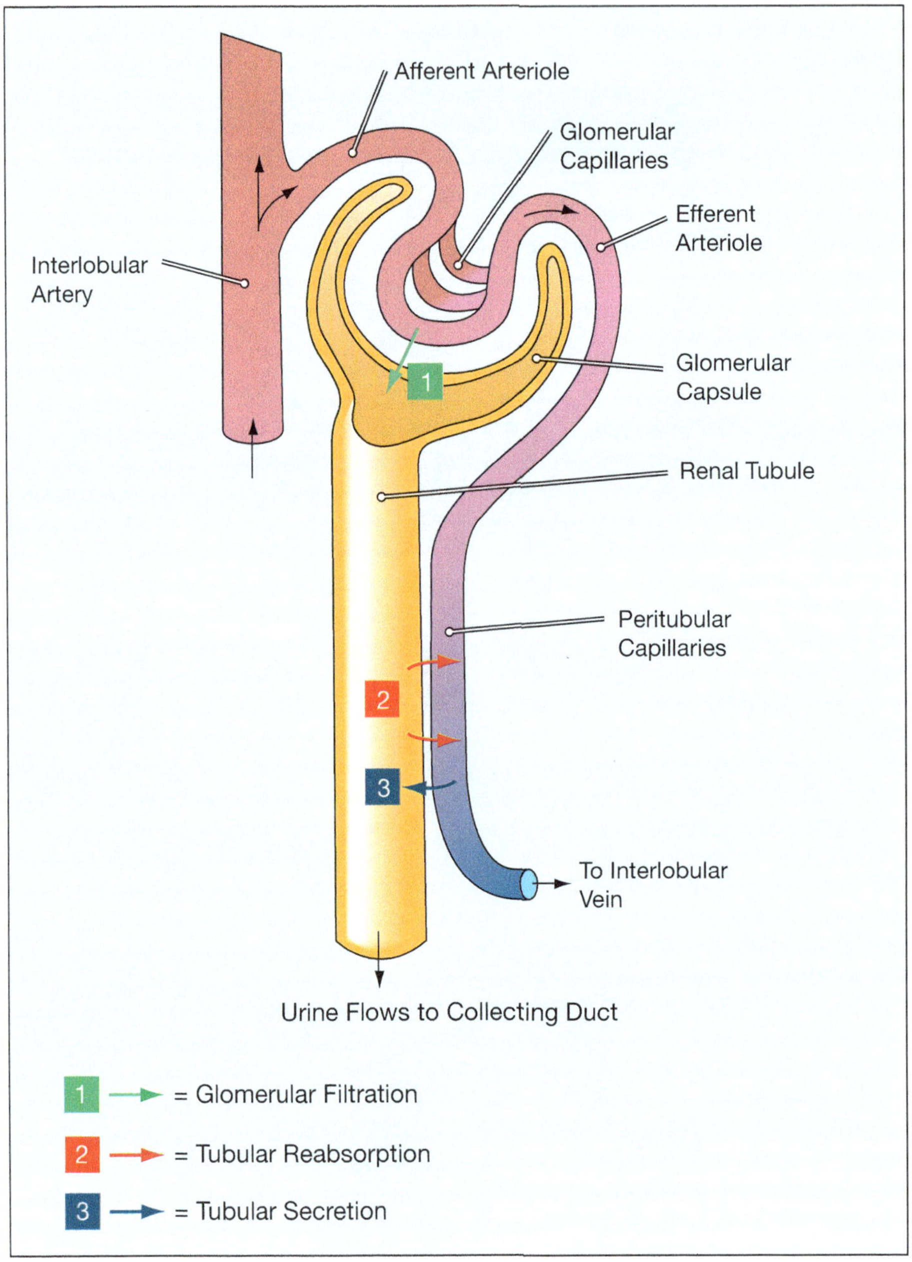

Figure 8.16: Stages of urine production.

1. Glomerular filtration — the rate blood is filtered in the glomerulus
2. Tubular reabsorption — process of moving solutes and water from the renal tubule into the blood
3. Tubular secretion — process of moving solutes and water from the blood into the renal tubule

Review: Labeling Activity

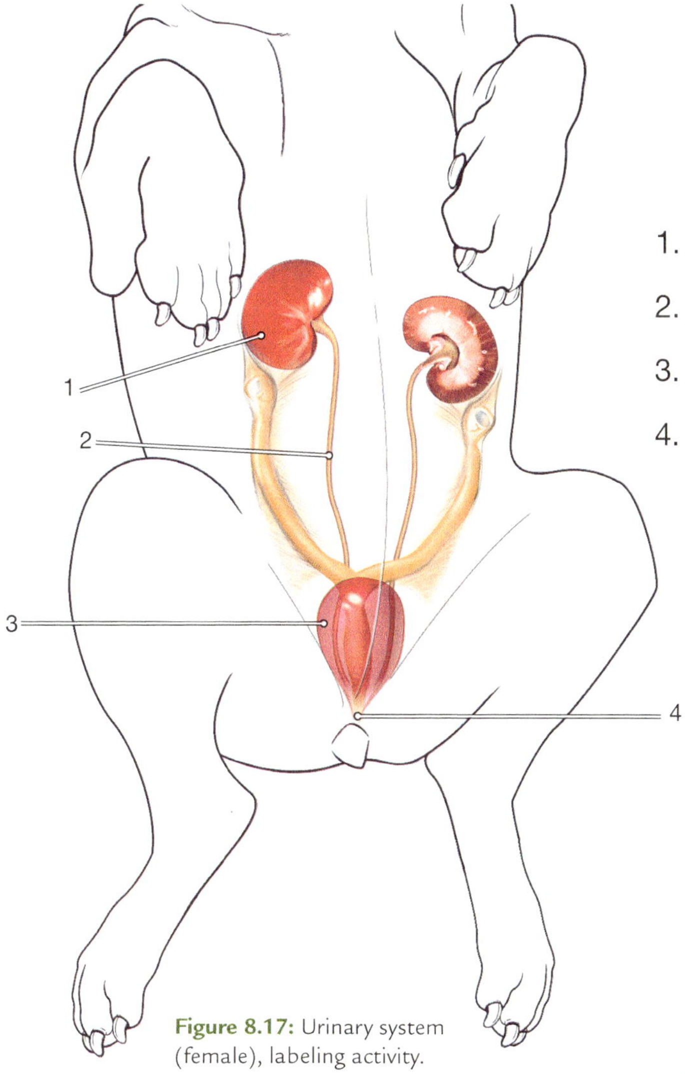

Figure 8.17: Urinary system (female), labeling activity.
© NCSU.

1. ______________________
2. ______________________
3. ______________________
4. ______________________

5. ______________________
6. ______________________
7. ______________________
8. ______________________
9. ______________________
10. ______________________
11. ______________________
12. ______________________

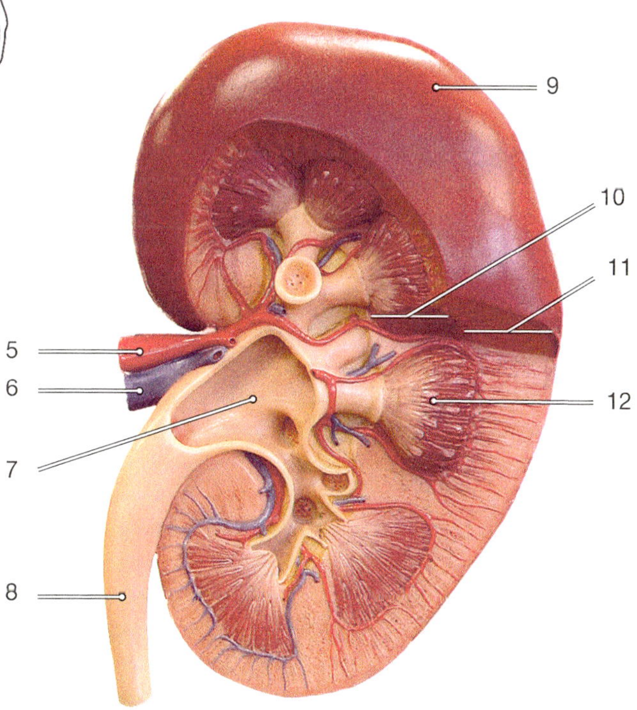

Figure 8.18: Kidney structure, labeling activity.
© David G. Ward.

Review: Crossword Puzzle

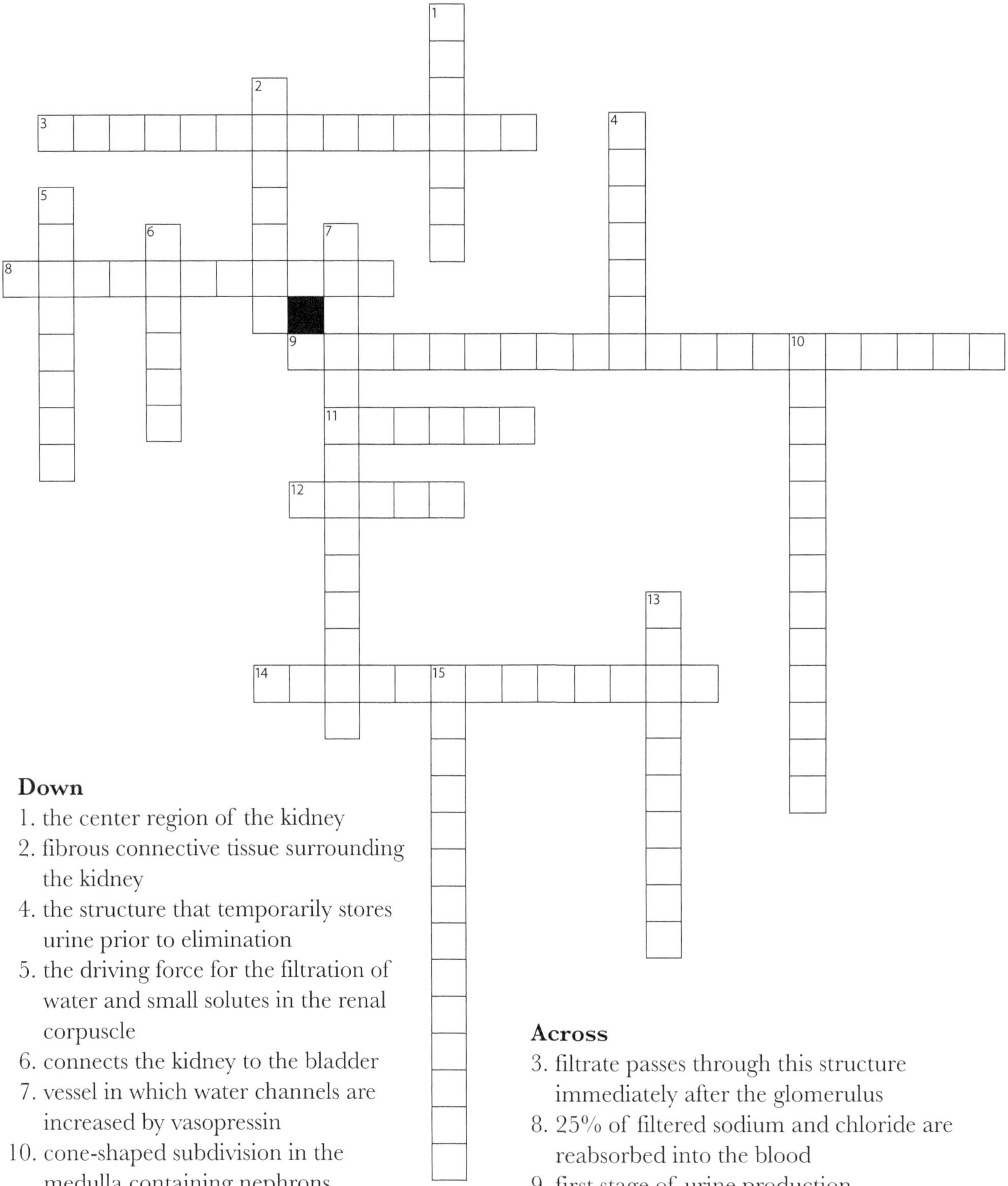

Down

1. the center region of the kidney
2. fibrous connective tissue surrounding the kidney
4. the structure that temporarily stores urine prior to elimination
5. the driving force for the filtration of water and small solutes in the renal corpuscle
6. connects the kidney to the bladder
7. vessel in which water channels are increased by vasopressin
10. cone-shaped subdivision in the medulla containing nephrons
13. a tuft of capillaries located in Glomerular capsule
15. where reabsorption of water by diffusion through open water channels takes place

Across

3. filtrate passes through this structure immediately after the glomerulus
8. 25% of filtered sodium and chloride are reabsorbed into the blood
9. first stage of urine production
11. the outermost region of the kidney
12. where blood vessels, nerves and ureters enter and leave the kidney
14. reabsorption of sodium and chloride by active transport

Review: Word Search

T	R	V	W	Y	W	N	L	E	S	X	N	T	S	Z	G	Z	W	G	N	G	L	Z	E
U	E	E	K	R	C	B	O	A	Z	P	M	X	X	T	O	M	A	B	S	R	M	T	T
B	N	L	X	X	Y	G	V	I	N	U	E	S	Y	X	M	T	Q	V	T	B	A	S	A
U	A	C	Y	R	P	D	N	N	T	E	O	C	M	M	R	L	N	C	D	R	F	F	G
L	L	S	N	I	E	V	L	A	N	E	R	L	Y	O	K	Z	O	D	N	V	F	Y	A
A	A	U	G	N	D	B	I	E	V	M	R	D	A	Q	O	B	Z	O	Z	E	M	P	M
R	R	P	K	R	U	Z	B	P	S	G	J	C	A	I	P	T	I	Q	R	G	X	N	C
R	T	R	A	N	S	I	T	I	O	N	A	L	E	P	I	T	H	E	L	I	U	M	N
E	E	O	G	N	D	P	G	X	A	P	S	L	T	S	A	W	N	M	Y	E	T	G	A
A	R	C	L	E	V	O	W	B	S	X	Z	C	L	R	R	T	S	U	U	Z	L	T	Q
B	Y	L	V	M	E	G	P	Z	V	X	U	U	T	E	A	A	G	A	R	S	B	D	W
S	K	A	B	U	M	D	K	G	N	D	I	L	F	R	C	N	L	Y	U	M	C	J	S
O	Z	N	O	L	X	H	W	D	G	A	I	Z	T	L	J	E	S	U	I	Z	Z	L	S
R	F	E	W	U	Z	K	I	N	L	F	R	E	X	H	E	X	F	U	B	S	J	O	E
P	Z	R	T	N	Q	A	I	D	R	Y	R	F	D	G	J	Z	Y	O	I	U	R	E	B
T	N	E	U	B	F	T	O	A	M	I	L	L	I	O	S	M	O	L	E	S	T	J	T
I	A	Z	I	D	C	S	L	C	O	L	D	N	T	H	I	V	O	V	A	E	M	H	F
O	E	D	V	E	T	U	H	L	E	O	N	T	G	X	T	I	A	V	V	Z	S	R	P
N	W	W	L	E	R	H	E	U	P	X	R	F	Q	C	L	I	K	S	Y	B	V	Q	Y
A	B	L	R	E	V	T	Q	P	O	A	V	J	J	I	W	M	Q	W	J	V	I	E	S
C	O	O	M	E	L	O	I	R	E	T	R	A	T	N	E	R	E	F	F	E	T	K	P
C	N	O	C	B	Q	T	I	L	O	F	Q	X	I	L	F	I	V	E	A	M	U	W	D
E	L	S	U	E	P	K	V	A	K	C	Q	H	V	P	B	N	W	M	Q	F	Z	W	M
G	X	G	E	O	Z	L	M	X	H	T	Z	I	K	J	P	K	U	P	Y	L	N	C	R

1. cell type lining the lumen of the ureter
2. made up of the glomerulus and Bowman's capsule
3. provides the blood supply to the kidney
4. distal tubule drains into this structure
5. provides blood supply to the glomerulus
6. measure of osmolarity in the kidney
7. the rate blood is filtered in the glomerulus (expressed mm/min and abbreviated GFR)
8. second stage of urine production
9. the open cavity in the ureter
10. the gland located superior to each kidney
11. the hormone affecting the DCT
12. drains the blood supply from the kidney
13. the outermost layer of the ureter
14. third stage of urine production
15. removes blood supply from the glomerulus

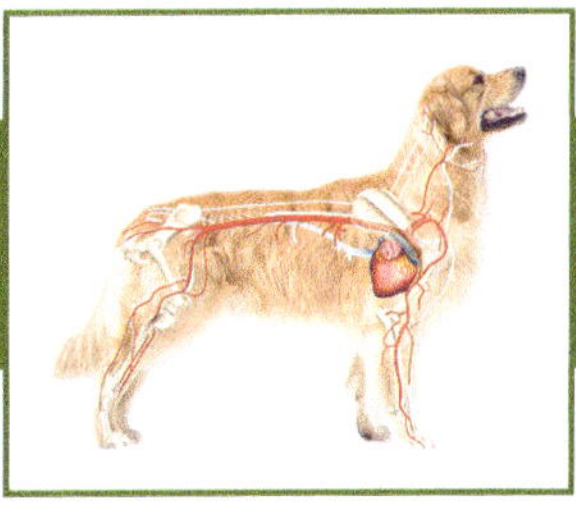

Chapter 9

Endocrine Glands and Reproductive Organs

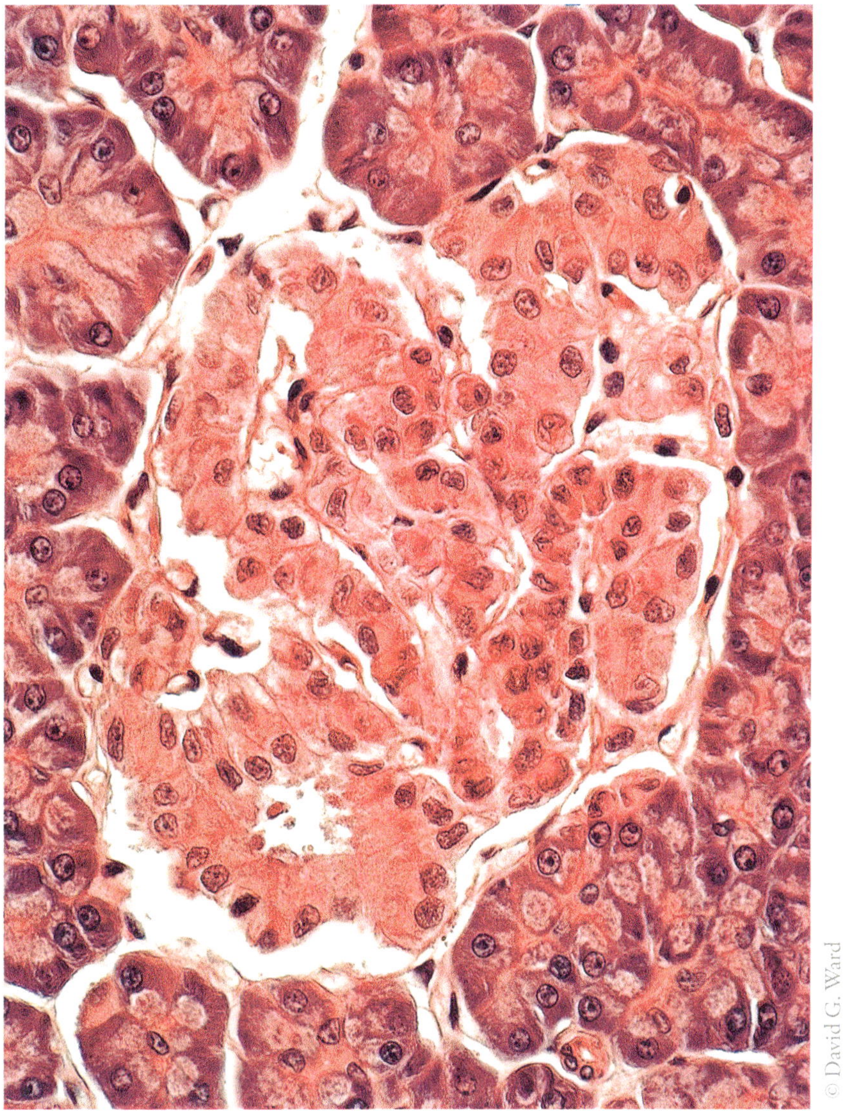

Pancreas Showing Pancreatic Islet (Brightfield, H&E; x430)

Overview of Endocrine Glands

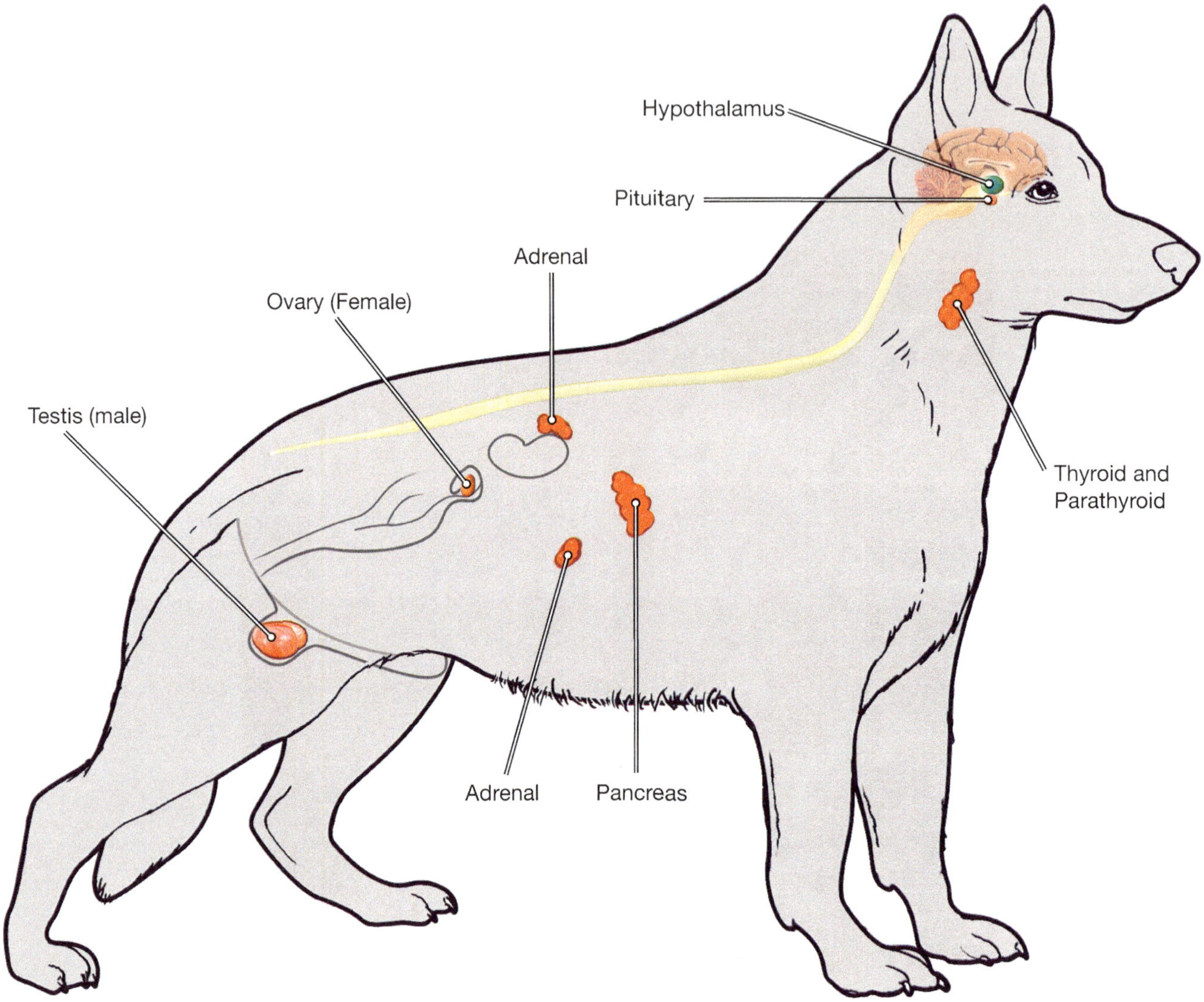

Figure 9.1: Endocrine glands.

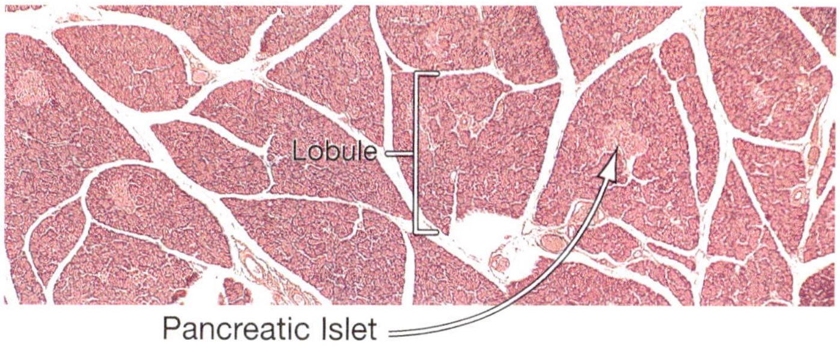

Figure 9.2: Pancreas.

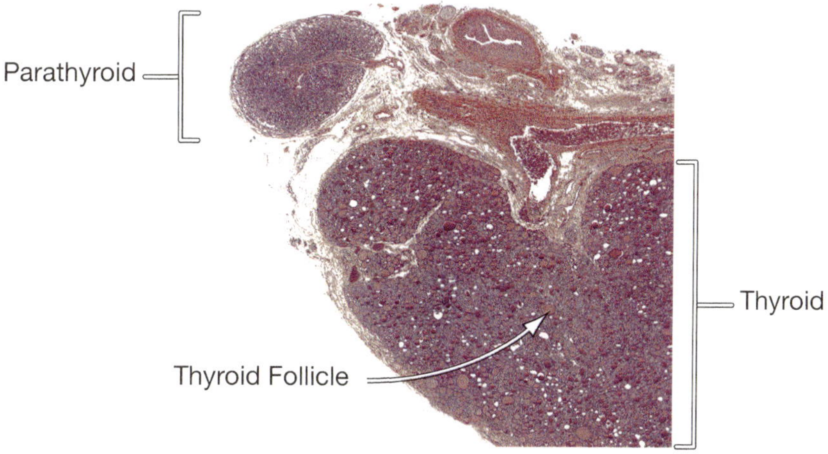

Figure 9.3: Parathyroid and Thyroid.

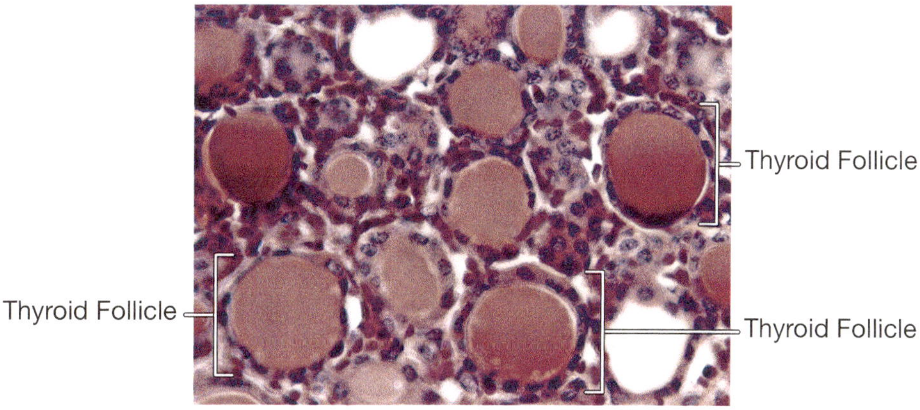

Figure 9.4: Thyroid.

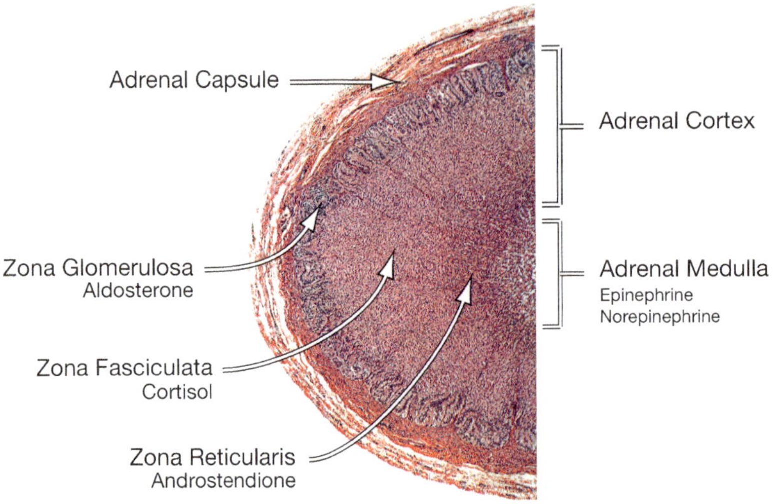

Figure 9.5: Adrenal gland.
© David G. Ward.

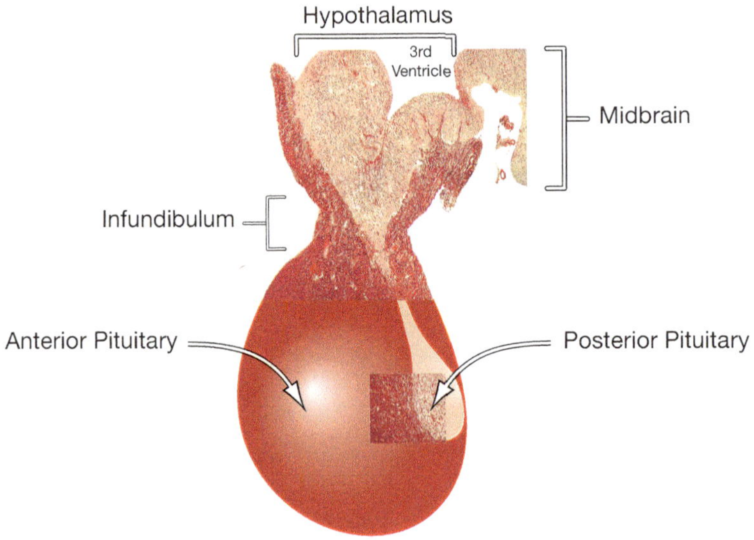

Figure 9.6: Pituitary gland.
© David G. Ward.

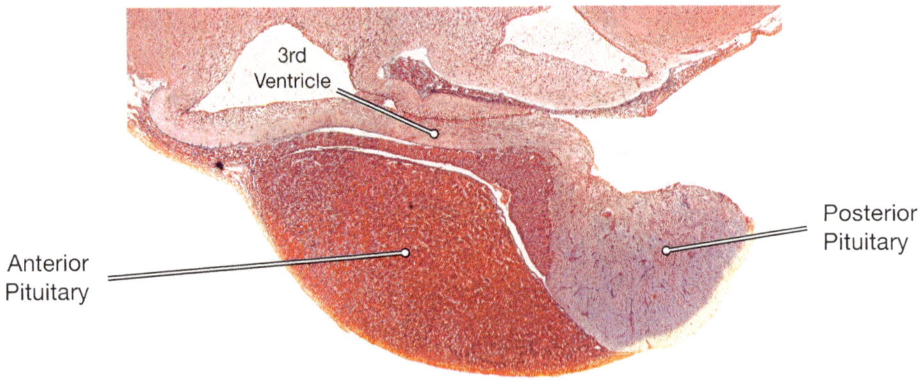

Figure 9.7: Pituitary gland.
© David G. Ward.

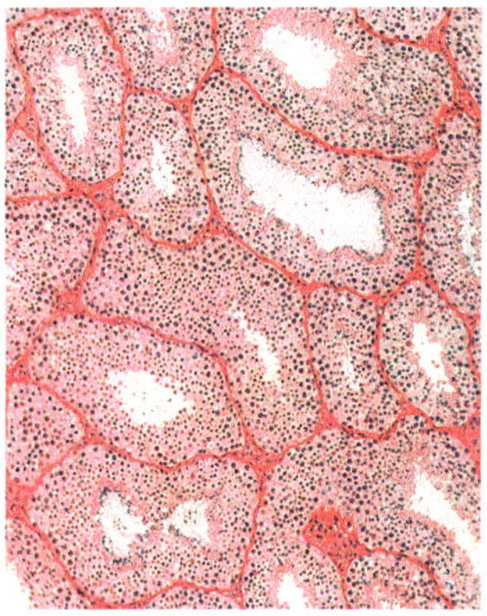

Figure 9.8: Sperm are produced in the seminiferous tubules.

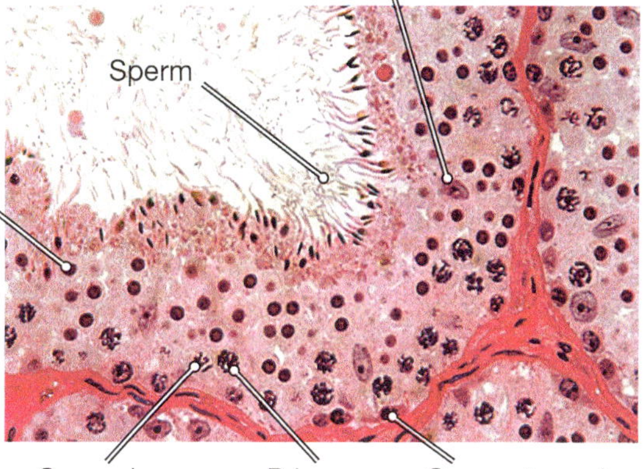

Figure 9.9: Seminiferous tubules.

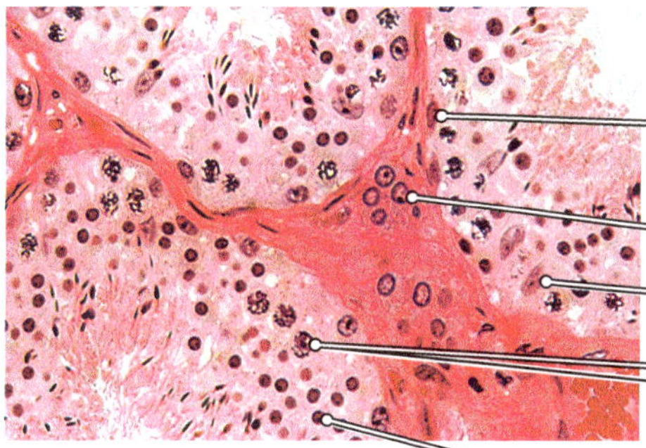

Figure 9.10: Interstitial cells produce testosterone.

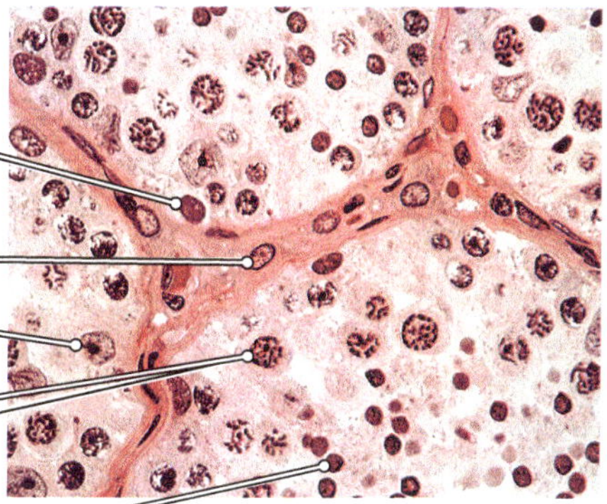

Figure 9.11: Seminiferous tubules.

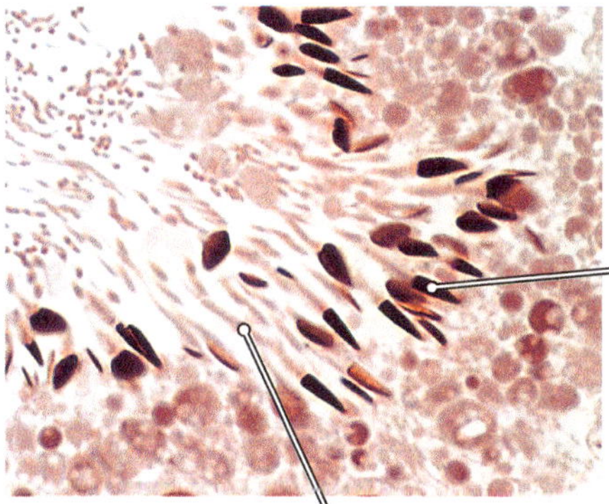

Figure 9.12: Sperm.

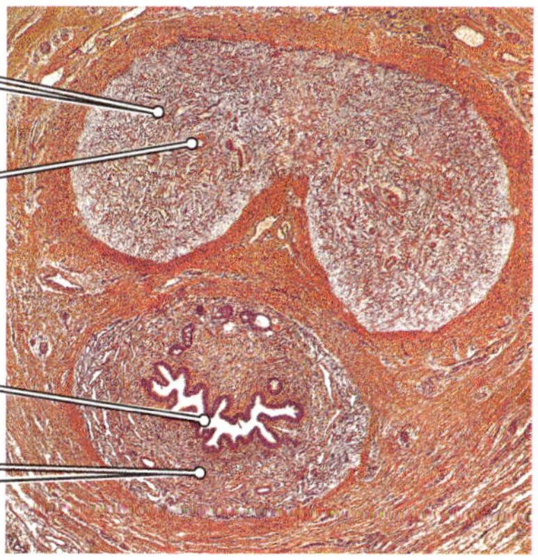

Figure 9.13: Penis, cross-section.

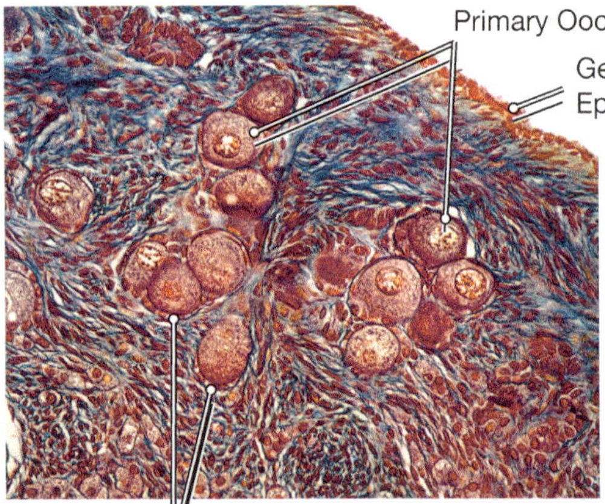

Figure 9.14: Primary Oocytes.

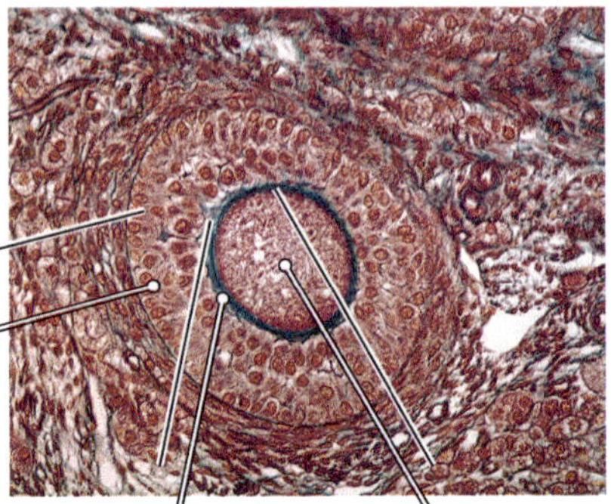

Figure 9.15: Developing follicle.

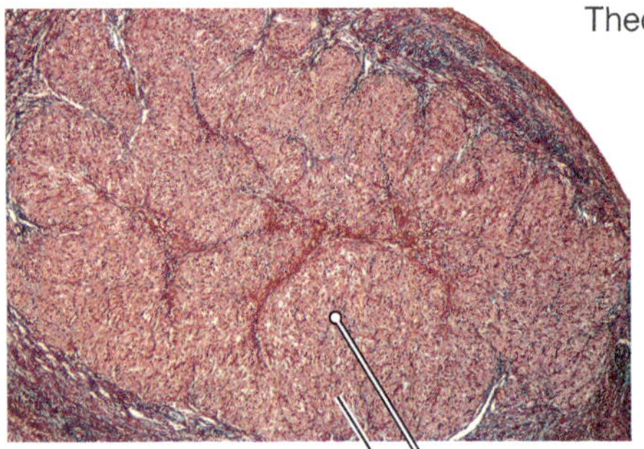

Figure 9.16: Corpus luteum.

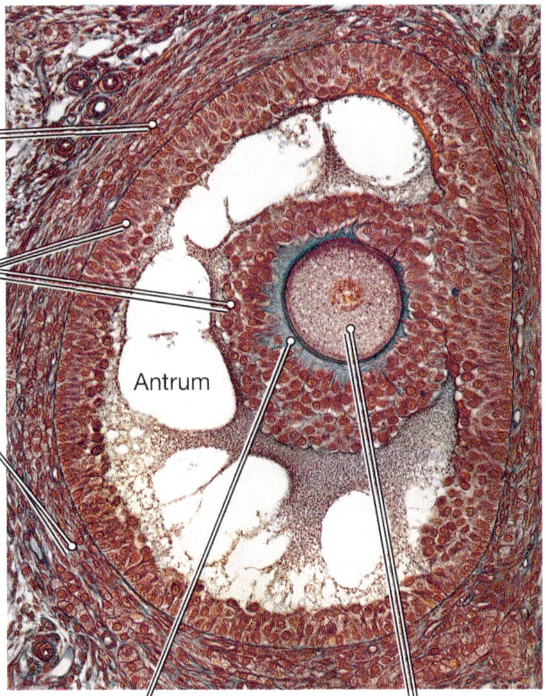

Figure 9.17: Mature follicle.

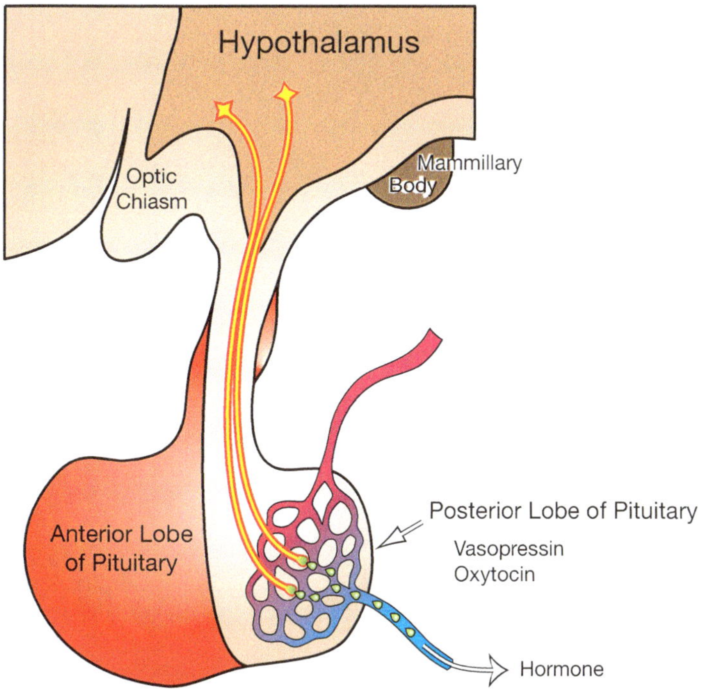

Figure 9.18: Posterior pituitary.
© David G. Ward.

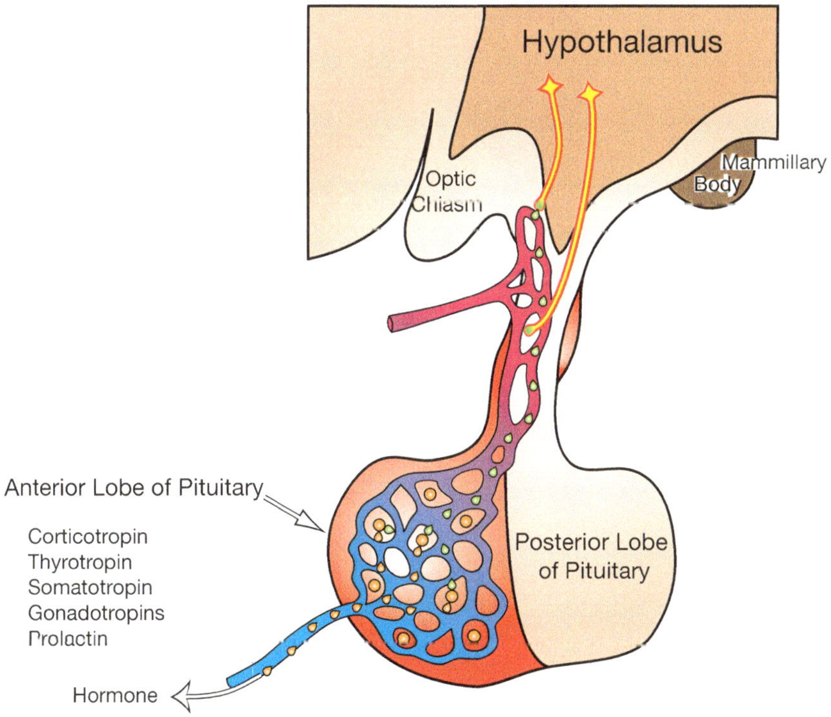

Figure 9.19: Anterior pituitary.
© David G. Ward.

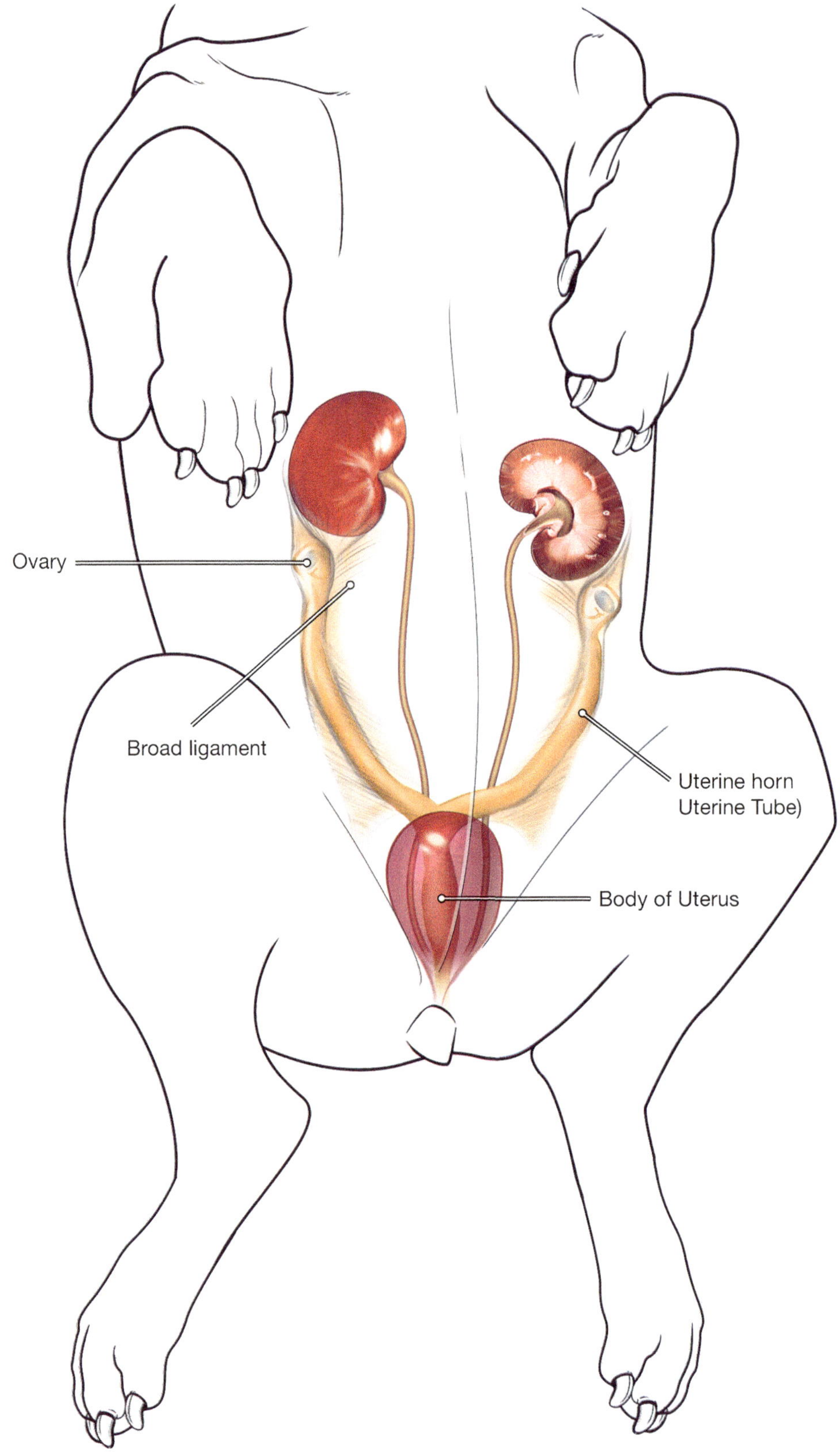

Figure 9.20: Female reproductive organs, ventral view.
© NCSU.

Female Reproductive Organs

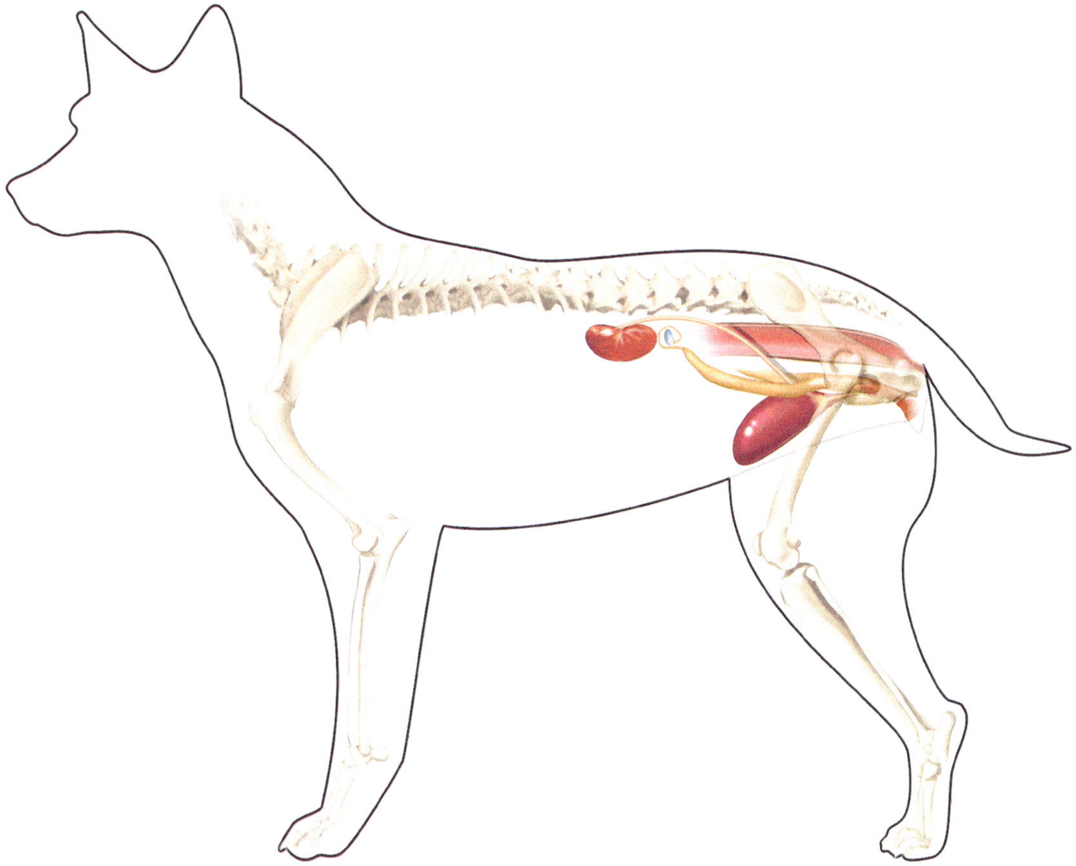

Figure 9.21: Female reproductive organs.
© NCSU.

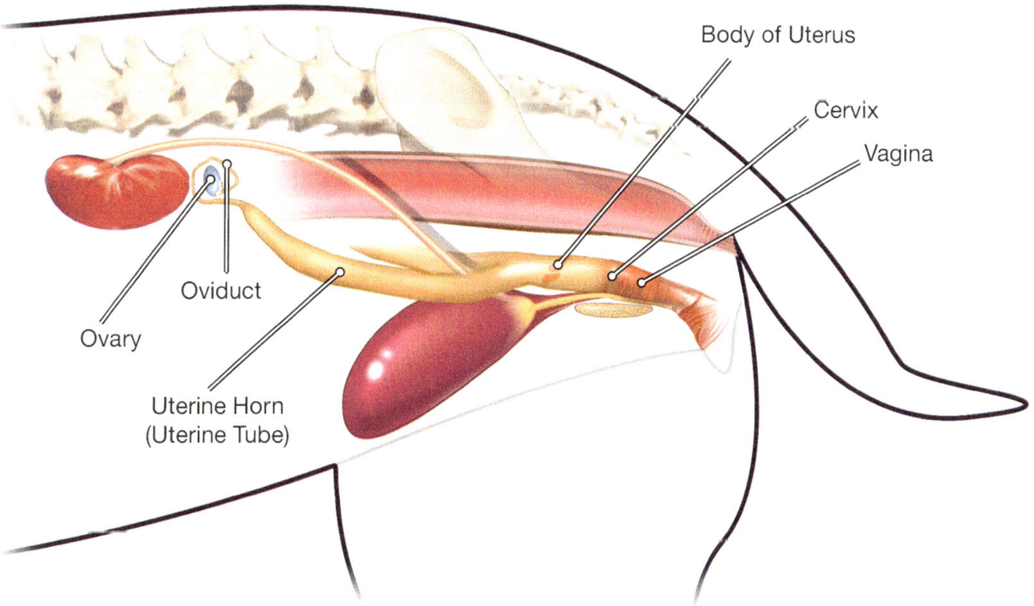

Figure 9.22: Female reproductive organs.
© NCSU.

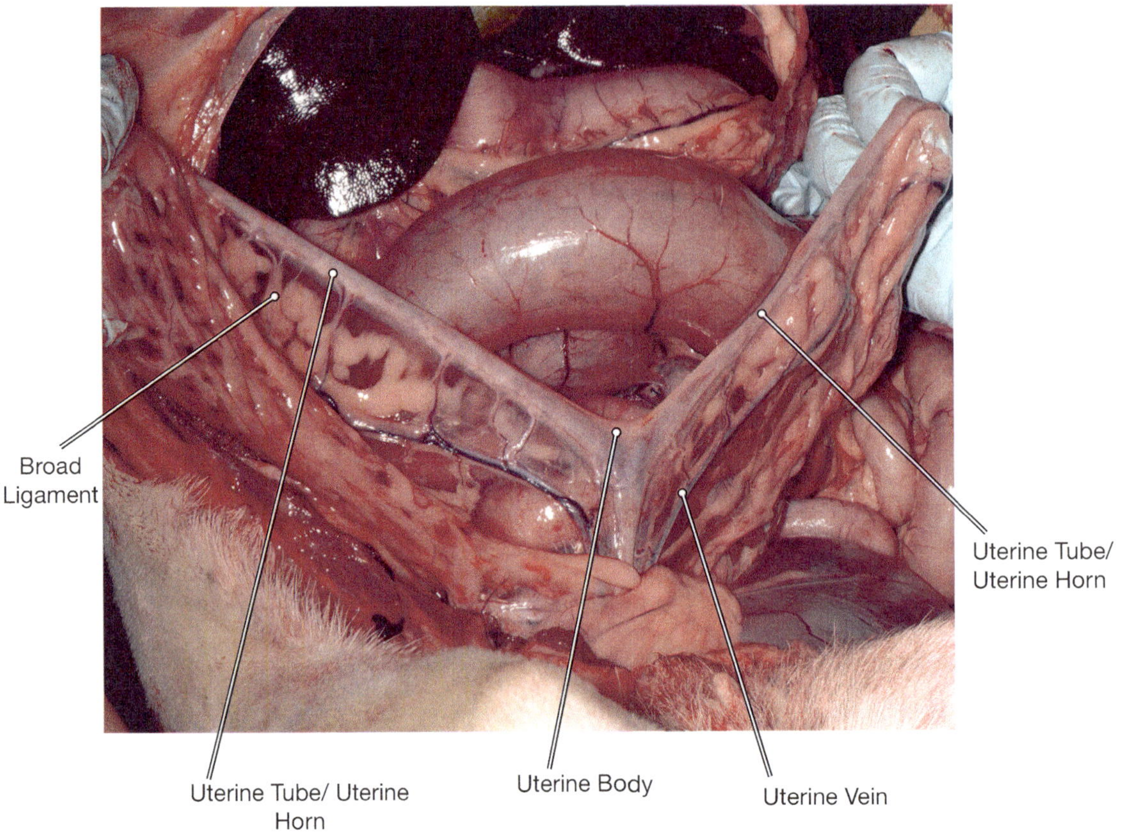

Figure 9.23: Non pregant female organs.
© NCSU.

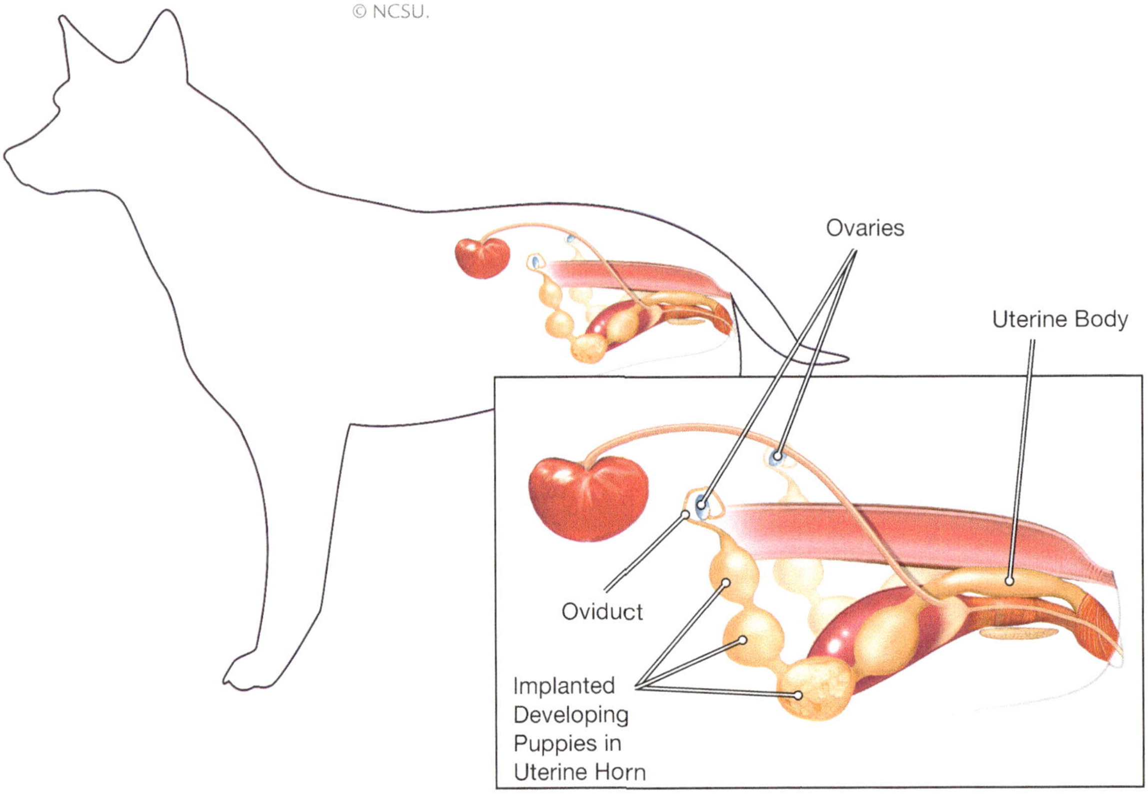

Figure 9.24: Pregant female organs.
© NCSU.

Male Reproductive Organs

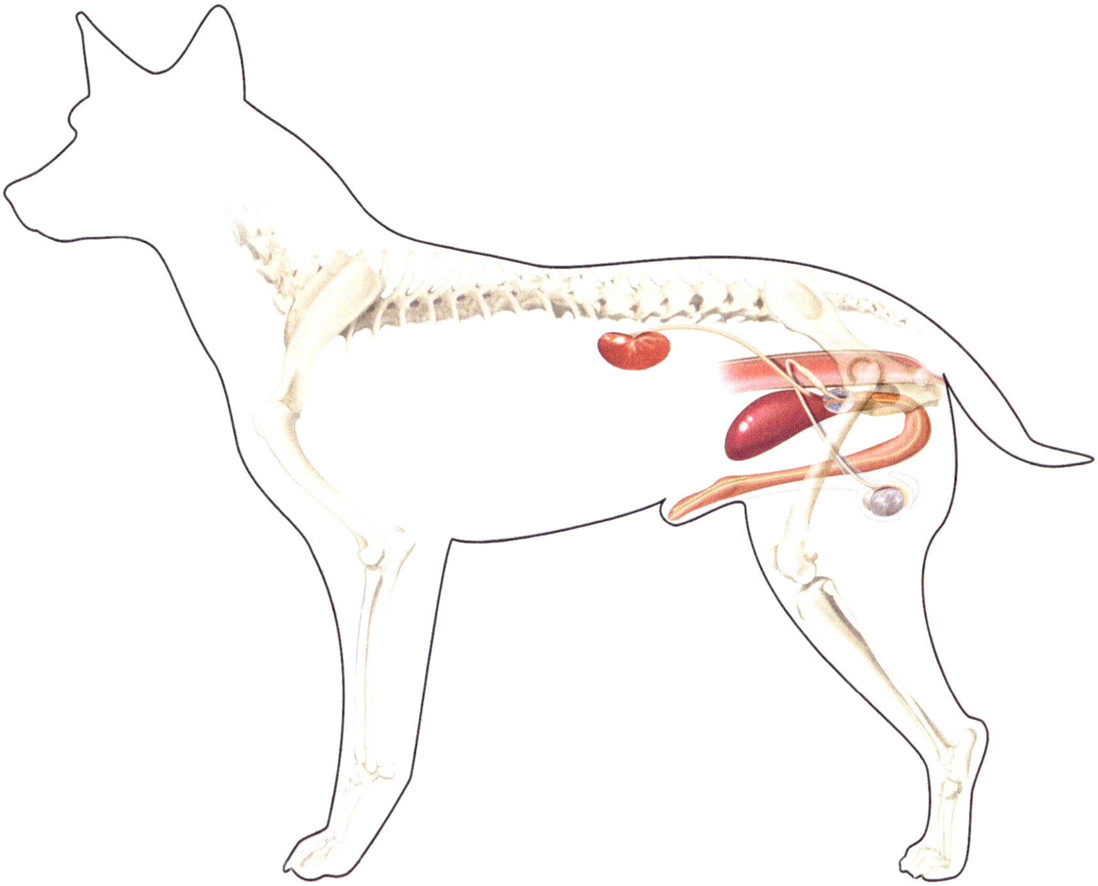

Figure 9.25: Male reproductive organs.

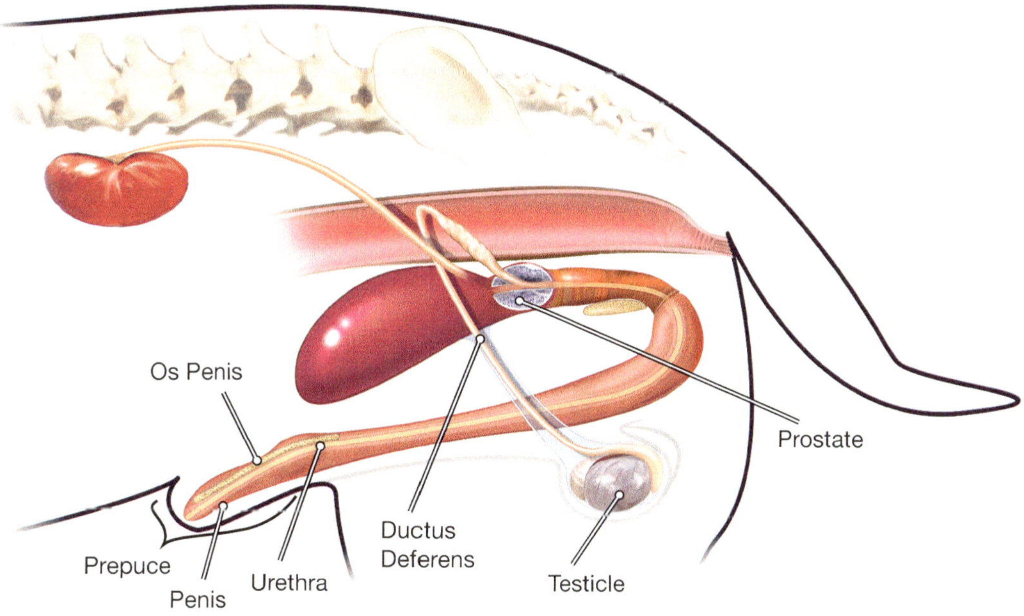

Figure 9.26: Male reproductive organs.

Review: Labeling Activity

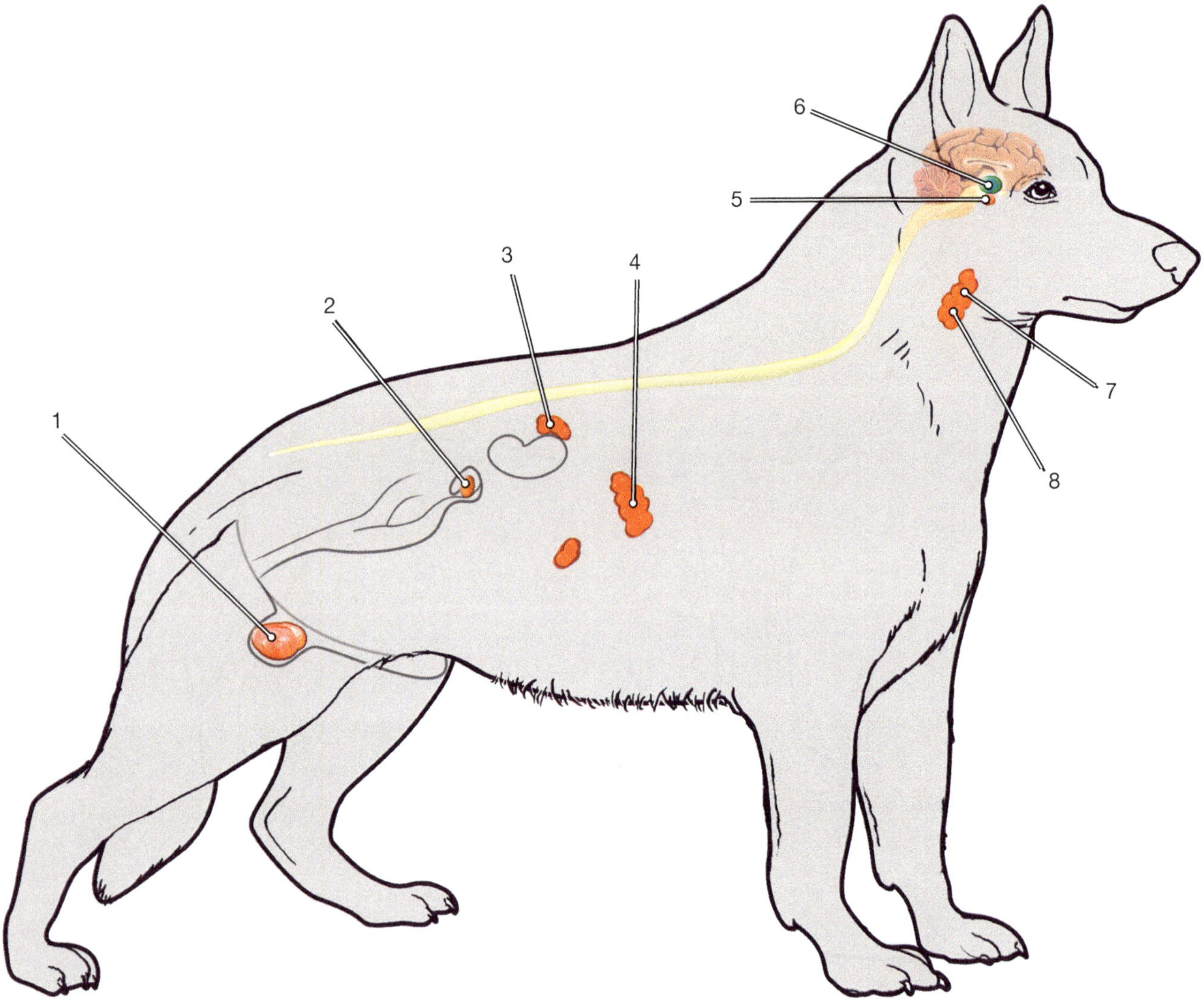

Figure 9.27: Endocrine glands, labeling activity.
© NCSU.

1. ______________________
2. ______________________
3. ______________________
4. ______________________
5. ______________________
6. ______________________
7. ______________________
8. ______________________

Review: Labeling Activity

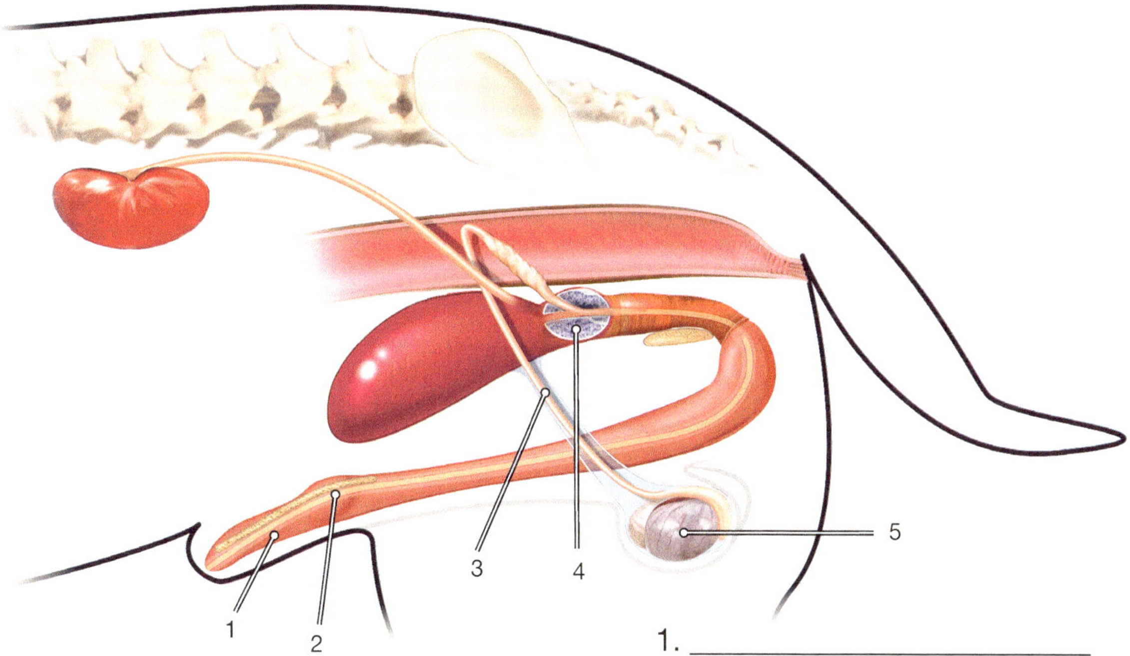

Figure 9.28: Male reproductive organs, labeling activity.
© NCSU.

1. ______________________________
2. ______________________________
3. ______________________________
4. ______________________________
5. ______________________________

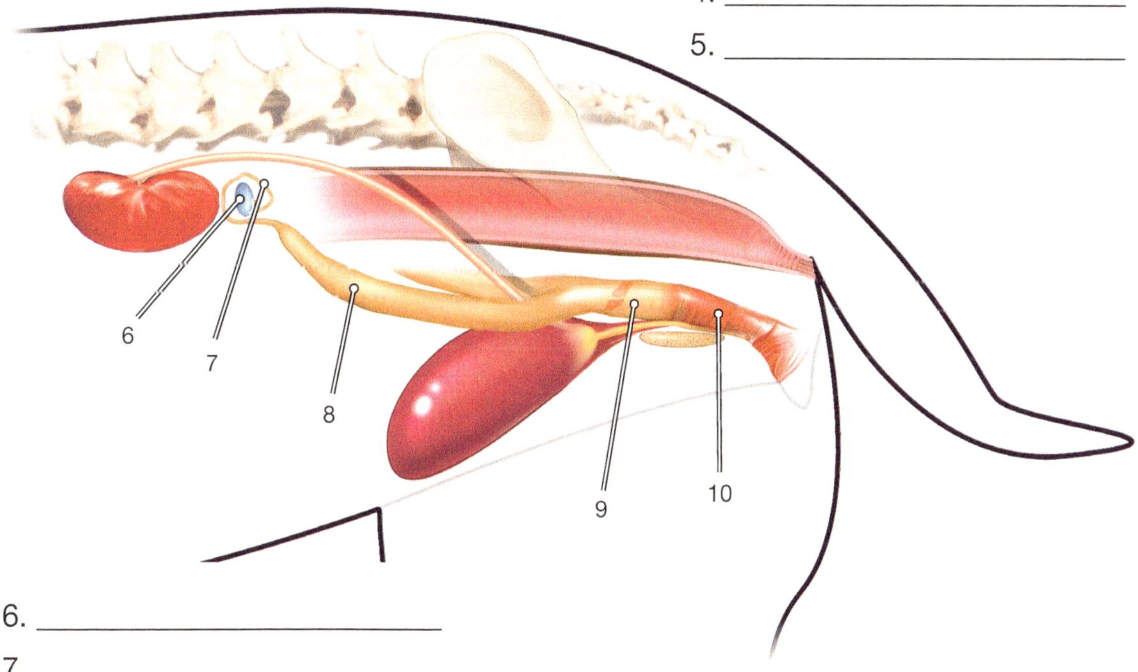

6. ______________________________
7. ______________________________
8. ______________________________
9. ______________________________
10. ______________________________

Figure 9.29: Female reproductive organs, labeling activty.
© NCSU.

Review: Crossword Puzzle

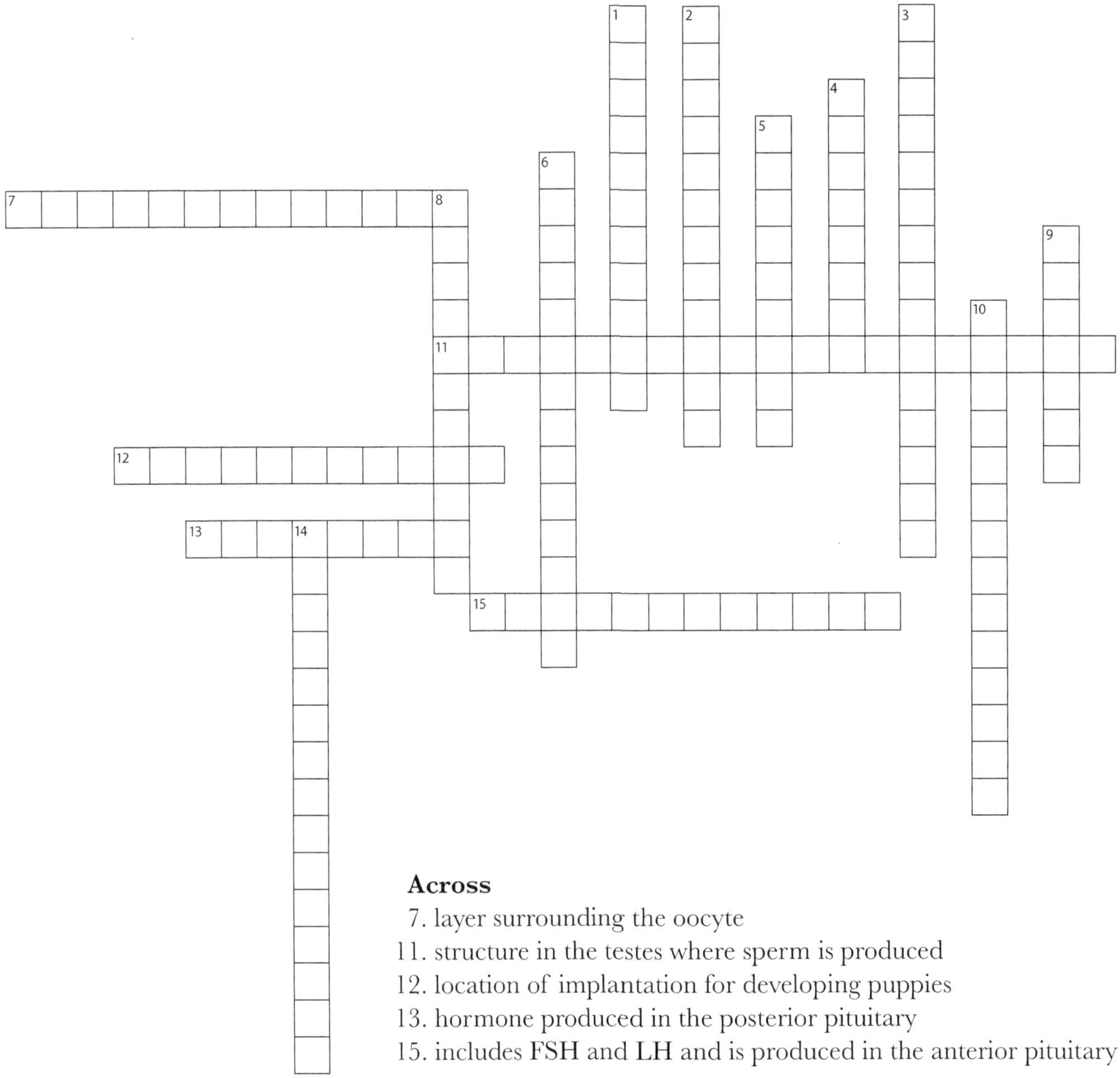

Across

7. layer surrounding the oocyte
11. structure in the testes where sperm is produced
12. location of implantation for developing puppies
13. hormone produced in the posterior pituitary
15. includes FSH and LH and is produced in the anterior pituitary

Down

1. hormone produced in the anterior pituitary
2. stucture on the ovary that produces progestin to maintain pregnancy
3. located in the adrenal cortex and produces androstendione
4. this endocrine gland aids in glucose regulation
5. cells surrounding the zona pellucida in the developing follicle
6. produced in the adrenal medulla
8. hormone produced in the zona glomerulosa
9. pair of endocrine glands located above the kidney
10. structure mature sperm move through between the testicle and the prostate
14. structure located in the thyroid gland